NEUTRON DIFFRACTION

Edited by **Irisali Khidirov**

Neutron Diffraction
Edited by Irisali Khidirov

Published by InTech

Janeza Trdine 9, 51000 Rijeka, Croatia

Publishing Process Manager Molly Kaliman

Technical Editor Teodora Smiljanic

Cover Designer InTech Design Team

Additional hard copies can be obtained from orders@intechopen.com

Neutron Diffraction, Edited by Irisali Khidirov
p. cm.
ISBN 978-953-51-0307-3

For EU product safety concerns:
IN TECH d.o.o., Prolaz Marije Krucifikse Kozulić 3, 51000 Rijeka, Croatia,
info@intechopen.com or visit our website at intechopen.com.

Meet the editor

Dr. Irisali Khidirov is the head of the Composition Materials Laboratory of the Institute of Nuclear Physics of the Uzbekistan Academy of Sciences (Tashkent), doctor of sciences in physics and mathematics, full professor of Solid State Physics, Honorary Doctor of Sciences in Physics of the Yorker International University in Italy. His research is concerned with the study of phase transformations and crystal structures of interstitial alloys, intermetallic compounds and their hydrides, superconducting ceramics and other ceramics by method of neutron diffraction. He has found various types of long-range orders, ordered antiphase domains, stable and metastable phases in interstitial alloys formed under the influence of temperature and γ - irradiation; the hydrogen thermoemission phenomenon from hydrogen containing powder crystals - full removal of hydrogen from a crystal lattice without change of elements symmetry. He showed that in the ordered layered hexagonal close-packed structure of interstitial solid solutions hydrogen isotopes H and D cannot be on the same plane. He has more than 150 scientific publications.

Contents

Preface

Presently the most effective method to study mutual disposition of atoms and magnetic structures is neutron diffraction. The appearance in last years of high flux atomic reactors, automatic neutron diffractometers controlled by computers and special software of processing neutron diffraction data extraordinarily expanded the possibilities of structural and magnetic neutronography. Now it's possible effectively to investigate internal stress and texture of materials and microsamples of nanocrystals by neutron diffraction. Protein structures with thousands of unique atomic positions can be determined with great precision. This all has caused sharp growth of interest to it from physicists, chemists, biologists, materials scientists and metallurgists. Present book comprises original results of world leading scientists recently in research of crystal state of various materials by neutron diffraction method. It has 13 chapters in various sphere of neutron diffraction application. The book may be useful to wide circle of researchers interested in structure of crystal state of various materials and neutron diffraction application.

Prof. Irisali Khidirov
Doctor of Sciences in Physics and Mathematics, Head of the Composition Materials Laboratory of the Institute of Nuclear Physics of Academy Sciences of Uzbekistan
Uzbekistan

1

Hydrides of Cu and Mg Intermetallic Systems: Characterization and Catalytic Function

M. Helena Braga, Michael J. Wolverton,
Maria H. de Sá and Jorge A. Ferreira
[1]*CEMUC, Engineering Physics Department, FEUP, Porto University*
[2]*LANSCE, Los Alamos National Laboratory*
[3]*LNEG*
[1,3]*Portugal*
[2]*USA*

1. Introduction

The worldwide demand for energy in the 21st century is growing at an alarming rate. The European "World Energy Technology and Climate Policy Outlook" [WETO] predicts an average growth rate of 1.8% per annum for the period 2000-2030 for the world energy demand (European Commission, 2003). The increased demand is being met largely by reserves of fossil fuel that emit both greenhouse gases and other pollutants. Since the rate of fossil fuel consumption is higher than the rate of fossil fuel production by nature, these reserves are diminishing and they will become increasingly expensive.

Against this background, the transition towards a sustainable, carbon-free and reliable energy system capable of meeting the increasing energy demands becomes imperative. Renewable energy resources, such as wind, solar, water, wave or geothermal, can offer clean alternatives to fossil fuels. Despite of their obvious advantages renewable energy sources have also some drawbacks in their use because they are unevenly distributed both over time and geographically. Most countries will need to integrate several different energy sources and an advanced energy storage system needs to be developed.

1.1 Hydrogen storage: A brief overview

Hydrogen has also attracted intensive attention as the most promising secure energy carrier of the future (Jain, 2009) due to its prominent advantages such as being:

1. Environmentally friendly. It is a "clean, green" fuel because when it burns in oxygen there is no pollutants release, only heat and water are generated:

$$2H_2\,(g) + O_2\,(g) \leftrightarrows 2H_2O\,(g)\,,\ \triangle H = 120\ kJ/g\ H_2 \qquad (1)$$

2. Easy to produce. Hydrogen is the most abundant element in the Universe and is found in great abundance in the world, allowing it to be produced locally and easily from a great variety of sources like water, biomass and organic matter;

3. Light. Hydrogen is the Nature's simplest and lightest element with only one proton and one electron with high energy per unit mass.

Nonetheless, opposing to the advantages of hydrogen as an energy carrier is the difficulty in storing it. Hydrogen storage remains a problem, in particular for mobile/vehicular applications (Felderhoff et al., 2007). High-pressure hydrogen gas requires very large volumes compared to petrol, for producing the same amount of energy. On the other hand, liquid hydrogen storage systems are not viable for vehicular applications due to safety concerns in addition to volumetric constraints. Thus, hydrogen storage viability has prompted an extensive effort to develop solid hydrogen storage systems but no fully satisfactory solutions have been achieved to date (Churchard et al., 2011).

The goal is to find a material capable of simultaneously absorbing hydrogen strongly enough to form a stable thermodynamic state, but weakly enough to release it on-demand with a small temperature rise (Jeon et al., 2011) in a safe, compact, robust, and efficient manner. There have been many materials under development for solid hydrogen storage, including metal hydrides (MH_x), via the chemical reaction of H_2 with a metal or metal alloy (M):

$$(x/2)\ H_2\ (g) + M\ (s) \leftrightarrows MH_x\ (s) \qquad (2)$$

Generally, a typical hydriding reaction is known to involve several steps: (1) gas permeation through the particle bed, (2) surface adsorption and hydrogen dissociation, (3) migration of hydrogen atoms from the surface into the bulk, (4) diffusion through the particle and finally (5) nucleation and growth of the hydride phase. Any delay in one of these steps will reduce the kinetic properties of the process (Schimmel et al., 2005).

1.2 Magnesium hydride

Magnesium-based hydrogen storage alloys have been considered most promising solid hydrogen storage materials due to their high gravimetric hydrogen storage densities and volumetric capacity (see Table 1 adapted from (Chen & Zhu, 2008) for comparison) associated to the fact that magnesium is abundant in the earth's crust, non toxic and cheap (Grochala & Edwards, 2004; Jain et al., 2010; Schlapbach & Züttel, 2001).

Metal	Hydrides	Structure	Mass %	P_{eq}, T
$LaNi_5$	$LaNi_5H_6$	Hexagonal	1.4	2 bar, 298 K
$CaNi_3$	$CaNi_3H_{4.4}$	Hexagonal	1.8	0.5 bar, 298 K
ZrV_2	$ZrV_2H_{5.5}$	Hexagonal	3.0	10^{-8} bar, 323 K
TiFe	$TiFeH_{1.8}$	Cubic	1.9	5 bar, 303 K
Mg_2Ni	Mg_2NiH_4	Monoclinic (LT) / Cubic (HT)	3.6	1 bar, 555 K
Ti-V-based	Ti-V-based-H_4	Cubic	2.6	1 bar, 298 K
Mg	MgH_2	Tetragonal	7.6	1 bar, 573 K

Table 1. Structure and hydrogen storage properties of typical metal hydrides

Magnesium can be transformed in a single step to MgH_2 hydride with up to 7.6 wt% of hydrogen with a volumetric storage efficiency of 110g H_2/l (Milanese et al., 2010a), according to:

$$Mg\ (s) + H_2\ (g) \leftrightarrows MgH_2\ (s) \tag{3}$$

Magnesium metal is hexagonal with $P6_3/mmc$ space group (α-structure) but the absorption of hydrogen induces a structural change into the tetragonal rutile-type structure α-MgH_2 ($P4_2/mnm$) (Aguey-Zinsou & Ares-Fernández, 2010) (see Fig. 1).

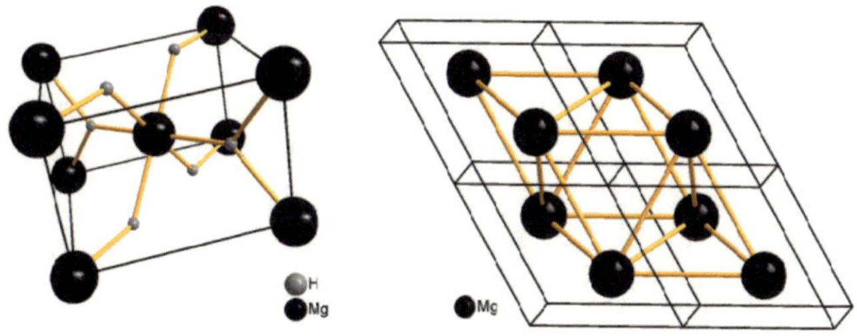

Fig. 1. Crystal structure of magnesium (left) and magnesium hydride (right) obtained with Materials Design® software

At high temperature and pressure, the latter phase undergoes polymorphic transformations to form two modifications: γ-MgH_2 and β-MgH_2, having an orthorhombic structure and a hexagonal structure, respectively (Schlapbach & Züttel, 2001). Other high-pressure metastable phases have also been reported (Cui et al., 2008; Ravindran et al., 2004). The charge density distribution in these materials has also been investigated and revealed a strong ionic character. The charge density determination of MgH_2 by means of synchrotron X-ray powder diffraction at room temperature, the maximum entropy method (MEM) and Rietveld refinement revealed that the ionic charge of Mg and H can be expressed by $Mg^{1.91+}$ and $H^{0.26-}$, respectively, denoting that Mg in MgH_2 is fully ionized, but the H atoms are in a weak ionic state (Noritake et al., 2003). The high strength of these bonds results however in an unacceptably high thermodynamic stability which diminishes the potentialities of using MgH_2 in practical applications. The hydrogen desorption temperature is well above 573 K, which is related to its high dissociation enthalpy (75 kJ/mol H_2) under standard conditions of pressure (Schlapbach & Züttel, 2001). In addition, the high directionality of the ionic bonds in this system leads to large activation barriers for atomic motion, resulting in slow hydrogen sorption kinetics (Vajo & Olson, 2007).

Several solutions were envisaged to circumvent these drawbacks. They can be accomplished to some extent by changing the microstructure of the hydride by ball-milling it (Huot et al., 1999; Zaluski et al., 1997). In this process the material is heavily deformed, and crystal defects such as dislocations, stacking faults, vacancies are introduced combined with an increased number of grain boundaries, which enhance the diffusivity of hydrogen into and out of the material (Suryanarayana, 2008). Alloying the system with other metallic additives, like 3d elements (Ti, Fe, Ni, Cu or Al), or $LaNi_5$, FeTi, Pd, V among others and oxides like V_2O_5 or Nb_2O_5 can also be a way of improving kinetic and/or thermodynamic properties by changing the chemical interaction between the atoms (Reule et al., 2000; Rude et al., 2011; Tan et al., 2011a). The use of a proper destabilization or catalyst element/alloy into the system has also been shown to improve adsorption/desorption kinetics and to lower the adsorption temperature (Beattie et al., 2011). Furthermore, substantial improvements in the hydriding-dehydriding properties can be achieved by nanoengineering approaches using nanosized reactants or by nanoconfinement of it (Jeon et al., 2011; Jurczyk et al., 2011; Vajo, 2011; Zaluska et al., 1999a; Zhao-Karger et al., 2010). The latter allows shorter diffusion distances and larger surface area, resulting in faster reaction kinetics. It can also introduce alternative mechanisms to hydrogen exchange modifying the thermodynamic stability of the process.

As previously referred, an alternative approach for altering the thermodynamics of hydrogenation-dehydrogenation is achieved by using additives that promote hydride destabilization by alloy or compound formation in the dehydrogenated state. This approach is known as chemical destabilization. The principle underlying this approach is that the additives are capable to form compounds or alloys in the dehydrogenated state that are energetically favourable with respect to the products of the reaction without additives. Destabilization occurs because the system can cycle between the hydride and the additive instead of the elemental metal. A generalized enthalpy diagram illustrating this approach - destabilization of the generic hydride AH_2 through alloy formation (AB_x) promoted by the presence of the alloying species B - was given by Vajo and Olson (Vajo & Olson, 2007), and is shown in Fig. 2.

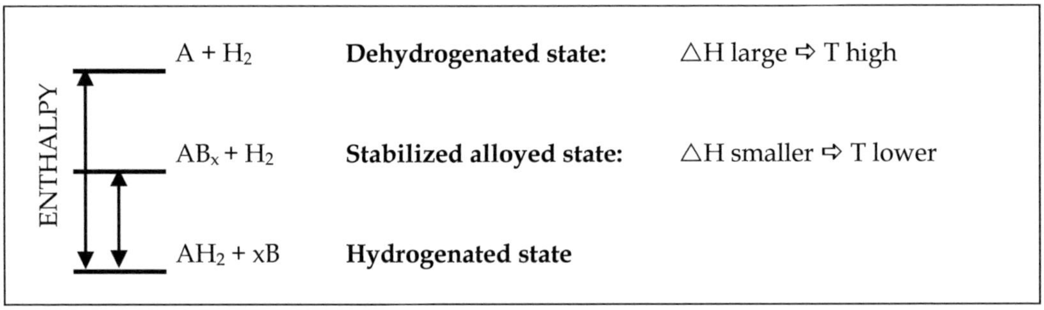

Fig. 2. Generalized enthalpy diagram illustrating destabilization through alloy formation upon dehydrogenation (adapted from Vajo & Olson, 2007)

1.3 Cu-Mg, Ni-Mg and other MgH_2 destabilizing systems

The work of Reilly and Wiswall provided the first evidences of this concept (Reilly & Wiswall, 1967, 1968). In their work, they showed that MgH_2 can be destabilized by Cu_2Mg. The formation of $CuMg_2$ occurs upon dehydrogenation at lower reaction temperatures than those obtained with just pure MgH_2. The compound $CuMg_2$ crystallizes in the orthorhombic structure (Braga et al., 2010c) and has a hydrogen capacity of 2.6 wt. % at 573 K (Jurczyk et al., 2007). The hydride formation enthalpy is approximately 5 kJ/mol H_2 lower than that of the hydrogenation of MgH_2 from Mg and this process obeys to the following scheme (Reilly & Wiswall, 1967):

$$2CuMg_2\,(s) + 3H_2\,(g) \leftrightarrows 3MgH_2\,(s) + Cu_2Mg\,(s) \qquad (4)$$

The intermetallic cubic compound Cu_2Mg does not hydrogenate under conventional hydrogenation conditions and seems to improve dehydrogenation kinetics (as compared to MgH_2) due to improve resistance towards oxygen contamination (Andreasen et al., 2006; Kalinichenka et al., 2011; Reilly & Wiswall, 1967). As to the hexagonal intermetallic compound $NiMg_2$, Reilly and Wiswall (Reilly & Wiswall, 1968) established that it reversibly reacts with hydrogen to form a ternary hydride Mg_2NiH_4, with a hydrogen content of 3.6 wt. %, according to the following scheme:

$$NiMg_2\,(s) + 2H_2\,(g) \leftrightarrows NiMg_2H_4\,(s) \tag{5}$$

Results obtained by A. Zaluska and co-workers (Zaluska et al., 1999b) showed that ball-milling the mixtures MgH_2 and $NiMg_2H_4$ results in a synergetic effect of desorption, allowing the mixture to operate at temperatures as low as 493 K - 513 K, with good absorption / desorption kinetics and with total hydrogen capacity exceeding 5 wt.%. They point out however that the ball-milled mixtures of the hydrides behave differently from two metal phases that are firstly ball-milled and then hydrogenated. In the latter case volume changes occur during hydrogenation with associated volume expansion of the material, in contrast to what happen in their study in which $NiMg_2H_4$ promoted the hydrogen release from an adjacent MgH_2 matrix since they undergo a significant volume contraction, which facilitates their dehydrogenation.

Many more studies have focused on changes in the hydriding/dehydriding properties of Ni-Mg binary alloys with compositional changes and changes in processing variables. Nonetheless, we highlight the study of C. D. Yim and collaborators (Yim et al., 2007) that showed that the $NiMg_2$ compound acted as a catalyst in the dissociation of the hydrogen molecule, which resulted in a faster nucleation of magnesium hydride compared to pure Mg. It also revealed that the capacity and kinetics of hydriding were larger and faster when the average size of the hydriding phase was smaller and the volume fraction of the phase boundary was larger, since phase boundaries between the eutectic α-Mg and $NiMg_2$ phases acted as a fast diffusion path for atomic hydrogen.

In the full hydrogenated state, the $NiMg_2H_4$ structure consists of tetrahedral $[NiH_4]^{4-}$ complexes in a framework of magnesium ions and two different forms exist, high-temperature (HT) and low-temperature (LT). Under the partial pressure of 1 atm of hydrogen, the HT cubic structure phase transforms into a LT monoclinically distorted structure between 518 and 483 K (Zhang et al., 2009). The LT phase has also two modifications the untwined (LT1) and micro-twinned (LT2), which depend on the thermomechanical history of the sample (Cermak & David, 2011). The hydride formation enthalpy for the $NiMg_2H_4$ has been determined experimentally for the HT form, and it is in the rangeof -64.3 to -69.3 kJ/mol H_2, for the LT form this value ranges from -68.6 to -81.0 kJ/mol H_2 (Tan et al., 2011b).

In the pioneer work of Reilly and Wiswall (Reilly & Wiswall, 1968) it was pointed out the catalytical effect of $NiMg_2$ on the hydrogen desorption characteristics of MgH_2. Recently, Cermak and David (Cermak & David, 2011) showed that $NiMg_2$, and more efficiently the LT1 phase of $NiMg_2H_4$, were responsible for the catalytic effect of Ni reported in the literature. The fact that $NiMg_2$ is a metal whereas $NiMg_2H_4$ behaves like a semiconductor has opened the way to the possibility of using this system also as a switchable mirror upon hydrogenation and dehydrogenation (Setten et al., 2007). A switchable mirror will switch from mirror to transparent material upon hydrogenation. A more detailed study of Ni-Mg-based hydrides can be found in (Orimo & Fujii, 2001).

Despite all the interest and extensive research on the above referred systems, a problem still remains; the hydrogen holding capacities of these materials are considerably less than that of MgH_2 (Sabitu et al., 2010). A way to overcome this limitation was found by combining MgH_2 with $LiBH_4$ (which involves the formation of MgB_2 and Li-Mg alloy (Yu et al., 2006)) since pure $LiBH_4$ has high gravimetric and volumetric hydrogen densities, 18.5 wt. % and

121 kg H_2/m^3, respectively (Bösenberg et al., 2010; Xia et al., 2011). However, although the reaction enthalpy is lowered and the hydrogen storage capacity increases (10.5 wt. %), the sorption and absorption processes occurs at high temperatures with relatively slow kinetic even though more additives are being tested in order to overcome this problem (Fernández et al., 2011; Xia et al., 2011). Alternatively, the study of the destabilization of MgH_2 with TiH_2 has also been taken experimentally (Choi et al., 2008; Sohn et al., 2011). Observations point to a substantially reduced apparent activation energy of 107-118 kJ/mol and significantly faster kinetics, compared with the 226 kJ/mol for the similarly milled MgH_2. The latter system constitutes a promising material to be used in practical applications for hydrogen storage.

The combined destabilization effect of Ni-Mg and Cu-Mg intermetallics towards MgH_2 was also tested and the Mg-rich ternary Cu-Ni-Mg alloys were recognized to have high potential for solid state hydrogen storage and have attracted many research interests. The study recently reported by Tan and co-workers (Tan et al., 2011b) elucidates about the influence of Cu substitution on the hydrogen sorption properties of magnesium rich Ni-Mg films. This study shows a two-step hydrogen absorption process. The first step is due to the absorption of Mg not alloyed in the form of $NiMg_2$ and/or $CuMg_2$, hereafter denoted as "free Mg" and is very quick, because it is mainly catalyzed by the intermetallic phase, $NiMg_2$. But the second step, due to the hydrogen absorption of intermetallic $NiMg_2$ and/or $CuMg_2$ ("bonded Mg") is significantly slow. The Cu substitution shows positive effects on desorption kinetics during full capacity hydrogen cycling, but shows strongly negative effects on absorption kinetics, particularly for the second absorption step, due to the segregation of $CuMg_2$ towards the grain boundaries of MgH_2, forming a closed shell that traps the hydrogen in MgH_2. The authors also reported that the Cu substitution has no Thermo-destabilization effect on MgH_2, but since a significant amount can be dissolved in $NiMg_2$, even at elevated temperatures, thermo-destabilization of $NiMg_2H_4$ and better desorption kinetics are observed. Hong and collaborators (Hong et al., 2011) on their study on the hydrogen storage properties of x wt.% Cu-23.5 wt.% Ni-Mg (x = 2.5, 5 and 7.5) prepared by rapid solidification process and crystallization heat treatment have also reported that the $NiMg_2$ phase has higher hydriding and dehydriding rates than Mg under similar conditions and that the addition of a smaller amount of Cu is considered favourable to the enhancement of the hydriding and dehydriding rates of the sample. The 2.5 wt.% Cu-23.5 wt.% Ni-Mg alloy had the highest hydriding and dehydriding rates. These observations are in line with the ones previously reported by the group of Milanese (Milanese et al., 2010b; 2008), who also observed the high sorption capacity and good sorption performance of Cu-Ni-Mg mixtures and proposed a two steps sorption process with different kinetics. The first step corresponds to the quick hydrogenation of "free Mg", according to reaction (3). After this step, absorption keeps on with a slower rate corresponding to the second step, hydrogenation of the "bonded Mg" phases, $NiMg_2$ and $CuMg_2$, according to reactions (4) and (5). They also showed that Ni is more effective than Cu in catalyzing the desorption reactions and that $NiMg_2H_4$ and Cu_2Mg phases destabilized each other with the beneficial effect of decreasing the dissociation temperature of about 50 K in comparison to the MgH_2, from "free Mg". The positive effect of Cu as a catalyst on the hydrogenation and thermodynamic properties of $NiMg_2$ mixed by ball milling technique was also studied and recently reported by Vyas and co-workers (Vyas et al., 2011) showing that hydrogen storage capacity and enthalpy of formation of $NiMg_2$ with 10 wt.% Cu reduces to 1.81 wt.% and 26.69 kJ $(mol\ H)^{-1}$ from 3.56 wt.% and 54.24 kJ $(mol\ H)^{-1}$ for pure $NiMg_2$ at 573 K, respectively. They attributed the decrement in the absorption capacity to the formation of the intermetallic phase Cu_2Mg, which does not absorb the hydrogen but itself behaves like a catalyst. However, in the case of nanocrystalline $Cu_xNi_{10-x}Mg_{20}$ (x = 0 - 4) alloys synthesized by melt-spinning technique, it was found (Zhang et al., 2010a, 2010b) that the substitution of Ni by Cu does not change the major phase $NiMg_2$ although it leads to a refinement of grains with increased cell volume and the formation of a secondary phase $CuMg_2$. This in turn leads to a decrease of the hydride stability with a clear improve of the hydrogen desorption capacity and kinetics of the alloys. The presence of $CuMg_2$ seems to act as a catalyst for the hydride-dehydride reactions of Mg and Mg-based alloys. Similar behaviour was found in $Cu_{0.25}Ni_{0.75}Mg_2$ and $Cu_{0.4}Ni_{0.6}Mg_2$ alloys that were prepared by mechanical alloying and subsequent thermal treatment (Simičić et al., 2006). The latter effect was also investigated on $Cu_{1-x}Ni_xMg_2$ (x = 0 - 1) alloys by Hsu and collaborators (Hsu et al., 2010). They observed that by substituting Cu by Ni in $CuMg_2$, the cell volume decreased (since the radius of Cu atom is slightly larger than Ni atom) and with increasing Ni content, the effect of Ni is actually effective in MgH_2 and Mg_2NiH_4 destabilization, leading to a decrease of desorption temperature in these two phases. They also showed that substituted nickel caused the hydriding reaction because absorption kinetics and hydrogen storage capacity increased with the rise of Ni-substitution contents.

1.4 Lithium hydride

An alternative route to be considered is to explore other hydrides besides MgH_2 for solid hydrogen storage. One of most interesting is lithium hydride, because it contains 12.5 wt.% hydrogen. Nonetheless, the desorption temperature is 1183 K for an equilibrium pressure of 1 bar (Vajo et al., 2004). However, it has been shown (Chen et al., 2003) that when LiH (see Fig. 3) reacts with lithium amide ($LiNH_2$) by thoroughly mixing the substances, hydrogen is released at temperatures around 423 K, with formation of lithium imide (Li_2NH) or Li-rich imide (Li_xNH_{3-x}) and lithium nitride (Li_3N) depending on the temperature and molar ratio of ($LiH/LiNH_2$) according to the following schemes:

- Below 593 K: $$LiH\ (s) + LiNH_2\ (s) \rightarrow 2H_2\ (g) + Li_2NH\ (s) \quad (6)$$

$$2LiH\ (s) + LiNH_2\ (s) \rightarrow (x-1)\ H_2\ (g) + Li_xNH_{3-x}\ (s) + (3-x)\ LiH(s) \quad (7)$$

- At higher temperatures: $$2LiH\ (s) + LiNH_2\ (s) \rightarrow H_2\ (g) + Li_3N\ (s) \quad (8)$$

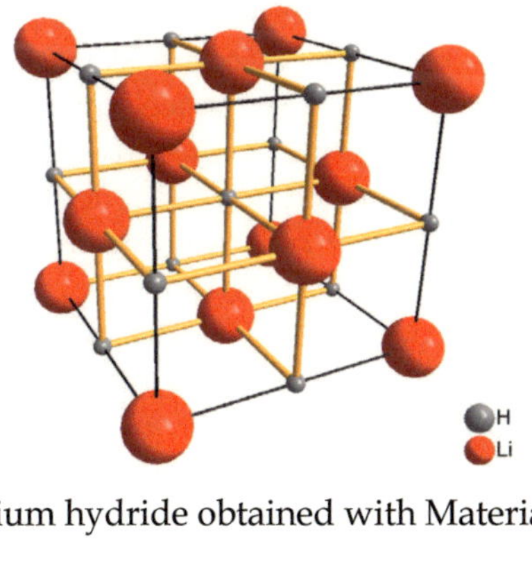

Fig. 3. Crystal structure of lithium hydride obtained with Materials Design® software

From a detailed analysis of high-resolution synchrotron x-ray diffraction data for the lithium amide ($LiNH_2$) - lithium imide (Li_2NH) hydrogen storage system (David et al., 2007), the authors were able to propose an alternative mechanism that does not need to have the materials mechanically milled to enhance mixing, as previously recognized by Chen and collaborators (Chen et al., 2003) as essential. The mechanism they propose for the transformation between lithium amide and lithium imide during hydrogen cycling is a bulk reversible reaction that occurs in a non-stoichiometric manner within the cubic anti-fluorite-like Li-N-H structure, based on both Li^+ and H^+ mobility within the cubic lithium imide. Concluding that increasing the Li^+ mobility and/or disorder it is likely to improve the hydrogen cycling in this and related Li-based systems. Recently, further systematical evaluation of the decompositions of $LiNH_2$ and Li_2NH was carried out by Zhang and Hu (Zhang & Hu, 2011), who also examined the effect of Cl^- anion on the decomposition process. Cl^- is widely employed as a promoter to improve various catalysts. As a result, decomposition mechanisms were established. The decomposition of $LiNH_2$ producing Li_2NH and NH_3 occurs in two steps at the temperature range of 573-723 K. $LiNH_2$ decomposes into a stable intermediate species ($Li_{1.5}NH_{1.5}$) and then into Li_2NH. Furthermore, Li_2NH is decomposed into Li, H_2, and N_2 without formation of Li_3N at the temperature range of 823-1023 K. The introduction of Cl^- can decrease the decomposition *temperature of Li_2NH by about 110 K.*

1.5 Neutron techniques associated with hydrogen solid storage

Though some progress have been made, the state-of-art materials are still far from meeting the aimed targets for hydrogen solid storage material (Churchard et al., 2011). This huge task can be facilitated by employing state-of-the-art techniques like, computational first-principles calculations to evaluate the thermodynamic properties of the potential materials (Alapati et al., 2007; Siegel et al., 2007; Yang et al., 2007). This allows a quick screen of a large number of potential candidates, searching for thermodynamically suitable ones (saving time and money). Once thermodynamic appropriate materials have been found other considerations such as structure and dynamics of the materials during hydrogenation/dehydrogenation will become crucial in order to understand the fundamental properties of hydrogen storage, in realistic conditions and hence design new hydrogen storage materials.

Neutron scattering techniques are highly suitable for structure and dynamics studies related to hydrogen in solids and bound on surfaces. The energy distribution of thermal neutrons is nearly ideal for the study of condensed matter in general because it is of the same order of magnitude as most molecular and lattice excitations and the de Broglie wavelengths of thermal neutrons match quite well with interatomic distances in most solids (Squires, 1978). Neutrons have some unique advantages over photons and electrons as scattering media which are of particular use for the analysis of hydrides. For these purposes the two most useful neutron scattering interactions are coherent elastic scattering for Neutron Diffraction (ND) and incoherent Inelastic Neutron Scattering (INS) to measure vibrational density of states. The distinction of coherent and incoherent scattering interactions is important to the unique advantages offered by ND and INS respectively. This is because the relative scattering intensity of a given interaction is dependent highly upon the nucleus involved, and as such is isotope dependant. Each isotope has a different scattering cross section for both coherent (σ_{coh}) and incoherent (σ_{inc})

interactions measured in barns (1 barn = 10^{-28} m^2). In general these scattering cross sections do not follow any specific trend regarding nucleus size.

INS has numerous advantages to other common techniques of obtaining vibrational spectra such as infrared (IR) and Raman spectroscopy. INS spectroscopy is hyper sensitive to the presence of hydrogen. The protium (^1H) nucleus has scattering cross sections of σ_{coh} = 1.8 and σ_{inc} = 80.2 barns respectively. This means neutron scattering in materials containing natural abundance hydrogen is largely inelastic. Additionally, the incoherent cross section of ^1H is one to two orders of magnitude higher than any other isotope (Ross, 2008). This means that in hydrides INS spectra are dominated by vibrational modes of hydrogen almost exclusively. This hyper sensitivity to hydrogen means that hydride phases are detectible even when present in relatively miniscule concentration. Another advantage of INS is the complete absence of selection rules for the excitation of vibrational modes. Lattice modes (i.e. phonons) are excited with equal opportunity to molecular vibrations. Because both IR and Raman spectroscopy rely upon different types of charge symmetry interactions, many materials have vibrational modes that cannot be excited by Raman or IR. In particular lattice modes are far more easily observable in INS spectra than any other type of vibrational spectroscopy. INS is also more useful for comparison with *ab initio* calculated density of states because relative excitation amplitudes are simply dependent upon the magnitude of motion and σ_{inc} of the excited nucleus (Squires, 1978; Ross, 2008). Free software, such as a-Climax is available to generate a theoretical INS spectrum from the density of states output files from numerous common *ab initio* packages such as Gaussian, AbInit and Dmol (Ramirez, 2004).

For these reasons INS is extremely useful in identifying the presence of different hydride phases which may not be structurally apparent (for example, due to structural disorder). A good example is the INS study of Schimmel et al. on MgH_2 produced from Mg processed by high energy ball milling. Ball milling of Mg to reduce particle size, and introduce fractures, defects, and faults has a beneficial effect of increasing hydride formation rate, and reducing the temperature required for absorption. Comparison of the INS spectra of the MgH_2 produced from ball milled Mg with well-ordered MgH_2 revealed a partial composition of γ-MgH_2, which is metastable and normally exists only at high temperatures (Schimmel et al, 2005). Presence of γ-MgH_2 was indicative of internal stress from mechanical processing. However after hydrogen sorption cycling, γ-MgH_2 was no longer observable in the INS spectrum of the ball milled material, while the fast kinetics and lower sorption temperature remained. In this way INS was indispensable in revealing that the particle size reduction is more significant in the role of lowering temperature and increasing sorption kinetics than the creation of faults and internal stresses after the high energy ball milling of Mg (Schimmel, 2005; Ross, 2008).

Neutron diffraction also provides some unique advantages versus more conventional diffraction methods such as X-ray diffraction (XRD). Elastic neutron and X-ray scattering are similar in that both result in interference patterns according to Bragg scattering conditions (Squires, 1978). In XRD the intensity of a given Bragg reflection varies with the atomic number (Z) of the atom at the lattice site. This means that the exact position of hydrogen in a structure is practically impossible to determine with XRD. In ND the relative intensities of reflections are independent of Z, and instead depend on the coherent scattering cross section (σ_{coh}). This means that deuterium (^2H; σ_{coh} = 5.6, σ_{inc} = 2.0) is just as readily observable as most metal atoms. This allows for the observation of hydride phase transitions which differ only by the hydrogen occupation sites, such as in interstitial hydrides. ND also allows metals with similar Z values such as Ni (Z=28, σ_{coh} = 13.4) and Cu (Z=29, σ_{coh} = 7.5) to be easily distinguished, unlike in XRD. A great deal of caution must be taken to ensure that ^2H is not displaced by ^1H during sample preparation and handling, as the large σ_{inc} of ^1H will create a substantial background signal. Another advantage of ND is that intensity does not diminish greatly with scattering angle as it does in XRD (Massa, 2004). Beyond these differences, crystal structure determination techniques are very similar for ND and XRD. Common approaches include a combination of a structure solution method and the Rietveld refinement method.

ND and INS carry some common advantages and disadvantages intrinsic with the use of neutrons as a scattering medium. Common advantages are associated with the highly penetrating quality of neutron radiation through most materials. This provides some possibilities for variable depth of measurements. If the neutron beam is directed at a relatively thin portion of the sample, a greater quantity of surface and shallow depth material is surveyed, whereas in relatively thick segments predominantly material deep within the sample is surveyed. The high penetration of neutrons also allows for relatively clear in-situ measurements in a wide range of sample environments such as high pressure gas cells, furnaces, cryogenic refrigerators, anvil cells and other environments requiring obtrusive equipment. This allows for detailed structure and dynamics studies of metastable hydride phases, and phase transitions which occur only in extreme conditions.

There are numerous inconveniences associated with neutrons as well. The most prevalent and obvious is the relative scarcity and cost of neutron sources, which typically take two forms: a research reactor or a spallation source (fed by a high energy proton accelerator). Another drawback is the time required to conduct a measurement, which can range from several hours to several days (per measurement). This is due to the short range of nuclear forces and relatively low probability of a scattering event, which is the same reason neutron radiation penetrates so effectively. Because of the long measurement time and high operational cost beam time is allocated very carefully at neutron sources, and flight paths are rarely left idle during neutron production. ND and INS require larger sample sizes, often multiple grams, to increase the scattering rate.

2. Hydrides of Cu and Mg intermetallic systems

We have studied the Cu-Li-Mg system as a hydrogen storage system and, at the same time, as a catalyst of the hydrogen storage process, namely for the Ti/TiH_2 system (Braga & Malheiros, 2007a, 2007b; Braga et al., 2010a, 2010b). The only ternary compound the Cu-Li-Mg system holds is $CuLi_xMg_{2-x}$ (x = 0.08) with hexagonal $P6_222$ structure (Braga et al., 2010c). Since the phase diagrams of Cu-Mg and Ni-Mg are similar (see Fig. 4), and Cu and Ni have similar electron affinities, it was thought in the sixties that $CuMg_2$ would store hydrogen, too.

However this is not the case (Reilly & Wiswall, 1967). $NiMg_2$ has a hexagonal structure ($P6_222$), but $CuMg_2$ has an orthorhombic structure (Fddd), and this structural difference is assumed to be the reason that $NiMg_2$ stores H_2 forming a hydride, but $CuMg_2$ does not. $CuMg_2$ decomposes into Cu_2Mg and MgH_2 (Reilly & Wiswall, 1967) upon hydrogen loading

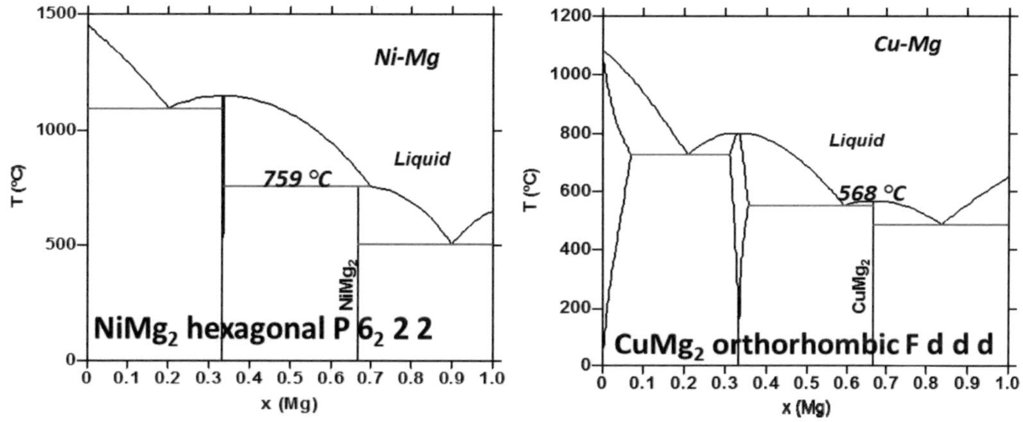

Fig. 4. Phase diagrams of Ni-Mg and Cu-Mg (Ansara et al., 1998)

as previously referred (4). As a result of this reaction and since $CuMg_2$ does not form a hydride, $CuMg_2$ was abandoned as a candidate material for hydrogen storage (Reilly & Wiswall, 1967; Schlapbach & Züttel, 2001) until the late destabilization studies that were previously cited. The hexagonal structure of $CuLi_xMg_{2-x}$, suggested the possibility of using this phase as a hydrogen storage material (Braga & Malheiros, 2007a, 2007b) because $CuLi_xMg_{2-x}$ has the same space group ($P6_222$) as $NiMg_2$ and $NiMg_2(H,D)_{0.3}$ (lattice parameters are almost identical: a = b = 5.250 Å and c = 13.621 Å (at 300 K) for $CuLi_xMg_{2-x}$ and a = b = 5.256 Å and c = 13.435 Å for $NiMg_2(H,D)_{0.3}$ (Senegas et al., 1984)). Therefore, we hypothesized that $CuLi_xMg_{2-x}$ (x = 0.08) would be a hydrogen storage material, just like $NiMg_2$ - a hypothesis that has been confirmed by now (Braga & Malheiros, 2007a, 2007b; Braga et al., 2010a).

The change of the $CuMg_2$ orthorhombic (Fddd) structure to a hexagonal structure ($P6_222$) upon addition of a small amount of Li has been firmly established (Braga et al., 2007). Isostructural phases to $CuLi_xMg_{2-x}$ are the hexagonal phase $NiMg_2$ and $NiMg_2H_{0.24-0.30}$ (Senegas et al., 1984). For the NiMg-hydrides, several hydrogen positions were reported: In $NiMg_2H_{0.29}$ the hydrogen atoms occupy Wyckoff 6f positions and could occupy the interstitial Wyckoff 6h position (Senegas et al., 1984). Other possibilities would be that the H atoms would just occupy interstitial Wyckoff 12k position (in $NiMg_2H_{0.26}$) or the Wyckoff 12k and 6j positions in $NiMg_2H_{0.24}$ (Senegas et al., 1984). This suggests a number of possible sites for Li in $CuLi_xMg_{2-x}$.

Interestingly V. Hlukhyy and collaborators (Hlukhyy et al., 2005) have reported a result closely related to our observations in the Sn-doped Ni-Mg system. These authors show that the synthesis of alloys in the Ni-Mg system is affected by the presence of small amounts of Sn (forming $NiMg_{2-x}Sn_x$ with x = 0.22 and 0.40). The replacement of Mg by Sn produces changes in the structure of $NiMg_2$, this time making the alloy change from the $NiMg_2$ type (hexagonal) to the $CuMg_2$ type (orthorhombic). While the structure of $NiMg_{1.85}Sn_{0.15}$ is still of $NiMg_2$ type, the structure of $NiMg_{1.78}Sn_{0.22}$ and $NiMg_{1.60}Sn_{0.40}$ is already of the $CuMg_2$ type. These results represent obviously the converse of our own observations in the $CuMg_2$ structure, and reaffirm our results with respect to $CuLi_xMg_{2-x}$.

2.1 The $CuLi_{0.08}Mg_{1.92}$ compound

We have used neutron diffraction to refine the composition of $CuLi_xMg_{2-x}$, site occupancies and lattice parameters at different temperatures. In Fig. 5, results from the Time-of-flight (TOF) Neutron Powder Diffractometer (NPDF) at the Los Alamos Neutron Science Center (LANSCE) are shown. It was analyzed a sample containing 37.5 at.% of $CuLi_{0.08}Mg_{1.92}$, 45.1 at.% of $CuMg_2$ and 17.4 at.% of Cu_2Mg. The structure was refined using the General Structure Analysis System (GSAS), a Rietveld profile analysis program developed by A. C. Larson and R. B. von Dreele (Larson & von Dreele, 2004).

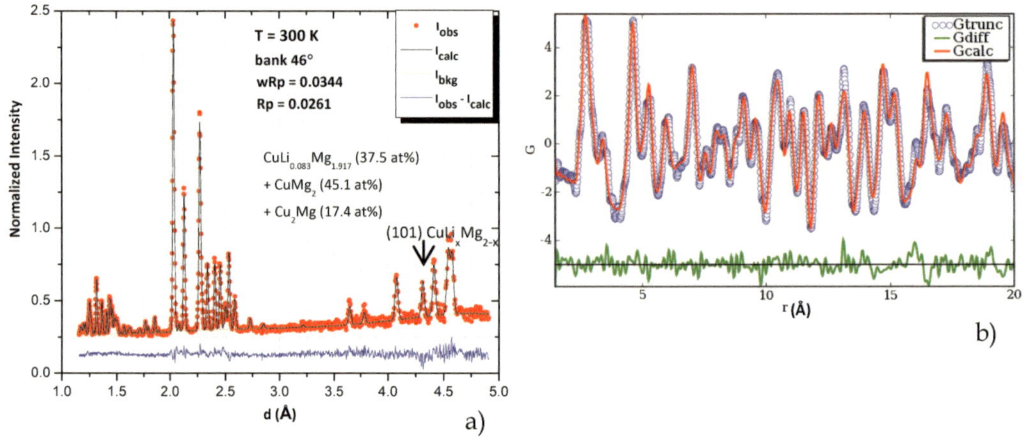

Fig. 5. a) Neutron diffraction pattern of a sample containing $CuLi_{0.08}Mg_{1.92}$, $CuMg_2$ and $CuMg_2$. The highlighted peak corresponds to the (101) reflection for the $CuLi_{0.08}Mg_{1.92}$ compound which is not overlapped by other phases. b) Pair Distribution Function (PDF) fitting for the same conditions of the pattern in a).

Furthermore, we've fitted the NPDF data using the Pair Distribution Function (PDF) in which G(r) was obtained via the Fourier Transform of the total diffraction pattern as indicated below:

$$G(r) = 4\pi r\left[\rho(r) - \rho_0\right] = \frac{2}{r}\int_0^\infty Q\left[S(Q) - 1\right]\sin(Qr)\,dQ \tag{9}$$

where $\rho(r)$ is the microscopic pair density, ρ_0 is the average atomic number density, and r the radial distance. Q is the momentum transfer $(Q = 4\pi\sin(\theta)/\lambda)$. $S(Q)$ is the normalized structure function determined from the experimental diffraction intensity (Egami & Billinge, 2003). PDF yields the probability of finding pairs of atoms separated by a distance r. PDF fittings were performed using the software PDFgui (Farrow et al., 2007).

Besides Neutron Diffraction, we have used theoretical complementary methods to determine the stoichiometry of the $CuLi_xMg_{2-x}$ compound. We relied on the Density Functional Theory (DFT) (Hohenberg & Kohn, 1964) to calculate the structure that minimized the Electronic Energy at 0 K, without accounting for the zero point energy. The latter energy gives us a good estimation of the Enthalpy of Formation at 0 K especially since we were relating data for stoichiometries that did not differ too much and for similar crystal structures. The results obtained are in close agreement with those obtained from ND after Rietveld refinement - $CuLi_xMg_{2-x}$ (x = 0.08). Nonetheless, no conclusions about Li site occupancies could be drawn from the use of the referred means. DFT shows that there isn't a clear preference, in terms of energy, for the different Li site occupancies. Then again, a technique that gives information about the average site occupancies - like the Rietveld refinement - is also inadequate to clarify this problem; therefore we have used PDF to determine Li preferred sites. With PDF fittings we were allowed to go further (see Fig. 5b). PDF does not see the average but the local structure and with PDF, all results but those in which Li would substitute Mg1 sites (1/2, 0, z), gave negative occupancies for Li. For Li substituting Mg1 we've obtained an average composition for $CuLi_xMg_{2-x}$ (x = 0.07) which is in agreement with the other obtained results. For further information please see (Braga et al., 2010c).

2.2 Hydrogen storage in the Cu-Li-Mg-H(D) system

To study the hydrogen storage in the Cu-Li-Mg system several techniques were used (Braga et al., 2010a). Besides absorption/desorption, Differential Scanning Calorimetry, Thermal Gravimetry Analysis (DSC/TGA), X-ray Diffraction (XRD) both at the laboratory and at the Synchrotron, we have used Neutron Diffraction and Inelastic Neutron Scattering. Owing to the low X-ray scattering power of hydrogen, neutron diffraction experiments on deuterides are necessary as previously highlighted in section 1.5.

Most atomic arrangements were determined on powders of different samples yet we have also used a bulk sample machined into a cylinder to obtain ND data in both the surface and the center of the sample during deuterium uptake.

The data were usually analyzed by the Rietveld method, yet in some cases in which the background was noisier we have used the biased method (Larson & von Dreele, 2004). For better convergence, the number of refined parameters in particular those of the atomic displacement amplitudes are reduced by constraints.

ND results obtained from the High-Intensity Powder Diffractometer HIPD at LANSCE, Los Alamos National Laboratory, for a sample initially containing 78 wt.% $CuLi_{0.08}Mg_{1.92}$ + 22 wt.%Cu_2Mg (from here on "initially containing" means before hydrogen/deuterium absorption) and that was deuterated *ex situ* at 473 K at P ≤ 50 atm in order to determine the crystal structure of the first deuteride phase formed in the sample (see Fig. 6 left). This pattern was refined using Rietveld's method.

The $CuLi_{0.08}Mg_{1.92}D_5$ crystal structure was determined to be monoclinic P121, with a = 15.14 Å, b = 6.88 Å, c = 5.55 Å and β = 91.73° according to the formula $CuLi_{0.08}Mg_{1.92}D_5$ = $0.5(Mg_3^{2+}.[CuD_4]_2^{3-}.MgD_2)$ corresponding to 4.4 wt% D per formula unit. $CuLi_{0.08}Mg_{1.92}D_5$ is the first deuteride/hydride to be formed. This result is interesting by itself, but the presence of MgD_2 in the diffraction pattern, highlights even further the possibilities of applications of this compound. According to these results, it can be obtained $MgH(D)_2$ from a sample that did not contain "free" Mg or $CuMg_2$. Furthermore, the deuteration process occurred at 473 K, which is considerably lower than the hydrogen absorption temperature reported for $CuMg_2$ (4) (Reilly & Wiswall, 1967).

The experiments with the bulk sample at SMARTS, LANSCE, Los Alamos National Laboratory, show that before MgD_2 is observed, $CuLi_{0.08}Mg_{1.92}D_5$ is already distinguishable at the surface even in a sample that initially contained $CuMg_2$ (see Fig. 7). Therefore, it

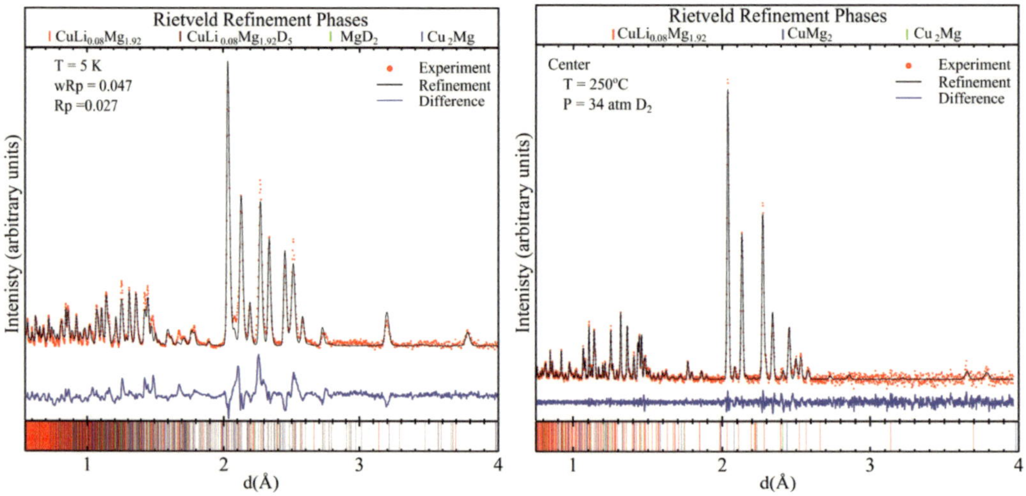

Fig. 6. (left) Rietveld refinement of a sample containing $CuLi_{0.08}Mg_{1.92}$, Cu_2Mg, MgD_2 and $CuLi_{0.08}Mg_{1.92}D_5$ obtained in HIPD. wRp and Rp are the reliability factors as defined in (Larson & von Dreele, 2004). (right) ND pattern of the center of a bulk cylinder sample containing $CuLi_{0.08}Mg_{1.92}$, Cu_2Mg, $CuMg_2$ obtained from the Spectrometer for Materials Research at Temperature and Stress SMARTS during an *in situ* reaction with D_2 at 523 K and ~34 atm. Both patterns show experimental, refined and difference between experimental and calculated intensities.

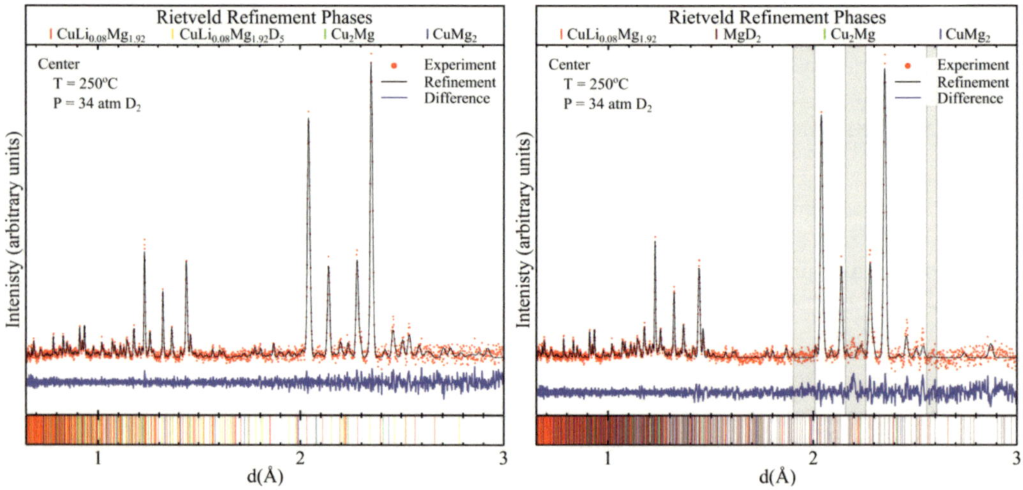

Fig. 7. (left and right) ND pattern of the surface of a bulk cylinder sample initially containing $CuLi_{0.08}Mg_{1.92}$, Cu_2Mg, and $CuMg_2$ obtained in SMARTS during an *in situ* reaction with D_2 at 523 K and ~34 atm. (right) it is highlighted that MgD_2 cannot justify some existing peaks. Both patterns show experimental, refined and difference between experimental and calculated intensities.

seems that $CuLi_{0.08}Mg_{1.92}D_5$ will have a catalytic and destabilizing roll that was additionally observed with the Ti/TiH_2 systems (Braga et al., 2010b).

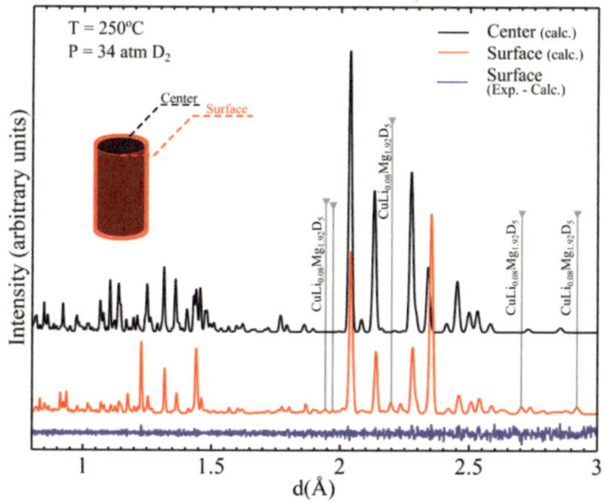

Fig. 8. ND refined pattern of the center and surface of a bulk cylinder sample initially containing $CuLi_{0.08}Mg_{1.92}$, Cu_2Mg, and $CuMg_2$ obtained in SMARTS during an *in situ* reaction with D_2 at 523 K and ~34 atm. Besides the texture effect that might be present in a bulk sample, it seems that the center initially contained more $CuLi_{0.08}Mg_{1.92}$ than the surface.

Experimental information about the metal-hydrogen interactions can be obtained by measuring lattice vibrations via INS, as previously highlighted in section 1.5. Because of the large difference between the masses of metal and H atoms in transition-metal-hydrogen systems, the acoustic dispersion branches of the phonon spectra can be attributed to the motion of the metal atoms, the optic branches to the vibrations of the light H atoms relative to the metal lattice. The densities of states of optic phonons typically show a pronounced maximum at the energy of the lattice vibrations at the Γ point in the centre of the phonon Brillouin zone (q = 0) e.g. in (Fukay, 1993). These phonon modes describe the vibration of the undistorted H sublattice relative to the rigid metal sublattice. Hence, they contain the metal-hydrogen interaction only. This is usually stronger than the H-H interaction, which leads to the dispersion of the optic branches. In the limit of very low H concentrations, the H vibrations can be imagined as independent vibrations of local Einstein oscillators at interstitial H sites. For both the lattice vibrations at the Γ point and the local vibrations, one can observe transitions from the ground state, the quantum-mechanical zero-point vibration of the H atoms, to the first excited states, e.g. by measuring optic phonon excitations, and transitions to higher excited states. Their energies, intensities and symmetry splittings yield an insight into the shapes of the potential and the wavefunctions for the vibrations of the light particles (Elsasser et al., 1998).

We have measured samples of the Cu-Li-Mg-H system by means of INS measured with the Filter Difference Spectrometer FDS, at LANSCE, Los Alamos National Laboratory. There is no doubt about the sequence of events; first there is the formation of $CuLi_{0.08}Mg_{1.92}H_5$ (see Fig. 9a) and then in subsequent cycles the formation of MgH_2 (see Fig. 9b), either for disproportionation of $CuLi_{0.08}Mg_{1.92}H_5$ or from hydrogenation of $CuMg_2$.

DSC/TGA experiments show that $CuLi_{0.08}Mg_{1.92}H_5$ starts desorbing hydrogen at 313 K to 328 K. In this range of temperatures the sample can release up to 1.3 wt.% (results for a isothermal experiment with a sample initially containing approximately 78 wt.% of $CuLi_{0.08}Mg_{1.92}$ and 22 wt.% of Cu_2Mg - which does not absorb hydrogen at the temperature and pressure that were used). In Fig. 10 it can be seen that 0.5 wt.% of a sample initially containing 61 wt.%

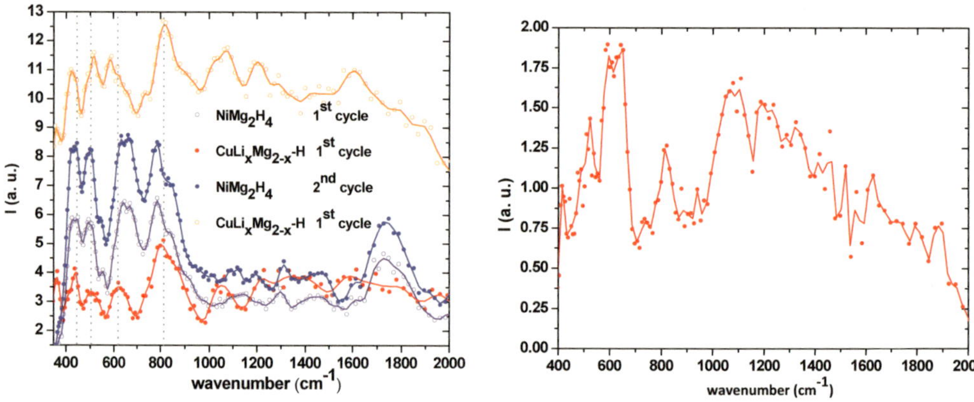

a) b)

Fig. 9. a) INS spectra for $NiMg_2H_4$ (1st and 2nd hydrogenation cycles) and for two samples containing $CuLi_{0.08}Mg_{1.92}$ (close circles below correspond to a sample that also contained Cu_2Mg and the open circles above correspond to sample that contained Cu_2Mg and $CuMg_2$ as well). All samples show the formation of a similar monoclinic structure. As in $NiMg_2H_4$, in which Ni is bonded to four atoms of H forming the tetrahedral complex $[NiH_4]^{4-}$, Cu is also bonded to four atoms of H forming the tetrahedral complex $[CuH_4]^{3-}$, which was previously referred on (Yvon & Renaudin, 2005). b) Sample initially containing approximately 61 wt.% of $CuLi_{0.08}Mg_{1.92}$, 23 wt.% of $CuMg_2$ and 16 wt.% of Cu_2Mg, after the 3rd hydrogenation cycle at 473 K, and ~50 atm. It is clear the formation of MgH_2 with a peak at ~620 cm^{-1}.

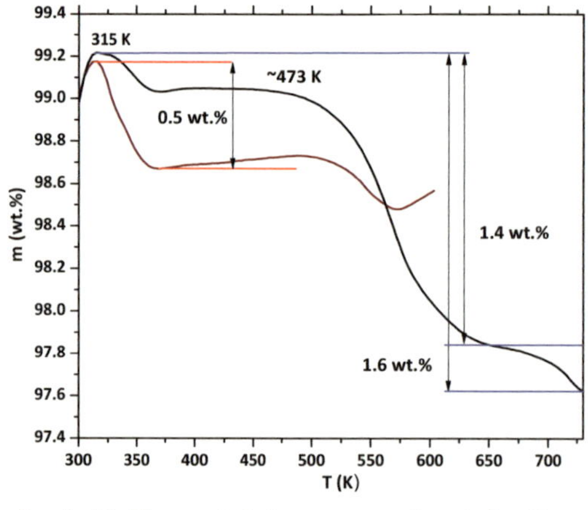

Fig. 10. TGA of two samples initially containing approximately 61 wt.% of $CuLi_{0.08}Mg_{1.92}$, 23 wt.% of $CuMg_2$ and 16 wt.% of Cu_2Mg. Samples were measured after hydrogenation but they are not expected to be saturated in hydrogen prior to the experiment.

of $CuLi_{0.08}Mg_{1.92}$, 23 wt.% of $CuMg_2$ and 16 wt.% of Cu_2Mg can be released at T < 350 K. In spite of the fact that there was some visible oxidation during this run, we think it was worth showing this initial desorption. This initial desorption seems to be due to $CuLi_{0.08}Mg_{1.92}H_5$. At ~473 K, hydrogen starts to be desorbed at a different rate, probably due to the disproportionation of $CuLi_{0.08}Mg_{1.92}H_5$, with the formation of MgH_2, which will start releasing hydrogen at 553-573 K. Additionally, MgH_2 can be formed upon hydrogenation of $CuMg_2$.

The DSC/TGA results show that the system containing $CuLi_{0.08}Mg_{1.92}$ and Cu_2Mg can destabilize MgH_2 in a more efficient way than Cu_2Mg by itself can. In fact, in a DSC experiment in which kinetics must be accounted for, MgH_2 will release hydrogen at 553-573 K, which can only be obtained when particles are reduced to nanopowders.

2.3 Hydrogen storage in the Cu-Li-Mg-H(D)+Ti system

A sample with 60.5 at% of $CuLi_{0.08}Mg_{1.92}$, 23.9 at% of $CuMg_2$ and 15.6 at% of Cu_2Mg was mechanically alloyed to titanium resulting into Cu-Li-Mg+Ti samples. The brittle CuLiMg alloy was mixed with Ti (99.9% purity, 325 mesh, Alfa Aesar) so that 68.2 at% / 47.3 wt% of the final mixture was Ti. The mixture was ball-milled for 3 h in a dry box under a He protective atmosphere. The Cu-Li-Mg+Ti mixture was then sealed inside a stainless steel crucible and kept at 473 K for 9h under D_2 at P = 34 bar. These samples were then cooled to 5 K (HIPD, neutron powder diffraction) over a period of 2 to 3 hours.

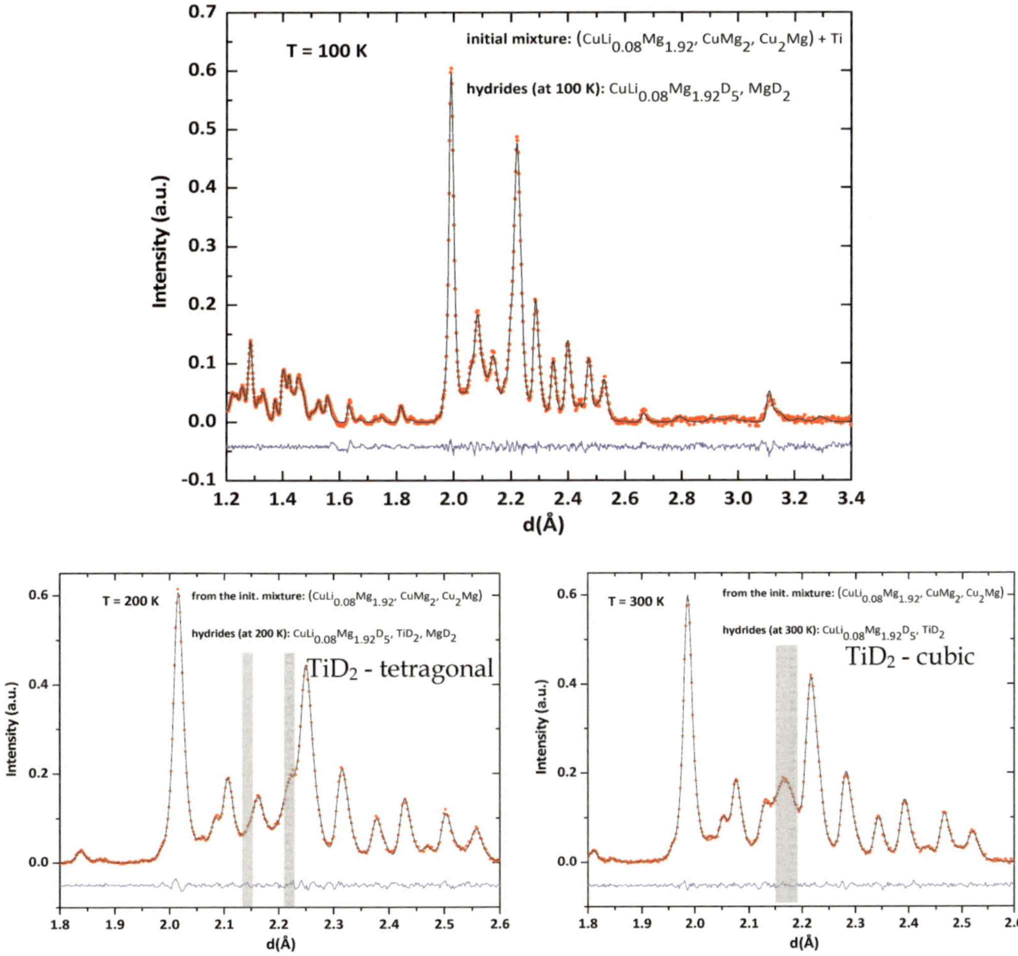

Fig. 11. Comparison of Cu-Li-Mg+Ti-D neutron diffraction pattern refinements from measurements taken at T= 100K, 200K, and 300K.

Cu-Li-Mg+Ti-D neutron diffraction samples were quenched and measured at the lowest temperature, 5K, first with 60K, 100K, 200K and 300K measurements taken subsequently. For diffraction patterns collected at 5K - 100K there is *no evidence* of any titanium deuteride phases. This is immediately obvious in Fig. 11 which shows a comparison of the 100, 200 and 300 K Cu-Li-Mg+Ti-D. Bragg peaks belonging to tetragonal TiD_2 begin to appear in the 200K pattern of the Cu-Li-Mg+Ti-D sample. The refinement at 100 K includes the 5 phases: $CuMg_2$, Cu_2Mg, $CuLi_{0.08}Mg_{1.92}$, $CuLi_{0.08}Mg_{1.92}D_5$ and α-Ti; at 200 K: $CuMg_2$, Cu_2Mg, $CuLi_{0.08}Mg_{1.92}$, $CuLi_{0.08}Mg_{1.92}D_5$ and MgD_2 plus tetragonal TiD_2 and at 300 K: $CuMg_2$, Cu_2Mg, $CuLi_{0.08}Mg_{1.92}$, $CuLi_{0.08}Mg_{1.92}D_5$ plus cubic TiD_2 and represents an excellent fit to the data as confirmed by the residual. At 300K a structural phase transition (tetragonal to cubic) occurs in TiD_2 (Yakel, 1958). The appearance of the TiD_2 cubic phase in the 300K diffraction pattern is confirmed by refinement as shown in Fig. 11.

Fig. 11 shows the changes in the Cu-Li-Mg+Ti-D sample as the temperature is increased from 100K to 300K: The $CuLi_{0.08}Mg_{1.92}D_5$, MgD_2 and α-Ti phases are progressively reduced in intensity while TiD_2 appears at 200K transforming from tetragonal to cubic at 300K. Since there was no further exposure of the sample to deuterium, the formation of TiD_2 must have been facilitated through *solid state diffusion* of deuterium from a separate phase. The decrease in intensity of Bragg peaks belonging to the Cu-Li-Mg-D phase implicates that the mechanism involves solid state transfer of deuterium from Cu-Li-Mg-D to Ti.

DSC/TGA measurements show the dehydrogenation of $CuLi_{0.08}Mg_{1.92}H_5$ accounts for approximately third of the total mass loss. Given that no other hydride phases were present in any significant quantity, and that MgH_2 began dehydrogenation at 553 K in the Cu-Li-Mg-H samples, the mass loss beginning at 590 K is due to the release of hydrogen from TiH_2. This demonstrates a significant catalytic effect for desorption as well given that TiH_2 ordinarily does not dissociate until well above 723 K (Gibb & Kruschwitz, 1950). For further information please check (Braga et al., 2010b).

3. Conclusion

The hydrogen storage world still offers a considerable amount of challenges since no universal solution has been found. Eventually, different solutions will be proposed to suite different applications.

The Cu-Li-Mg system provides other possibilities for catalytic and destabilization effects yet not fully explored.

There are several techniques that can be employed to study systems containing hydrogen. Nonetheless, Neutron Scattering is a very useful resource, in particular, Neutron Diffraction. In the latter, crystal structure of deuteride phases are directly studied since deuterium can be detected by ND and accurate results can be obtained either in *ex situ* or *in situ* experiments as shown previously.

4. Acknowledgments

The authors would like to acknowledge FCT - Portugal and FEDER - EU, for the PTDC/CTM/099461/2008 project. This work has benefited from the use of HIPD, NPDF, SMARTS and FDS at LANSCE, LANL, funded by DOE, DE-AC52-06NA25396. The authors would like to acknowledge the Lujan Center's, LANSCE, instrument scientists for their support and helpful discussions.

5. References

Aguey-Zinsou, K. & Ares-Fernández, J. (2010). Hydrogen in magnesium: new perspectives toward functional stores. *Energy & Environmental Science,* Vol. 3, No. 5, (February 2010), pp. 526-543, ISSN 1754-5706

Alapati, S., Johnson, J. & Sholl, D. (2007). Using first principles calculations to identify new destabilized metal hydride reactions for reversible hydrogen storage. *Physical Chemistry Chemical Physics,* Vol. 9, No. 12, (February 2007), pp. 1438-1452, ISSN 1463-9084

Andreasen, A., Sørensen, M., Burkarl, R., Møller, B., Molenbroek, A., Pedersen, A., Vegge, T. & Jensen, T. (2006). Dehydrogenation kinetics of air-exposed MgH_2/Mg_2Cu and $MgH_2/MgCu_2$ studied with in situ X-ray powder diffraction. *Applied Physics A,* Vol. 82, No. 3, (February 2006), pp. 515-521, ISSN 1432-0630

Ansara, I., Dinsdale, A. & Rand, M. (Ed(s)). (1998). *COST 507, Thermochemical database for light metal alloys,* Vol. 2, pp. 170-174, Office for Official Publications of the European Communities, ISBN 92-828-3902-8, Luxembourg

Beattie, S., Setthanan, U. & McGrady, G. (2011). Thermal desorption of hydrogen from magnesium hydride (MgH2): An in situ microscopy study by environmental SEM and TEM. *International Journal of Hydrogen Energy,* Vol. 36, No. 10, (May 2011), pp. 6014-6021, ISSN 0360-3199

Bösenberg, U., Ravnsbæk, D., Hagemann, H., D'Anna, V., Minella, B., Pistidda, C., Beek, W., Jensen, T., Bormann, R. & Dornheim, M. (2010). Pressure and Temperature Influence on the Desorption Pathway of the $LiBH_4-MgH_2$ Composite System. *The Journal of Physical Chemistry C,* Vol. 114, No. 35, (August 2010), pp. 15212-15217, ISSN 1932-7455

Braga, M., Acatrinei, A., Hartl, M., Vogel, S., Proffen, T. & Daemen, L. (2010a). New Promising Hydride Based on the Cu-Li-Mg system. *Journal of Physics: Conference Series,* Vol. 251, (December 2010), pp. 012040 [4 pages], ISSN 1742-6596

Braga, M., Wolverton, M., Llobet, A. & Daemen, L. (2010b). Neutron Scattering to Characterize Cu/Mg(Li) Destabilized Hydrogen Storage Materials. *Materials Research Society Symposium Proceedings,* Vol. 1262, pp. 1262-W03-05 (April 2010), ISSN 0272-9172

Braga, M., Ferreira, J., Siewenie, J., Proffen, Th., Vogel, S. & Daemen, L. (2010c). Neutron powder diffraction and first-principles computational studies of $CuLi_xMg_{2-x}$ ($x \cong 0{,}08$), $CuMg_2$, and Cu_2Mg. *Journal of Solid State Chemistry,* Vol. 183, No. 1, (January 2010), pp. 10-19, ISSN 0022-4596

Braga, M. & Malheiros, L. (2007a). $CuMg_{2-Y}Li_X$ alloy for hydrogen storage. International patent, WO 2007046017 (A1)

Braga, M. & Malheiros, L. (2007b). $CuMg_{2-Y}Li_X$ alloy for hydrogen storage. National patent, PT 103368 (A)

Braga, M., Ferreira, J. & Malheiros, L. (2007). A ternary phase in Cu–Li–Mg system. *Journal of Alloys and Compounds,* Vol. 436, No. 1-2, (June 2007), pp. 278-284, ISSN 0925-8388

Cermak, J. & David, B. (2011). Catalytic effect of Ni, Mg_2Ni and Mg_2NiH_4 upon hydrogen desorption from MgH_2. *International Journal of Hydrogen Energy,* Vol. 36, No. 21, (October 2011), pp. 13614-13620, ISSN 0360-3199

Chen, P. & Zhu, M. (2008). Recent progress in hydrogen storage. *Materials Today,* Vol. 11, No. 12, (December 2008), pp. 36-43, ISSN 1369-7021

Chen, P., Xiong, Z., Luo, J., Lin, J. & Tan, K. (2003). Interaction between lithium amide and lithium hydride. *The Journal of Physical Chemistry B,* Vol. 107, No. 37, (September 2003), pp. 10967-10970, ISSN 1520-5207

Choi, Y., Hu, J., Sohn, H. & Fang, Z. (2008). Hydrogen storage properties of the Mg–Ti–H system prepared by high-energy–high-pressure reactive milling. *Journal of Power Sources,* Vol. 180, No. 1, (May 2008), pp. 491-497, ISSN 0378-7753

Churchard, A., Banach, E., Borgschulte, A., Caputo, R., Chen, J., Clary, D., Fijalkowski, K., Geerlings, H., Genova, R., Grochala, W., Jaroń, T., Juanes-Marcos, J., Kasemo, B., Kroes, G., Ljubić, I., Naujoks, N., Nørskov, J., Olsen, R., Pendolino, F., Remhof, A., Románszki, L., Tekin, A., Vegge, T., Zäch, M., & Züttelc, A. (2011). A multifaceted approach to hydrogen storage. *Physical Chemistry Chemical Physics,* Vol. 13, No. 38, (September 2011), pp. 16955-16972, ISSN 1463-9084

Cui, S., Feng, W., Hu, H., Feng, Z. & Wang, Y. (2008). Structural phase transitions in MgH_2 under high pressure. *Solid State Communications,* Vol. 148, No. 9-10, (December 2008), pp. 403-405, ISSN 0038-1098

David, W., Jones, M., Gregory, D., Jewell, C., Walton, A. & Edwards, P. (2007). A mechanism for non-stoichiometry in the lithium amide/lithium imide hydrogen storage reaction. *Journal of the American Chemical Society,* Vol. 129, No. 6, (February 2007), pp. 1594-1601, ISSN 0002-7863

Egami, T. & Billinge, S. (2003). *Underneath the Bragg-Peaks: Structural Analysis of Complex Materials* (First edition), Pergamon Press, Elsevier Ltd, ISBN 0-08-042698-0, Amsterdam

Elsasser, C., Krimmel, H., Fahnle, M., Louie, S. & Chan, C. (1998). *Ab initio* study of iron and iron hydride: III. Vibrational states of H isotopes in Fe, Cr and Ni. *Journal of Physics: Condensed Matter,* Vol. 10, No. 23, (June 1998), pp. 5131 [16 pages], ISSN 0953-8984

European Commission (2003). World energy, technology and climate policy outlook 2030 – WETO, in: *Directorate-General for Research Energy, EUR 20366,* Available from: http://ec.europa.eu/research/energy/pdf/weto_final_report.pdf

Farrow, C., Juhas, P., Liu, J., Bryndin, D., Božin, E., Bloch, J., Proffen, T. & Billinge, S. (2007). PDFfit2 and PDFgui: computer programs for studying nanostructure in crystals. *Journal of Physics: Condensed Matter,* Vol. 19, No. 33, (July 2007), pp. 335219 [7 pages], ISSN 0953-8984

Felderhoff, M., Weidenthaler, C., von Helmoltb, R. & Eberleb, U. (2009). Hydrogen storage: the remaining scientific and technological challenges. *Physical Chemistry Chemical Physics,* Vol. 9, No. 21, (May 2007), pp. 2643-2653, ISSN 1463-9084

Fernández, A., Deprez, E. & Friedrichs, O. (2011). A comparative study of the role of additive in the MgH_2 vs. the $LiBH_4$–MgH_2 hydrogen storage system. *International Journal of Hydrogen Energy,* Vol. 36, No. 6, (March 2011), pp. 3932-3940, ISSN 0360-3199

Fukai, Y. (1993). *The Metal-Hydrogen System: Basic Bulk Properties* (First edition), Springer-Verlag, ISBN 3540-556370, Berlin

Gibb, T. & Kruschwitz, H. (1950). The Titanium-Hydrogen System and Titanium Hydride. I. Low-Pressure Studies. *Journal of American Chemical Society*, Vol. 72, No. 12 pp. 5365-5369

Grochala, W. & Edwards, P. (2004). Thermal Decomposition of the Non-Interstitial Hydrides for the Storage and Production of Hydrogen. *Chemical Reviews,* Vol. 104, No. 3, (March 2004), pp. 1283-1315, ISSN 0009-2665

Hlukhyy, V., Rodewald, U. & Pöttgen, R. (2005). Magnesium-Tin Substitution in $NiMg_2$. *Zeitschrift für anorganische und allgemeine Chemie,* Vol. 631, No. 15, (November 2005), pp. 2997–3001, ISSN 1521-3749

Hohenberg, P. & Kohn, W. (1964). Inhomogeneous Electron Gas. *Physical Review B*, Vol. 136, No. 3B, (November 1964) pp. B864-B871, ISSN 1098-0121

Hong, S., Bae, J., Kwon, S. & Song, M. (2011). Hydrogen storage properties of Mg-23.5Ni-xCu prepared by rapid solidification process and crystallization heat treatment. *International Journal of Hydrogen Energy,* Vol. 36, No. 3, (February 2011), pp. 2170-2176, ISSN 0360-3199

Hsu, F., Hsu, C., Chang, J., Lin, C., Lee, S. & Jiang, C. (2010). Structure and hydrogen storage properties of $Mg_2Cu_{1-x}Ni_x$ (x = 0–1) alloys. *International Journal of Hydrogen Energy,* Vol. 35, No. 24, (December 2010), pp. 13247-13254, ISSN 0360-3199

Huot, J., Liang, G., Boily, S., Neste, A. & Schulz, R. (1999). Structural study and hydrogen sorption kinetics of ball-milled magnesium hydride. *Journal of Alloys and Compounds,* Vol. 293-295, (December 1999), pp. 495-500, ISSN 0925-8388

Jain, I., Lal, C. & Jain, A. (2010). Hydrogen storage in Mg: A most promising material. *International Journal of Hydrogen Energy,* Vol. 35, No. 10, (May 2010), pp. 5133-5144, ISSN 0360-3199

Jain, I. (2009). Hydrogen the fuel for 21st century. *International Journal of Hydrogen Energy,* Vol. 34, No. 17, (September 2009), pp. 7368-7378, ISSN 0360-3199

Jeon, K., Moon, H., Ruminski, A., Jiang, B., Kisielowski, C., Bardhan, R. & Urban, J. (2011). Air-stable magnesium nanocomposites provide rapid and high-capacity hydrogen storage without using heavy-metal catalysts. *Nature Materials,* Vol. 10, No. 4, (April 2011), pp. 286-290, ISSN 1476-4660

Jurczyk, M., Nowak, M., Szajek, A. & Jezierski, A. (2011). Hydrogen storage by Mg-based nanocomposites. *International Journal of Hydrogen Energy,* (In press - available online 27 April 2011), doi:10.1016/j.ijhydene.2011.04.012, ISSN 0360-3199

Jurczyk, M., Okonska, I., Iwasieczko, W., Jankowska, E. & Drulis H. (2007). Thermodynamic and electrochemical properties of nanocrystalline Mg_2Cu-type hydrogen storage materials. *Journal of Alloys and Compounds,* Vol. 429, No. 1-2, (February 2007), pp. 316-320, ISSN 0925-8388

Kalinichenka, S., Röntzsch, L., Riedl, T., Gemming, T., Weißgärber, T. & Kieback, B. (2011). Microstructure and hydrogen storage properties of melt-spun Mg–Cu–Ni–Y alloys. *International Journal of Hydrogen Energy,* Vol. 36, No. 2, (January 2011), pp. 1592-1600, ISSN 0360-3199

Larson, A., von Dreele, R. (2004). GSAS Generalized Structure Analysis System, LANSCE, Los Alamos.

Massa, W. (2008). *Crystal Structure determination.* Springer-Verlag, Berlin Heidelberg. ISBN 978-3540206446

Milanese, J., Girella, A., Garroni, S., Bruni, G., Berbenni, V., Matteazzi, P. & Marini, A. (2010a). Effect of C (graphite) doping on the H_2 sorption performance of the Mg–Ni storage system. *International Journal of Hydrogen Energy,* Vol. 35, No. 3, (February 2010), pp. 1285-1295, ISSN 0360-3199

Milanese, C., Girella, A., Bruni, G., Cofrancesco, P., Berbenni, V., Matteazzi, P. & Marini, A. (2010b). Mg-Ni-Cu mixtures for hydrogen storage: A kinetic study. *Intermetallics,* Vol. 18, No. 2, (February 2010), pp. 203-211, ISSN 0966-9795

Milanese, C., Girella, A., Bruni, G., Cofrancesco, P., Berbenni, V., Villa, M., Matteazzi, P. & Marini, A. (2008). Reactivity and hydrogen storage performances of magnesium–nickel–copper ternary mixtures prepared by reactive mechanical grinding. *International Journal of Hydrogen Energy,* Vol. 33, No. 17, (September 2008), pp. 4593-4606, ISSN 0360-3199

Noritake, T., Towata, S., Aoki, M., Seno, Y., Hirose, Y., Nishibori, E., Takata, M. & Sakata, M. (2003). Charge density measurement in MgH_2 by synchrotron X-ray diffraction. *Journal of Alloys and Compounds,* Vol. 356-357, (August 2003), pp. 84-86, ISSN 0925-8388

Orimo, S. & Fujii, H. (2001). Materials science of Mg-Ni-based new hydrides. *Applied Physics A,* Vol. 72, No. 2, (April 2001), pp. 167-186, ISSN 1432-0630

Ramirez-Cuesta, A. (2004). aCLIMAX 4.0. 1, The new version of the software for analyzing and interpreting INS spectra. *Computer Physics Communications.* Vol. 157, No. 3, pp. 226-238.

Ravindran, P., Vajeeston, P., Fjellvåg, H. & Kjekshus, A. (2004). Chemical-bonding and high-pressure studies on hydrogen-storage materials. *Computational Materials Science,* Vol. 30, No. 3-4, (August 2004), pp. 349-357, ISSN 0927-0256

Reilly, J. & Wiswall, R. (1967). The Reaction of Hydrogen with Alloys of Magnesium and Copper. *Inorganic chemistry,* Vol. 6, No. 12, (December 1967), pp. 2220-2223, ISSN 0020-1669

Reilly, J. & Wiswall, R. (1968). The Reaction of Hydrogen with Alloys of Magnesium and Nickel and the Formation of Mg_2NiH_4. *Inorganic chemistry,* Vol. 7, No. 11, (November 1968), pp. 2254-2256, ISSN 0020-1669

Reule, H., Hirscher, M., Weißhardt, A. & Kronmüller, H. (2000). Hydrogen desorption properties of mechanically alloyed MgH_2 composite materials. *Journal of Alloys and Compounds,* Vol. 305, No. 1-2, (June 2000), pp. 246-252, ISSN 0925-8388

Ross, D. (2008). Neutron scattering studies for analysing solid state hydrogen storage, in: *Solid State Hydrogen Storage Materials and Chemistry* Walker, G. Ed. CRC Woodhead Publishing: Cambridge, England, pp. 135-172. ISBN 9781845692704

Rude, L., Nielsen, T., Ravnsbæk, D., Bösenberg, U., Ley, M., Richter, B., Arnbjerg, L., Dornheim, M., Filinchuk, Y., Besenbacher, F. & Jensen, T. (2011). Tailoring properties of borohydrides for hydrogen storage: A review. *Physica Status Solidi A,* Vol. 208, No. 8, (July 2011), pp. 1754-1773, ISSN 1862-6300

Sabitu, S., Gallo, G. & Goudy, A. (2010). Effect of TiH_2 and Mg_2Ni additives on the hydrogen storage properties of magnesium hydride. *Journal of Alloys and Compounds,* Vol. 499, No. 1, (June 2010), pp. 35-38, ISSN 0925-8388

Schimmel, H., Huot, J., Chapon, L., Tichelaar, F. & Mulder, F. (2005). Hydrogen Cycling of Niobium and Vanadium Catalyzed Nanostructured Magnesium. *Journal of the*

American Chemical Society, Vol. 127, No. 41, (September 2005), pp. 1438-14354, ISSN 0002-7863

Schlapbach, L. & Züttel, A. (2001). Hydrogen-storage materials for mobile applications. *Nature,* Vol. 414, (November 2001), pp. 353-358, ISSN 0028-0836

Senegas, J., Mikou, A., Pezat, M. & Darriet, B. (1984). Localisation et diffusion de l'hydrogene dans le systeme Mg_2Ni-H_2: Etude par RMN de $Mg_2NiH_{0,3}$ et Mg_2NiH_4. *Journal of Solid State Chemistry,* Vol. 52, No. 1, (March 1984), pp. 1-11, ISSN 0022-4596

Setten, M., Wijs, G. & Brocks, G. (2007). Ab initio study of the effects of transition metal doping of Mg_2NiH_4. *Physical Review B,* Vol. 76, (August 2007), pp. 075125 [8 pages], ISSN 1098-0121

Siegel, D., Wolverton, C. & Ozoliņš, V. (2007). Thermodynamic guidelines for the prediction of hydrogen storage reactions and their application to destabilized hydride mixtures. *Physical Review B,* Vol. 76, No. 13, (October 2007), pp. 134102 [6 pages], ISSN 1098-0121

Simičić, M., Zdujić, M., Dimitrijević, R., Nikolić-Bujanović, L. & Popović N.(2006). Hydrogen absorption and electrochemical properties of Mg_2Ni-type alloys synthesized by mechanical alloying. *Journal of Power Sources,* Vol. 158, No. 1, (July 2006), pp. 730-734, ISSN 0378-7753

Sohn, H. & Emami, S. (2011). Kinetics of dehydrogenation of the Mg–Ti–H hydrogen storage system. *International Journal of Hydrogen Energy,* Vol. 36, No. 14, (July 2011), pp. 8344-8350, ISSN 0360-3199

Squires, G. (1978). *Introduction to the theory of Thermal Neutron Scattering.* Dover Publications Inc., Mineola, New York, ISBN 978-0486694474

Suryanarayana, C. (2008). Recent developments in mechanical alloying. *Reviews on Advanced Materials Science,* Vol. 18, No. 3, (August 2008), pp. 203-211, ISSN 1605-8127

Tan, Z., Chiu, C., Heilweil, E. & Bendersky, L. (2011a). Thermodynamics, kinetics and microstructural evolution during hydrogenation of iron-doped magnesium thin films. *International Journal of Hydrogen Energy,* Vol. 36, No. 16, (August 2011), pp. 9702-9713, ISSN 0360-3199

Tan, X., Danaie, M., Kalisvaart, W. & Mitlin, D. (2011b). The influence of Cu substitution on the hydrogen sorption properties of magnesium rich Mg-Ni films. *International Journal of Hydrogen Energy,* Vol. 36, No. 3, (February 2011), pp. 2154-2164, ISSN 0360-3199

Vajo, J. (2011). Influence of nano-confinement on the thermodynamics and dehydrogenation kinetics of metal hydrides. *Current Opinion in Solid State & Materials Science,* Vol. 15, No. 2, (April 2011), pp. 52-61, ISSN 1359-0286

Vajo, J. & Olson, G. (2007). Hydrogen storage in destabilized chemical systems. *Scripta Materialia,* Vol. 56, No. 10, (May 2007), pp. 829-834, ISSN 1359-6462

Vajo, J., Mertens, F., Ahn, C., Bowman Jr., R. & Fultz, B. (2004). Altering Hydrogen Storage Properties by Hydride Destabilization through Alloy Formation: LiH and MgH_2 Destabilized with Si. *The Journal of Physical Chemistry B,* Vol. 108, No. 37, (August 2004), pp. 13977-13983, ISSN 1520-5207

Vyas, D., Jain, P., Khan, J., Kulshrestha, V., Jain, A. & Jain, I. (2011). Effect of Cu catalyst on the hydrogenation and thermodynamic properties of Mg_2Ni. *International Journal of*

Hydrogen Energy, (In press - available online 20 July 2011), doi:10.1016/j.ijhydene.2011.05.143, ISSN 0360-3199

Xia, G., Leng, H., Xu, N., Li, Z., Wu, Z., Du, J. & Yu, X. (2011). Enhanced hydrogen storage properties of $LiBH_4$—MgH_2 composite by the catalytic effect of $MoCl_3$. *International Journal of Hydrogen Energy,* Vol. 36, No. 12, (June 2011), pp. 7128-725, ISSN 0360-3199

Yakel, H. (1958). Thermocrystallography of higher hydrides of Titanium and Zirconium. *Acta Crystallographica,* Vol. 11, pp. 45-51

Yang, J., Sudik, A. & Wolverton, C. (2007). Destabilizing $LiBH_4$ with a Metal (M) Mg, Al, Ti, V, Cr, or Sc) or Metal Hydride (MH_2) MgH_2, TiH_2, or CaH_2). *The Journal of Physical Chemistry C,* Vol. 111, No. 51, (November 2007), pp. 19134-19140, ISSN 1932-7455

Yim, C., You, B., Na, Y. & Bae, J. (2007). Hydriding properties of Mg–xNi alloys with different microstructures. *Catalysis Today,* Vol. 120, No. 3-4, (February 2007), pp. 276-280, ISSN 0920-586

Yvon, K. & Renaudin, G. (2005). Hydrides: Solid State Transition Metal Complexes, In: *Encyclopedia of Inorganic Chemistry,* R. Bruce King (Editor-in-Chief), pp. 1814–1846, John Wiley & Sons Ltd, ISBN 0-470-86078-2, Chichester

Yu, X., Grant, D. & Walker, G. (2006). A new dehydrogenation mechanism for reversible multicomponent borohydride systems—The role of Li–Mg alloys. *Chemical Communications,* No. 37, (October 2006), pp. 3906-3908, ISSN 1359-7345

Zaluska, A., Zaluski, L., & Strom-Olsen, J. (1999a). Nanocrystalline magnesium for hydrogen storage. *Journal of Alloys and Compounds,* Vol. 288, No. 1-2, (June 1999), pp. 217-225, ISSN 0925-8388

Zaluska, A., Zaluski, L., & Strom-Olsen, J. (1999b). Synergy of hydrogen sorption in ball-milled hydrides of Mg and Mg_2Ni. *Journal of Alloys and Compounds,* Vol. 289, No. 1-2, (July 1999), pp. 197-206, ISSN 0925-8388

Zaluski, L., Zaluska, A. & Strom-Olsen, J. (1997). Nanocrystalline metal hydrides. *Journal of Alloys and Compounds,* Vol. 253-254, (May 1997), pp. 70-79, ISSN 0925-8388

Zhang, J. & Hu, Y. (2011). Decomposition of Lithium Amide and Lithium Imide with and without Anion Promoter. *Industrial & Engineering Chemistry Research,* Vol. 50, No. 13, (May 2011), pp. 8058–8064, ISSN 0888-5885

Zhang, Y., Li, B., Ren, H., Guo, S., Zhao, D. & Wang, X. (2010a). Microstructure and hydrogen storage characteristics of melt-spun nanocrystalline $Mg_{20}Ni_{10-x}Cu_x$ (x=0-4) alloys. *Materials Chemistry and Physics,* Vol. 124, No. 1, (November 2010), pp. 795-802, ISSN 0254-0584

Zhang, Y., Li, B., Ren, H., Guo, S., Zhao, D. & Wang, X. (2010b). Hydrogenation and dehydrogenation behaviours of nanocrystalline $Mg_{20}Ni_{10-x}Cu_x$ (x = 0–4) alloys prepared by melt spinning. *International Journal of Hydrogen Energy,* Vol. 35, No. 5, (March 2010), pp. 2040-2047, ISSN 0360-3199

Zhang, J., Zhou, D., He, L., Peng, P. & Liu J. (2009). First-principles investigation of Mg_2Ni phase and high/low temperature Mg_2NiH_4 complex hydrides. *Journal of Physics and Chemistry of Solids,* Vol. 70, No. 1, (January 2009), pp. 32-39, ISSN 0022-3697

Zhao-Karger, Z., Hu, J., Roth, A., Wang, D., Kübel, C., Lohstroh, W. & Fichtner, M. (2010). Altered thermodynamic and kinetic properties of MgH_2 infiltrated in microporous scaffold. *Chemical Communications,* Vol. 46, No. 44, (November 2010), pp. 8353-8355, ISSN 1359-7345

2

Neutron Diffraction Measurements for Residual Stresses in AL-6XN Stainless Steel Welded Beams

Xiaohua Cheng[1,*], Henry J. Prask[2,3], Thomas Gnaeupel-Herold[2,3], Vladimir Luzin[4,**] and John W. Fisher[5]
[1]New Jersey Department of Transportation, Trenton, NJ
[2]NIST Center for Neutron Research, Gaithersburg, MD
[3]University of Maryland, College Park, MD
[4]Australian Nuclear Science & Technology Organization
[5]ATLSS Research Center, Lehigh University, Bethlehem, PA
[1,2,3,5]USA
[4]Australia

1. Introduction

The primary objective of the current study was to quantify the magnitude and distribution of residual stresses in various welded details in non-magnetic super-austenitic AL-6XN stainless steel I-beams through saw cutting and neutron diffraction methods. The welded details included transverse groove welds and simulated bulkhead attachment welded details, both of which duplicated details in previous fatigue test specimens [Fisher et al., 2001; Cheng et al., 2003a]. It was expected that the results of a study of residual stresses could be used to analyze the effect of residual stresses on fatigue strength. This Chapter presents a description of specimen materials, specimen preparation, neutron diffraction measurement plan, method and results.

2. Background, materials and testing methods

2.1 Background

Residual stresses are introduced into structural steel components during manufacturing and fabricating processes. For welded structures, weld process, weld sequence, component size and setup restraint, temperature or cooling rate difference, and material composition and properties are primary factors that affect residual stresses. Residual stresses can have significant impact on ultimate strength, stability, fatigue strength and toughness depending on their magnitude and distribution with respect to stresses from applied external loads and dead loads.

---

* formerly with ATLSS Reserach Center, Lehigh University, USA
** formerly with NIST Center for Neutron Research, USA

Since a loaded structure is subjected to both internal residual stress and stress from externally applied loads, the resultant stress should be their superposition [Cheng et al., 2003b]. Tensile residual stresses can cancel out compressive loading stresses that are favourable for fatigue strength. Once the resultant stress exceeds the yield stress, plastic flow occurs in a mild steel material (such as carbon-manganese (C-Mn) steel). Consequently, stress range is one controlling factor for fatigue strength rather than the maximum applied stress. Combined with applied stresses, for example, the residual stress distribution along a lateral weld in a beam may indicate where a fatigue crack first develops, whereas the through-thickness residual stress gradient may affect fatigue crack propagation. Therefore, studies of the magnitude and distribution of residual stresses are needed.

The AL-6XN steel in this study is a non-magnetic super-austenitic stainless steel that has potential application in ship structures which use the advanced double hull (ADH) concept due to its high strength, superior crevice corrosion resistance, low magnetic signature, excellent fracture toughness and great ductility. It is a stable, single-phase austenitic (face-centered cubic) alloy. Standard tension and compression tests showed that the material is isotropic and homogeneous. The stress-strain relationship is approximately bilinear without the yielding flow plateau that C-Mn mild steel has. Because of its greater nickel content, AL-6XN steel has about a 20% higher coefficient of thermal expansion than carbon steel (15.20 μm/m/°K vs. 12.06 μm/m/°K or 8.44 μ in./in./°F vs. 6.7 μ in./in./°F), but only 1/4 the thermal conductivity of carbon steel (12.9 W/(m·K) vs. 51.9 W/(m·K) or 89.5 Btu·in./(ft^2·hr.·°F) vs. 360 Btu·in./(ft^2·hr·°F)) [INCO, 1964; Lamb, 1999; Dudt, 2000]. Consequently, residual stresses in welded components of AL-6XN steel may be more localized and greater than those observed in carbon steels. A well-understood residual stress state is desired as a part of material characterization to evaluate fatigue strength.

The magnitude of residual stresses introduced into a material is associated with the yield strength and ultimate tensile strength of the material. For large welded structural members, the higher the material yield strength, the greater the residual stresses. For AL-6XN steel, the minimum specified 0.2% offset yield strength and tensile strength are 45 ksi (310 MPa) and 95~100 ksi (655~690 MPa), respectively, depending on the plate thickness for the AL-6XN alloy [Rolled Alloys, 1997; Lamb, 1999]. The coupon test by Lehigh University [Lu *et al.*, 2002] and the mill report provided by Rolled Alloys showed that the 0.2% offset uniaxial nominal yield strength and ultimate tensile strength were 48.1~55.8 ksi (332~385 MPa) and 108~111 ksi (745~766 MPa), respectively, for ½" (12.7 mm) thick flange plates, and 50.9~62 ksi (351~428 MPa) and 105.1~113 ksi (725~780 MPa), respectively, for 3/8" (9.5 mm) thick web plates.

For comparison, the Inconel 625 filler metal used for welding the specimens in the current study typically has a yield strength of 72.5~85 ksi (500~590 Mpa) and a tensile strength of 114~116 ksi (790~800 Mpa). The mechanical properties of the IN625 filler metal (ERNiCrMo3) should comply with AWS A5.14 and ASME SFA5.14.

The measurements in this study focused on the magnitude and on lateral and through-thickness distributions of residual stresses in the beam's longitudinal direction, which was the same as the fatigue stress direction, for a transverse groove weld and an attachment fillet weld detail in AL-6XN stainless steel I-beams tested in the fatigue study program [Fisher, et al., 2001].

2.2 Types of residual stresses

There are three types of residual stresses: 1) Macroscopic residual stress, which extends over several grains and usually many more. It is in self-equilibrium and can be relieved elastically when the member is cut or sectioned; 2) Structural micro-stress, occurring in one grain or part of a grain. It can occur between different phases or between particles, and strongly depends on micro-structure characteristics, such as grain size and grain orientation. It arises because of different thermal contractions in different crystallographic directions and is not diminished by cutting the sample. It is a grain interaction stress that promotes deleterious processes, such as stress corrosion cracking and hydride cracking; 3) Intra-granular stress, even more microscopic, ranging over several atomic distances within a grain and equilibrated over this small part of a grain. In this study, only the macroscopic residual stress was addressed.

Macroscopic residual stresses produced by welding can be decomposed into local welding stress and global welding stress [Campus, 1954]. Local weld stresses are developed in every case, even when welded pieces are small and completely free. These stresses are localized near the weld, most concentrated in the heat affected zone (HAZ), and decrease rapidly away from the weld. When welded pieces are restrained, that is, the thermal deformations are restrained, stresses will be produced everywhere in the pieces, which are called global welding stresses. Local and global stresses exist together and are in equilibrium. Once restraints are removed, the elastic part of global stresses should disappear, but plastic deformation due to global stress yielding and local stresses remain. In most cases, it is hard to separate them in a large welded assembly. Nevertheless, understanding the concept of local and global stresses is important and helpful in explaining many engineering fractures and phenomena in large welded structures.

2.3 Measuring techniques and selection of measuring methods

Since Mathar's pioneering work using a hole drilling method in 1934, various techniques for measuring residual stresses have been developed and applied in industry and research laboratories [e.g. SEM, 1996; HYTEC, 2001; Ritchie et al., 1987; Sherman, 1969]. These techniques include (1) hole drilling; (2) layer removal; (3) sectioning; (4) X-ray diffraction; (5) neutron diffraction; (6) ultrasonic; and (7) electro-magnetic methods. Since each measuring method is limited to a certain use and precision, no one method is ideal to measure a large-scale specimen with a rather complex geometry, such as welded beams with attachments. In most cases, two or more methods are combined to meet industrial needs and achieve a required precision.

In this study, two measurement methods, saw cutting (sectioning) and neutron diffraction, were selected and employed. Neutron diffraction measurements were conducted on groove weld and bulkhead attachment details in fabricated AL-6XN stainless steel beams because of advantages over traditional mechanical methods: it is non-destructive; it has triaxial through-thickness measurement capacity; it has flexibility with respect to sample geometry and material properties; and the measurement is localized in a region. Segment cutting from large-scale welded beams was needed because of limitations on the size of the neutron diffraction equipment, as discussed in Section 4 below.

3. Test specimens and measurement plan

Two welded I-beams of AL-6XN steel identical to those for fatigue tests [Fisher et al., 2001] were used for residual stress measurements, one with transverse groove welds in flange and web (Figure 1), and the other with one AL-6XN steel attachment detail welded to a flange (Figure 2). The weld conditions and sequence can be found in the report [Cheng et al., 2003a]. Residual stresses were measured at three weld details in these two beams: the transverse groove weld in the first beam, and the longitudinal flange-web weld and attachment fillet weld in the second beam. Table 1 summarizes the measurement plan for the two beams, including saw cut segment, sectioning and neutron diffraction methods. The results from different methods are superposed to obtain the final results [Cheng et al., 2003a]. However, only the neutron diffraction method and results are presented in this Chapter.

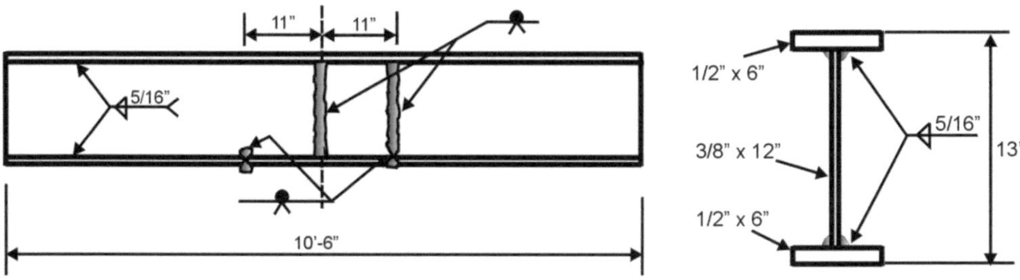

Fig. 1. Groove welded beam for residual stress measurement (unit: inch; 1 inch=25.4mm)

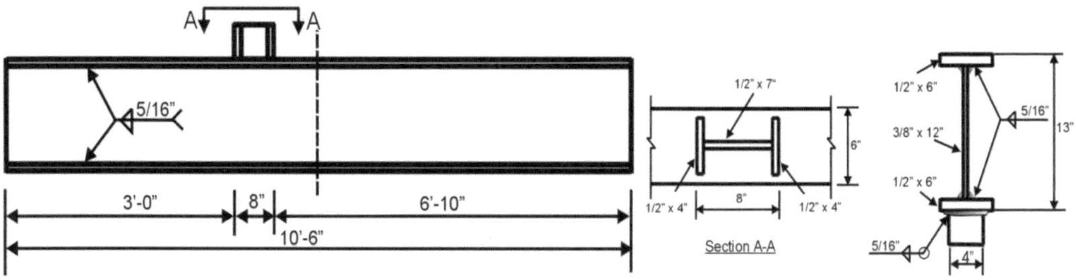

Fig. 2. Attachment Beam for Residual Stress Measurement (unit: inch; 1 inch=25.4mm)

Welded Beam	Weld Detail	Measuring Method	
Groove Welded Beam	Groove Weld	Two Steps	Saw Cut Segment
			Neutron Diffraction
Attachemnt Beam	Attachment Fillet Welds	Two Steps	Saw Cut Segment
			Neutron Diffraction
	Longitudinal Fillet Weld	-	Sectioning

Table 1. Summary of measurement plan

3.1 Saw cut segments from welded beams

Figures 3 and 4 show the locations, dimensions and most relevant portion of test segments for neutron diffraction meaurements for the groove-weld and attachment-weld details, respectively. Figure 5 shows a segment saw cut for the attachment beam. The stress release

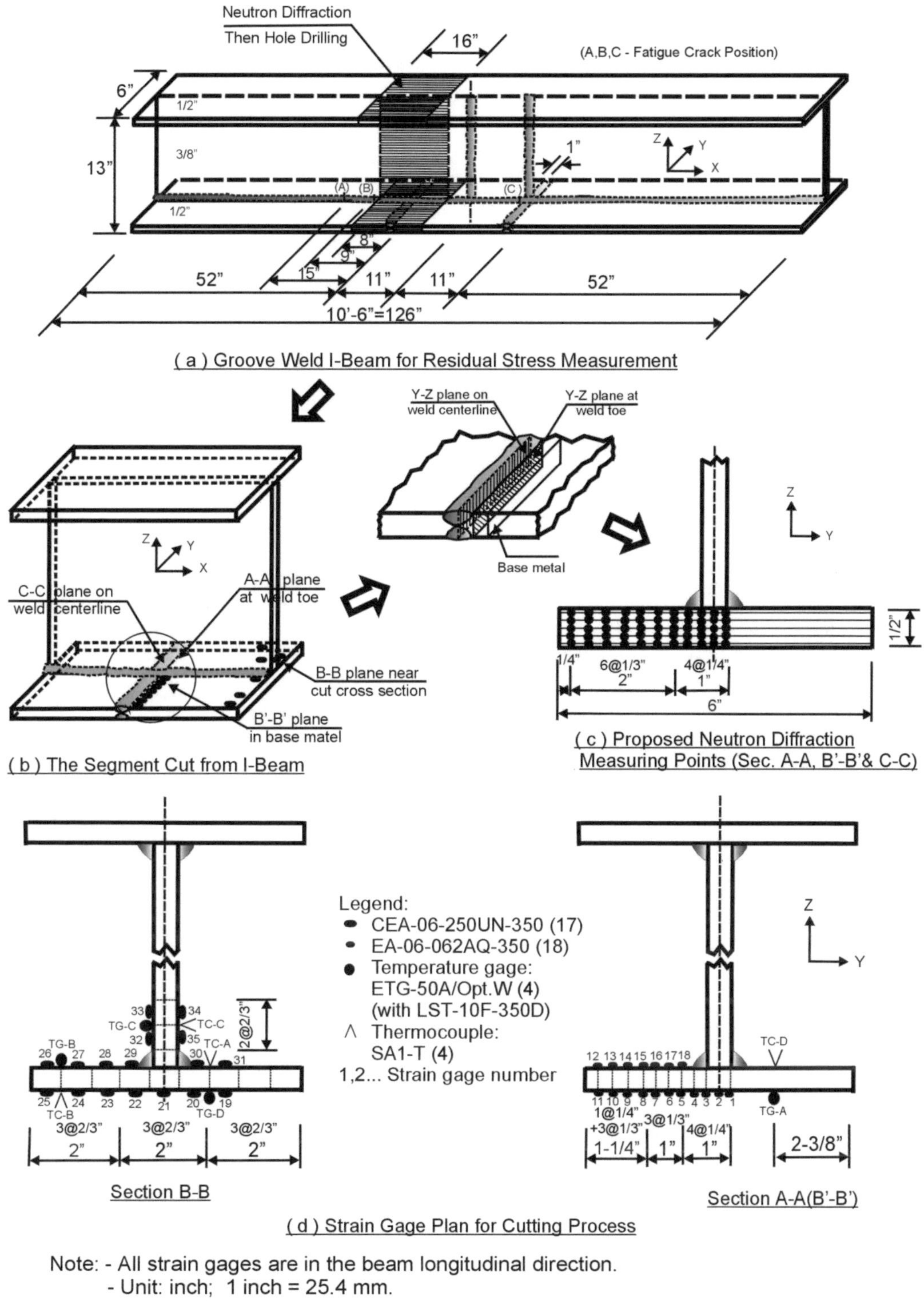

Fig. 3. Two-step residual stress measurement of groove welded beam

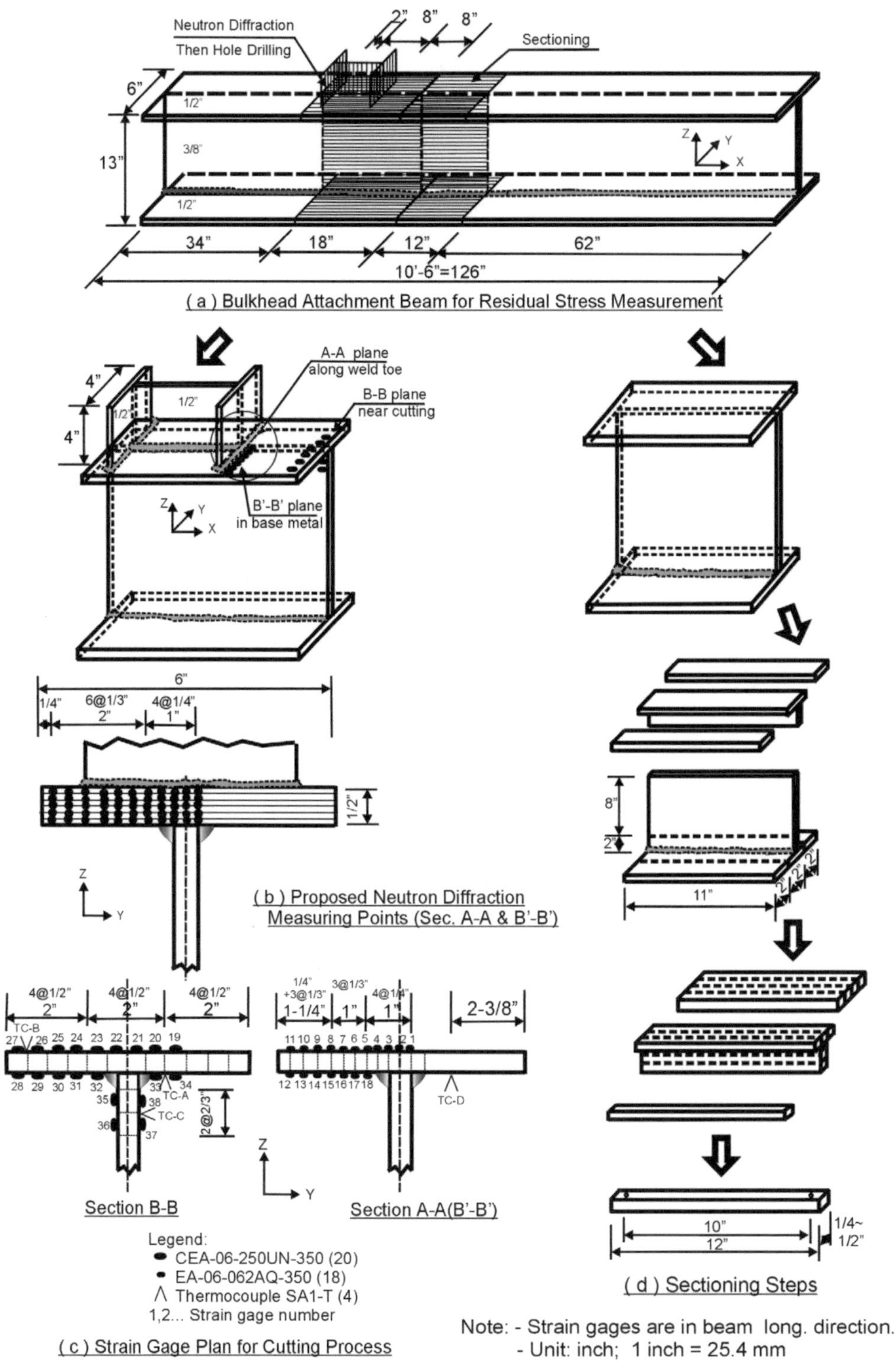

Fig. 4. Two-step residual stress measurement of attachment beam

Fig. 5. Saw cut for attachment beam

due to saw cutting was recorded by strain gages. Strain gages were mounted near the weld toe and adjacent to the cross section of saw cut, shown in Figure 3(d) and Figure 4(c), to measure the residual stress release due to saw cutting. More details for saw cut procedure, data collection of stress release, as well as the segment of longitudinal weld used with the sectioning method can be found in the report [Cheng et al., 2003a; Tebedge et al., 1973; 1969].

3.2 Segments for neutron diffraction measurement

After saw cutting and stress release analysis, the remaining segments for groove weld and attachment weld details were shipped to NIST for neutron diffraction measurements. The groove weld segment was 406 mm long, 152 mm wide and 330 mm deep (16 x 6 x 13 inches) (Figure 3(a)), while the attachment segment was 457 mm (18 in.) long (Figure 4(a)).

4. Neutron diffraction measuring method

4.1 Principle

Neutron diffraction measurement is a physical and non-destructive measuring process that evaluates residual strain/stress through measuring the change in crystallographic lattice spacing (d-spacing) and utilizing the relationship between crystallographic parameters and residual stresses. The technique utilizes the penetrating power of neutrons that is ~10^3 greater than X-rays in the diffraction wave-length regime to map subsurface triaxial stress distributions [e.g. NATO ASI Series, 1992; Prask et al., 2001; SEM, 1996]. The crystallographic lattice spacing is quantified by observing the intensity of the diffracted neutron beam. As shown in Figure 6 the lattice spacing in a certain orientation, d_{hkl}, can be obtained by the Bragg Law where λ is neutron beam wavelength and θ is the Bragg angle. Consequently, residual strain locked in the material can be calculated from Eq.1 by knowing the change of lattice spacing before (d_0) and after (d) residual stress was introduced.

$$\varepsilon_{hkl} = \frac{d_{hkl} - d_{hkl}^0}{d_{hkl}^0} \tag{1}$$

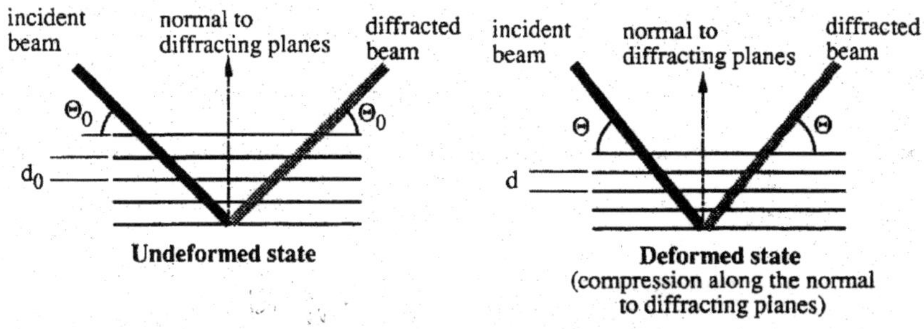

(a) Neutron Diffraction Measurement for Different Lattice Spacing

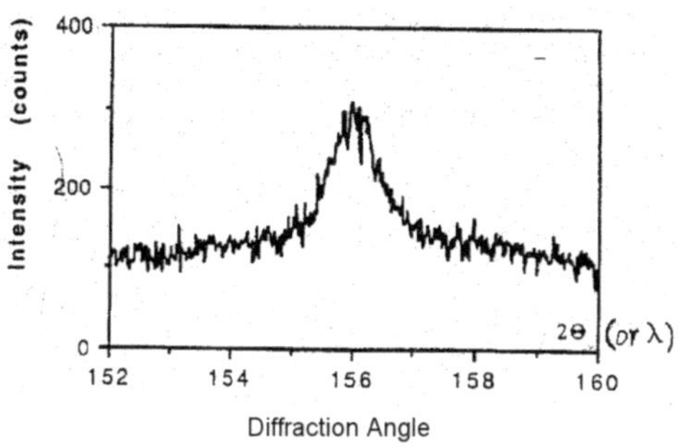

Bragg Law:

$$n\underline{\lambda} = 2d_{hkl}\sin\underline{\theta} \qquad (n=1)$$

(b) Neutron Diffraction Observation and Bragg Law for Calculating Lattice Spacing

Fig. 6. Principle of neutron diffraction measurement [SEM, 1996]

Two coordinate systems are utilized (Figure 7). One is the specimen-fixed coordinates (X, Y, Z) and the other is a laboratory coordinate system (L_i) corresponding to the scattering configuration. The measured lattice spacing is an average value over a number of grains covered in the neutron beam gage volume which is of the order of mm^3. The residual stress result is a kind of macro-stress that can be described by tensor components ε_{ij} or σ_{ij} in the specimen coordinate system (X,Y,Z). For a neutron beam with wavevector $Q(\phi,\psi)$ on the axis L_3, the measured strain along L_3 in the laboratory system ($\varepsilon_{\phi\varphi}{}'$ or $\varepsilon_{33}{}'$) that satisfies the Bragg Law, can be converted to stresses in specimen-fixed axes as shown in Section 5.2. Eq.2 shows how the measured strains in the laboratory system are related to the the actual strains in the sample:

$$\varepsilon'_{\varphi,\psi} = [\varepsilon_{xx}\cos^2\phi + \varepsilon_{yy}\sin^2\phi + \varepsilon_{xy}\sin 2\phi]\sin^2\psi +$$

$$+\varepsilon_{xz}\cos\phi\sin 2\psi + \varepsilon_{yz}\sin\phi\sin 2\psi + \varepsilon_{zz}\cos^2\psi \qquad (2)$$

The superscript ′ indicates strain in the laboratory coordinate system. Measuring strains at a point for at least six distinct (ϕ,ψ) orientations yields all six residual strain components at that point. These strains can be converted to strains in the specimen axes, including the beam longitudinal direction (fatigue stress direction in fatigue tests). Further details for neutron diffraction can be found elsewhere [Society for Experimental Mechanics (SEM) 1996; NATO 1992; Prask *et al.* 1996].

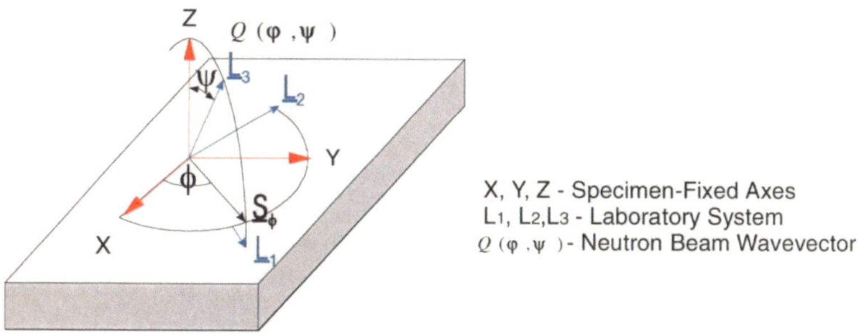

Fig. 7. Two coordinate systems for neutron diffraction measurement

4.2 Neutron diffraction measurement for segments

Due to limited space on the neutron diffraction table, from each 3.2 m (10.5 ft) long beam, a segment was cut to fit the instrument's X-Y-Z table and keep the residual stress relaxation at the location of interest as small as possible.

Segment saw cutting was done at Lehigh University (Bethlehem, PA), and neutron diffraction measurements were conducted at the NIST Center for Neutron Research (Gaithersburg, MD). At NIST, residual stress measurements on the segments were focused on weld toe (A-A plane), base metal (B′-B′ plane on the centerline of mounted strain gages) and weld metal (C-C plane) (Figures 3(b), 4(b)). In each plane, measurements included both points near the surface for distribution along the weld bead and points through the plate thickness. To reduce the total number of measurement points, it was assumed that the residual stress distribution was symmetrical about the beam web centerline. The measurement point mesh (resolution) depends on the expected residual stress gradient and gage volume of the neutron beam, shown in Figure 3(c) for groove weld and in Figure 4(b) for attachment fillet weld. After evaluating the grain size of virgin AL-6XN material and needed path lengths, a gage volume 3 x 3 x 3 mm^3 was chosen.

4.3 Test setup and measurement procedure at NIST facilities

The neutron diffraction measurements were made with a constant wavelength from a steady state reactor. The diffractometer (BT-8) is specially designed and well-suited for strain measurement of various shape-complexities and compositions [NIST, 2011]. Figure 8 shows a schematic of the measurement method. The (311) reflection of face-centered cubic (FCC) iron (γ-iron) with corresponding d-spacing about 1.095 Å (1Å=0.1nm (10^{-7}mm)) was chosen for the strain measurements. A wavelength λ=1.518 Å was used at a scattering angles of 88°. A position sensitive detector (PSD) centered at 88° covered an angular range from 84° to 91°, which corresponds to 1.133 to 1.064 Å in d-spacing scale. Figure 9 shows the groove weld segment placed on the X-Y-Z translator table of the neutron diffraction equipment. The incident beam is from the left tube, while the aperture of the neutron detector in the back of the photo receives the diffracted (scattered) beam. The X-Y-Z table is under computer control with a specimen weight limit of 50 kg (110 lbs). The part of the greatest interest for the groove weld and attachment fillet weld, is shown by the circles in Figure 3 and Figure 4, respectively.

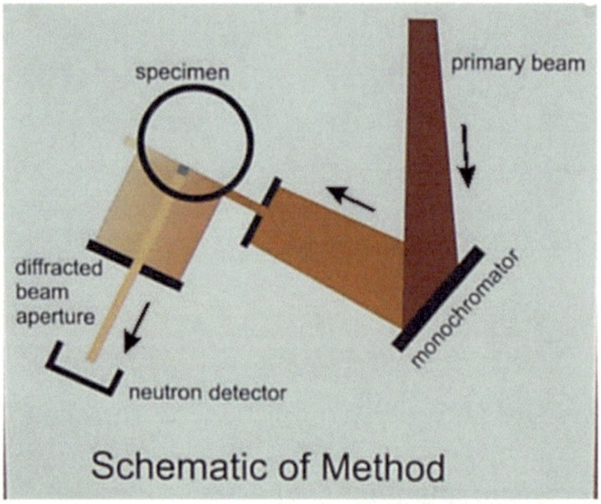

(Courtesy of NIST Center for Neutron Research)

Fig. 8. Schematic of neutron diffraction method

(Courtesy of NIST Center for Neutron Research)

Fig. 9. Groove welded beam segment on neutron diffraction X-Y-Z translator table

For a given specimen, d_0-spacing varies across the specimen (from base metal to weld metal). To measure the d_0-spacing, a smaller piece with base metal and weld cut from a similar AL-6XN beam was used to provide a stress-free reference sample. From this piece, one columnar shape coupon was cut from the base metal region and three were cut from the weld metal [NIST, 2003]. The d_0-spacing measurements were made on these coupons with gage volume of 2 mm x 2 mm x 2 mm. The average values of d_0 ($d_0(x)$, $d_0(y)$, $d_0(z)$) in three different directions were used for strain calculation (Eq.1).

5. Results from neutron diffraction measurements for groove weld segment

The groove weld joint was approximately in the middle of the segment specimen. It was a one-side GMAW weld without bevels on each plate and was back gouged. Groove welds were first made to connect two plates forming a flange or web member, then two welded flanges and a web were welded by longitudinal welds to assemble an I-beam. The maximum width of the weld zone was about 20mm (25/32 in.), as seen in Figure 10.

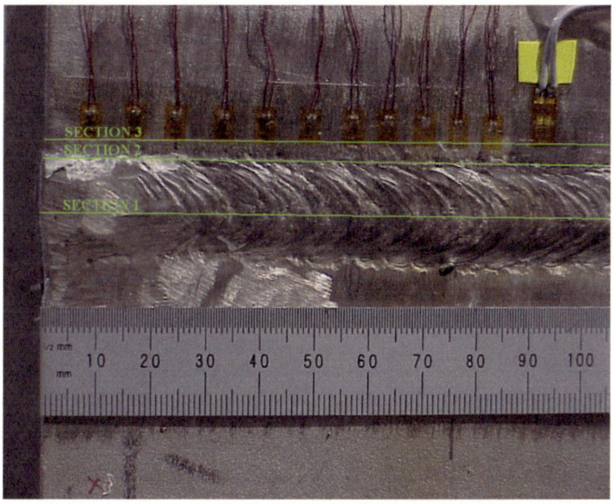

Fig. 10. Groove weld area on flange of the groove welded beam (outer surface)

5.1 Measurement locations

As shown in Figure 11, neutron diffraction measurements were made at three different sections: at the weld toe (Section 2 or A-A plane in Figure 3(b)), weld centerline (Section 1 or C-C plane) and base metal (Section 3 or B′-B′ plane). At each section the measuring mesh was basically the same (Figure 11). Measurement of the (311) lattice spacing was made in four different directions for each location. The specimen-fixed coordinate system is X-Y-Z, shown in Figure 11. The X-component which is in the fatigue stress direction was obtained by measuring two oblique directions corresponding to ϕ =±41.7° (≈±42°) with respect to the X-axis (see Figure 7) in the X-Y plane because of geometric complexity and high neutron beam absorption. The nominal increment step in flange thickness (Z) direction was 2.25 mm (0.09 in.) and was dictated mainly by spatial resolution of the experimental setup of the neutron diffractometer. The mesh in the Y direction (along the groove weld bead) was the same as the strain gage spacing used during segment saw cutting, so that straightforward superposition of the results from the neutron diffraction measurements and the saw cutting measurement could be made.

5.2 Measurement results

From the change of measured lattice spacing *d* and d_0 (d_0 & *d*: before and after residual stress is introduced), strain was obtained at each measurement point using Eq.1. The original strain data for the weld metal, weld toe and base metal are shown in Figures 12(a), 12(b), and 12(c), respectively. Figure 13 shows the stresses at the weld toe converted from the obtained strains using Eq.3:

$$\sigma_{ij} = \frac{1}{\frac{1}{2}S_2(hkl)}\{\varepsilon_{ij} - \delta_{ij}\frac{S_1(hkl)}{\frac{1}{2}S_2(hkl)+3S_1(hkl)}(\varepsilon_{xx}+\varepsilon_{yy}+\varepsilon_{zz})\} \quad (3)$$

where i, j denotes one of the X, Y, Z specimen-fixed coordinates; S_1 and S_2 are the two diffraction elastic constants; $S_1(311)$ = 6.64 TPa^{-1} and $1/2S_2(311)$ = -1.61 TPa^{-1} for AL-6XN

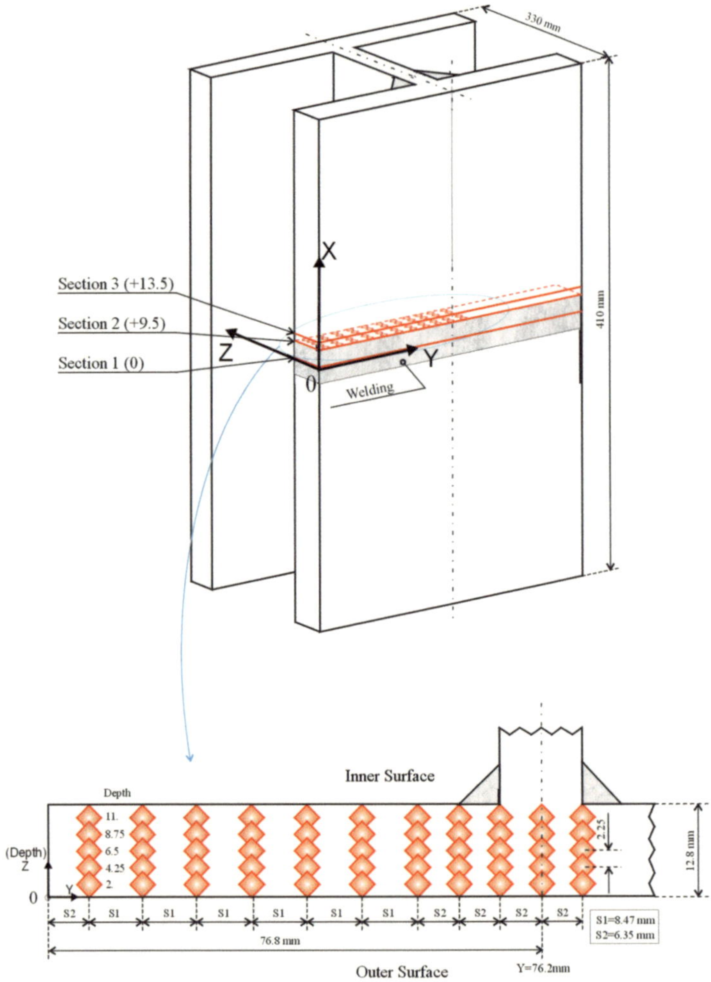

Fig. 11. Schematic measurement grid in specimen-fixed coordinate system (X-Y-Z), for groove weld segment (Unit: mm; 25.4mm=1 inch; gage volume 3x3x3mm^3)

stainless steel. S_1 and S_2 were obtained using Young's modulus, E=1.97x10^5 MPa (28,550 x10 ksi) and Poisson's ratio, v=0.33, for AL-6XN stainless steel, and single crystal constants given in [Danilkin et al., 2001]. δ_{ij} is the Kronecker delta which equals one when $i = j$ and zero when $i \neq j$. The uncertainties on measured d-spacings are determined by counting statistics and least-squares fits to the Gaussian peak shapes. The uncertainties on strains and stresses are determined from standard error propagation methods. Additional systematic errors are discussed below. Figure 14 shows the stress contour plots (for the weld toe) from the same measurements.

At Section 2 (at the weld toe), the measurement near the outer surface (nominal depth Z=2mm) included both base metal and weld metal, and would be sensitive to the weld content. The presence of weld metal complicated the measurement because of the different grain size and d_0 value for the base metal and weld metal. This was most obvious near the flange tip (Y=6.3mm) where more weld metal was included due to the start/stop end of

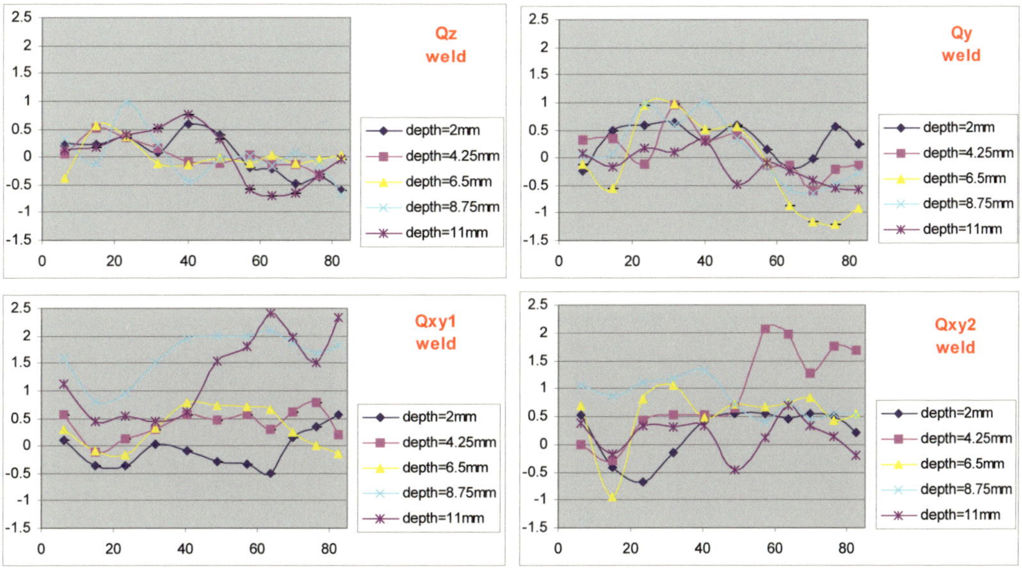

Fig. 12(a). Original strain data (x10³μ) for weld centerline in groove-weld segment, (Qxy1 and Qxy2 correspond to φ = ±42°; counting-statistic uncertainties are typically about ±0.035; abscissa is Y-distance from flange edge (mm; beam centerline is 76.8mm)

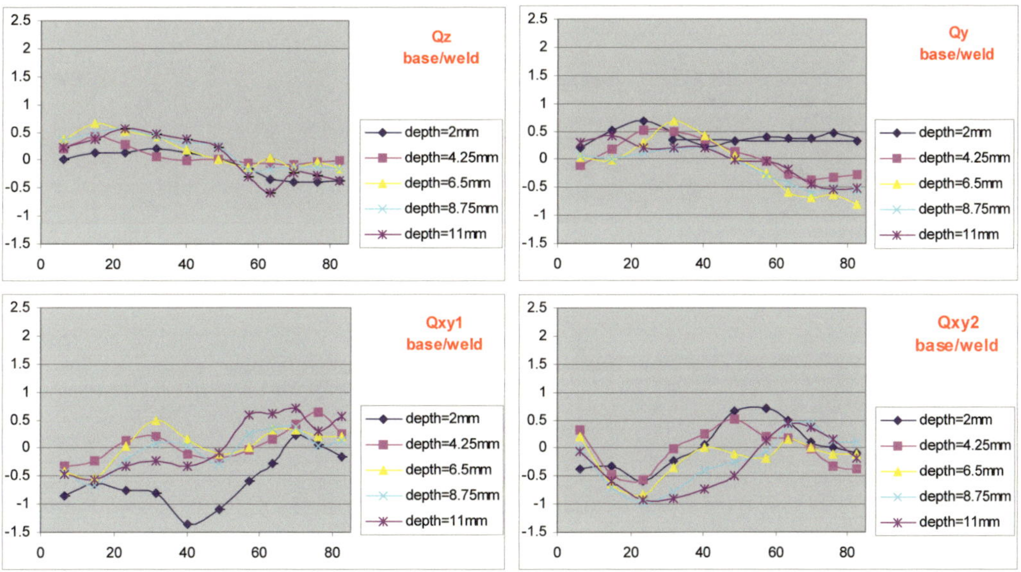

Fig. 12(b). Original strain data (x10³ micro-strain) for weld toe in groove-weld segment, (Qxy1 and Qxy2 correspond to φ = ±42°; counting-statistic uncertainties are typically about ±0.035; abscissa is Y-distance from flange edge (mm; beam centerline is 76.8mm)

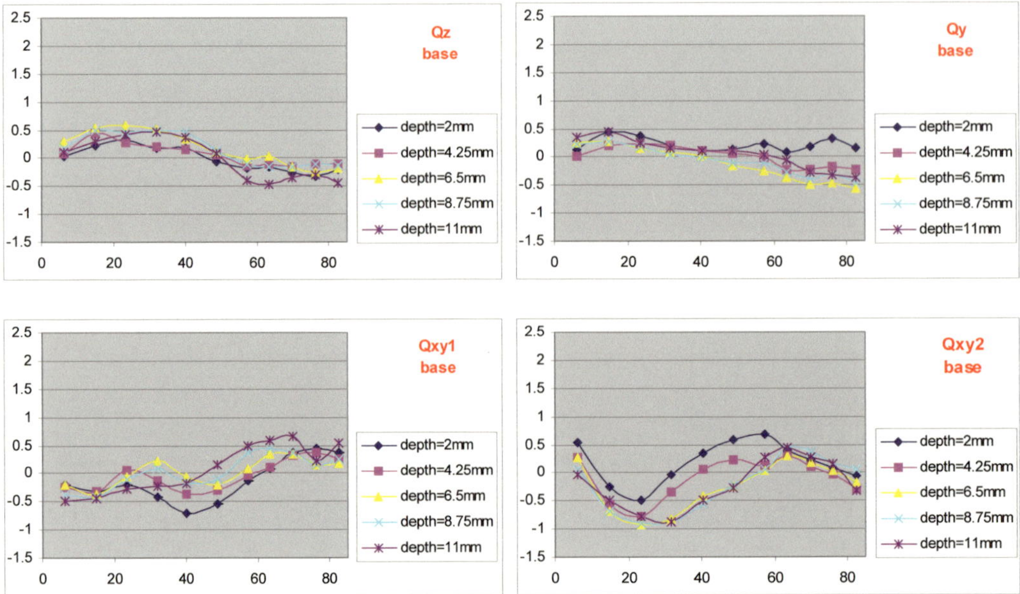

Fig. 12(c). Original strain data (x10³ micro-strain) for base metal in groove-weld segment (Qxy1 and Qxy2 correspond to φ = ±42°); counting- statistic uncertainties are typically about ±0.035; abscissa is Y-distance from flange edge (mm; beam centerline is 76.8mm)

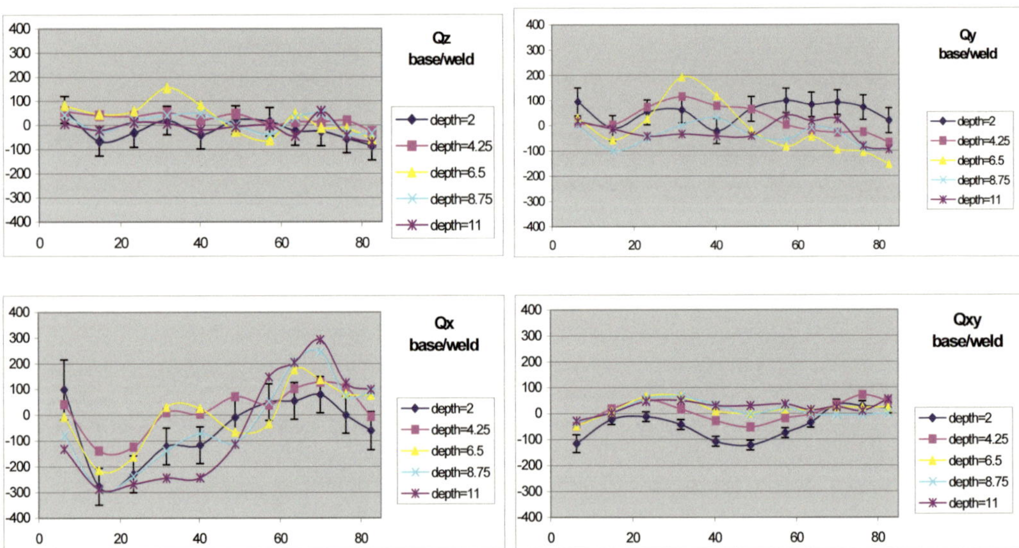

Fig. 13. Converted stress data (unit: MPa; 6.895 MPa=1ksi) for weld toe in groove-weld segment (Qxy Corresponds to stress shear component. Depth unit is mm; abscissa is Y-distance from flange edge (mm; beam centerline is 76.8mm)

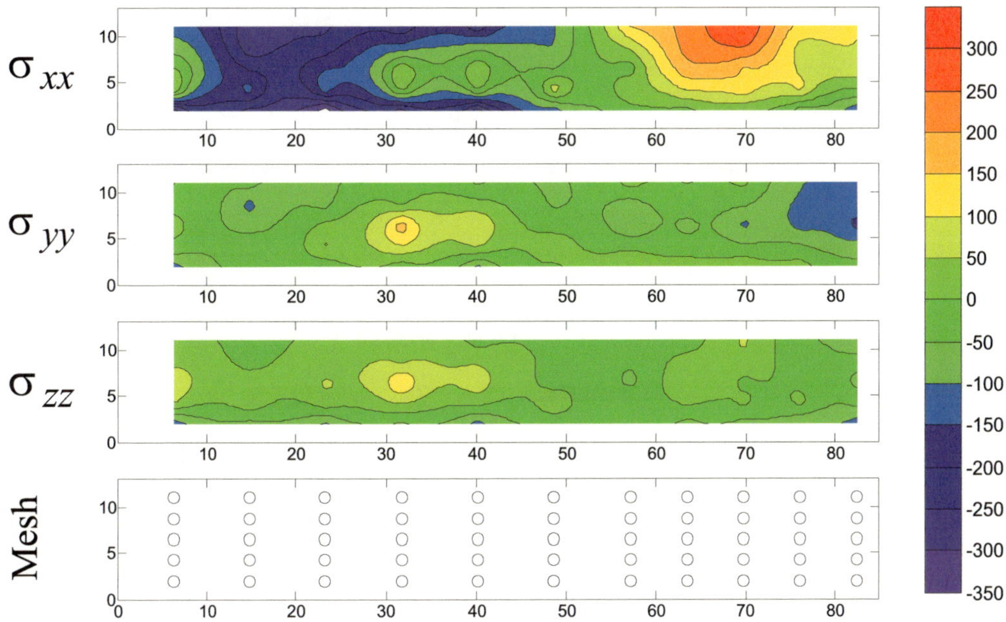

Fig. 14. Contour plots of stress data in Figure 13 for weld toe (MPa; 6.895 MPa=1ksi)

groove weld. Appropriate corrections were made for the measurement points at nominal depth of 2mm based on the FWHM (full-width at half-maximum) values that were used as an indicator of metal differences (base or weld). Relevant details are available elsewhere [NIST, 2003]. The converted stresses in Figure 13 are the data after the corrections. Since the values of d_0 were not obtained directly from the segment itself but from small pieces of corresponding materials (base metal or weld metal) in a similar weld detail, they were given as a range (the upper and lower bounds) for the uncertainty due to the slight material difference and counting statistics in neutron diffraction measurements. Therefore, the values of stresses were also presented as a range which is actually larger than the individual stress-value uncertainties. For the points at nominal depth of Z=2 mm (near outer surface), upper and lower bounds presenting the stress data range are shown as error bars. For the stresses at points on the outer surface near the flange tip (Z=2mm, Y=6.3 mm) where more weld metal was included, the range (difference between the maximum and the minimum values) was greater than other points. Other measurment points in Figure 13 are the average stresses of the maximum and minimum values.

As fatigue stress in the fatigue-tested beams was in the X-direction (longitudinal direction in I-beams), X-direction residual stresses are of the greatest interest. The distributions of near-surface X- stresses along the groove weld bead were re-plotted in Figure 15 for the weld metal, groove weld toe and base metal, respectively. These stress distributions are at nominal depth of about 1.5 to 2mm beneath the outer and inner flange plate surfaces (nominal Z=2 mm and Z=11 mm). Examination of these distributions show that residual stresses near the inner surface (web side) were high in tension near the flange-web connection and high in compression near the flange tip. The stress distribution pattern was similar to the residual stress distribution for longitudinal web-to-flange weld detail [Cheng et al. 2003a], but rather complex since the flanges and web were welded into an I-beam by

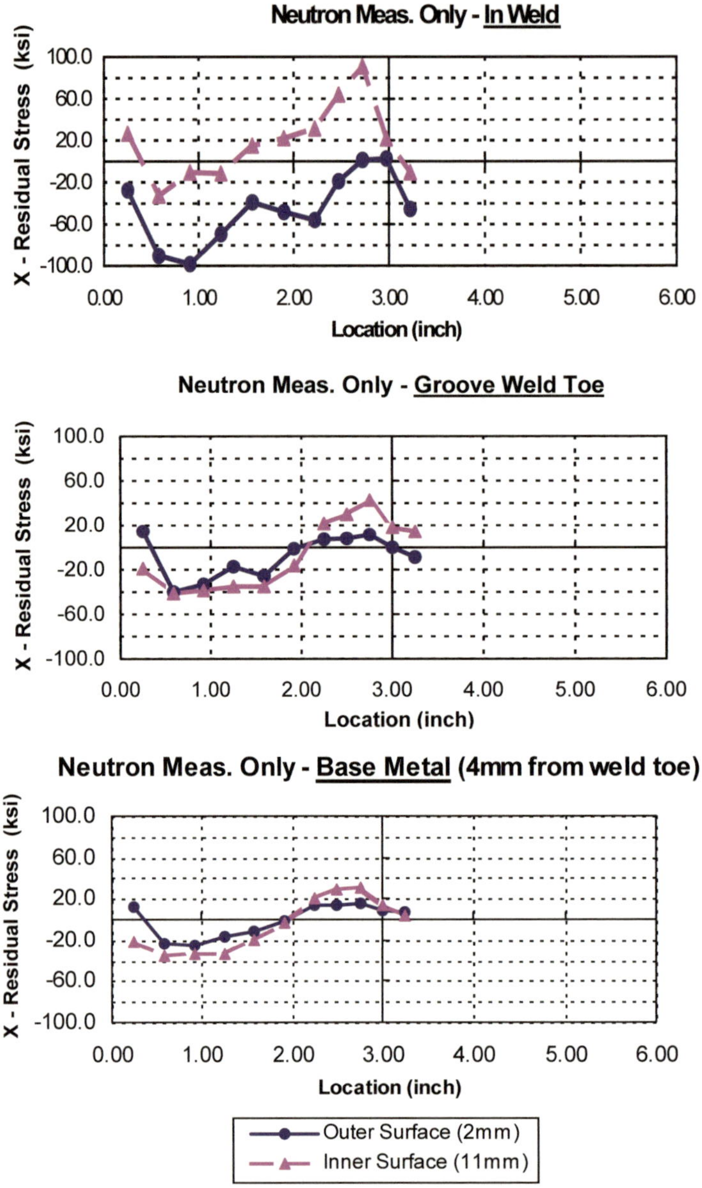

Fig. 15. X-direction stress distributions in groove welded beam (1 in.=25.4 mm; 1 ksi=6.895 MPa; abscissa is Y-distance from flange edge (beam centerline is 76.8mm (~3"))

longitudinal welds after the flange groove weld was completed. Weld residual stresses self-adjusted when single members were assembled into an I-beam to maintain self-equilibrium. The tensile stress near the inner surface reached about 90.6 ksi (625 MPa) in the weld, 42.5 ksi (293 MPa) at the weld toe and 31.3 ksi (216 MPa) in the base metal. These residual stresses in the weld and at the weld toe exceeded or were close to the nominal yield stress, 48~56 ksi, of AL-6XN steel. In fatigue tests, some fatigue cracks have developed from the interior region of the groove weld at the intersection of transverse groove weld and longitudinal fillet weld, as well as at the weld toe, as shown in Figure 16. On the other hand, residual stresses on outer subsurface (free surface side) are primarily in compression. The magnitudes of the maximum compressive stresses were -98.4 ksi (-679 MPa) in the weld and -40.1 ksi (-277 MPa) at the weld toe, and were comparable to the tensile residual stresses on the inner subsurface. In general, the magnitude of residual stresses decreased when the location was away from the groove weld centerline.

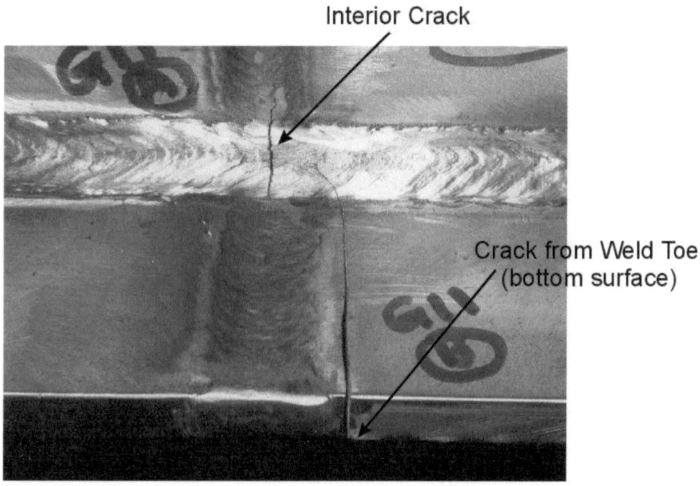

Fig. 16. Typical fatigue cracks from groove weld in AL-6XN beam

It is noted that the data across Section 3 (base metal) and Section 2 (plane at the weld toe, mainly base metal) were of better quality because of the stable conditions in d_0 and d measurements, and fairly good balance was obtained in X-direction equilibrium. For weld metal, grain size variability had an impact on the d_0 measurement due to the much larger grain size than base metal and the irregularity of grain size/micro-structural orientation in the weld material, which caused a d_0 and d incompatibility in calculating strain. Therefore the data scatter for weld material was much larger than for base metal.

6. Results from neutron diffraction measurements for attachment segment

6.1 Measurement locations

Residual Stresses in the attachment-weld segment were also determined by neutron diffraction. The relevant details regarding the use of neutron diffraction were the same as described earilier for the groove-welded segment.

The weld detail studied was an attachment fillet weld between the attachment and the beam flange located near the middle of the segment. Measurements were carried out in a plane along weld toe (A-A Plane in Figure 4) and a plane 7.1 mm away from the weld toe where the strain gages for segment saw cut were located (B′-B′ Plane in Figure 4). Figure 17(a) shows a photo of the attachment weld area. The mesh mapping the nominal measurement points for both A-A and B′-B′ planes is shown in Figure 17(b). The mesh spacing in the Y-direction parallel to fillet weld bead was the same as the strain gage spacing for the segment saw cut.

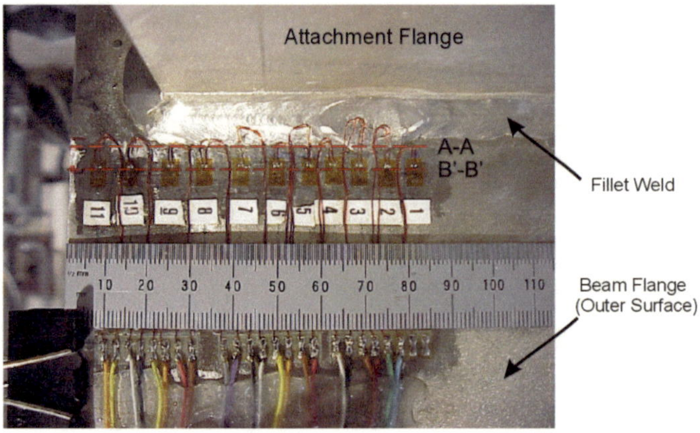

(a)

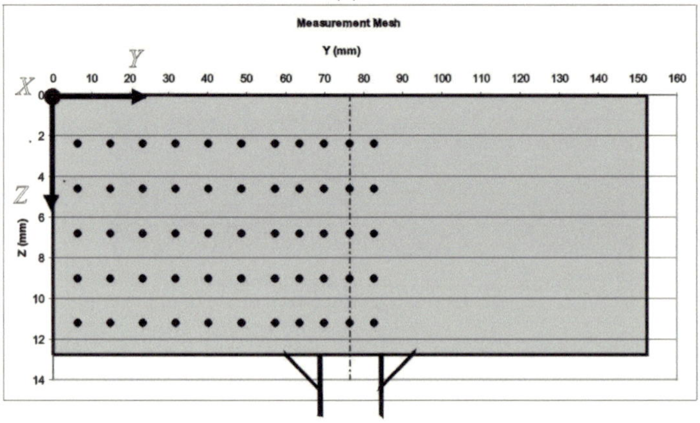

(b)

Fig. 17. Attachment welded beam segment measurement: (a)Close-up view of fillet weld for attachment welded segment; A-A plane at weld toe and B′-B′ plane at 7.1 mm away from weld toe. (b) Complete measurement mesh for beam flange planes at weld toe (A-A) and 7.1 mm away from weld toe (B′-B′). Beam web centerline is at 76.2 mm (3 inch; 1 inch =25.4 mm)

6.2 Measurement results

Figures 18(a) shows the measurement results of the 3-D stresses in the A-A plane (along the weld toe). Both X-stresses (longitudinal direction) and Y-stresses (transverse direction) in the A-A plane near the beam centerline were very high near the outer surface, whereas Z-stresses (depth direction) were very low. The magnitude of the tensile residual stresses in the X-direction was 324 MPa (47.0 ksi) at a nominal depth of 2.4mm and 405 MPa (58.7 ksi) at 4.6 mm, which exceeded the nominal yield stress of AL-6XN steel (48~56 ksi). The maximum tensile stresses in Y-direction were 406 MPa (58.8 ksi) at 2.4 mm depth and 466 MPa (67.5 ksi) at 4.6 mm depth, both of which also exceeded the nominal yield stress. The measurement results for both X-stresses and Y-stresses of the A-A plane, that the stress magnitude was lower at d=2.4 mm than at d=4.6 mm, were not expected. The highest residual tensile stresses were found near the intersection of attachment flange and attachment web welds.

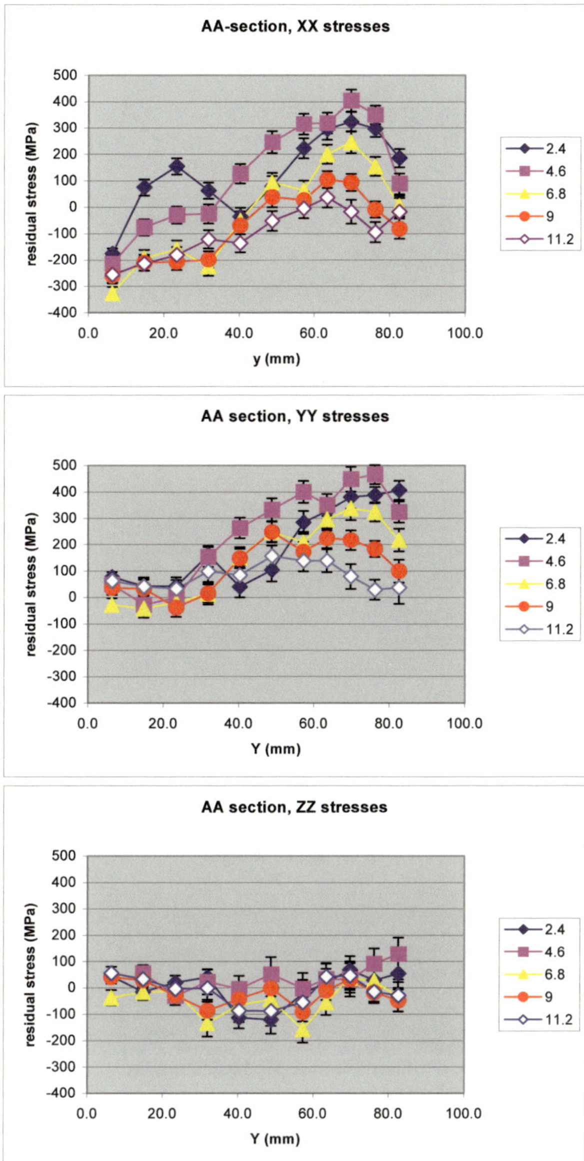

Fig. 18(a). 3-D stress data in A-A plane (weld toe) in attachment weld segment (1 ksi=6.895 MPa; beam centerline at 76.2 mm (1 inch=25.4mm))

Figure 18(b) shows the measurement results of the 3-D stresses in B′-B′ plane (7.1 mm away from A-A) for base metal. The stresses for the B′-B′ plane were less. Further, the dip in stress magnitude with Z=2.4 mm was not seen in the B′-B′ plane. The dip was probably the result of either weld metal or HAZ (heat affected zone) metal being present in the gage volume for 30mm$\leq$Y$\leq$60mm in the scan length along the weld toe, and was manifested in the raw data by peak widths which were clearly larger than other peak widths for the A-A plane and all of the B′-B′ plane peak widths. In the absence of sufficient data to make a reliable

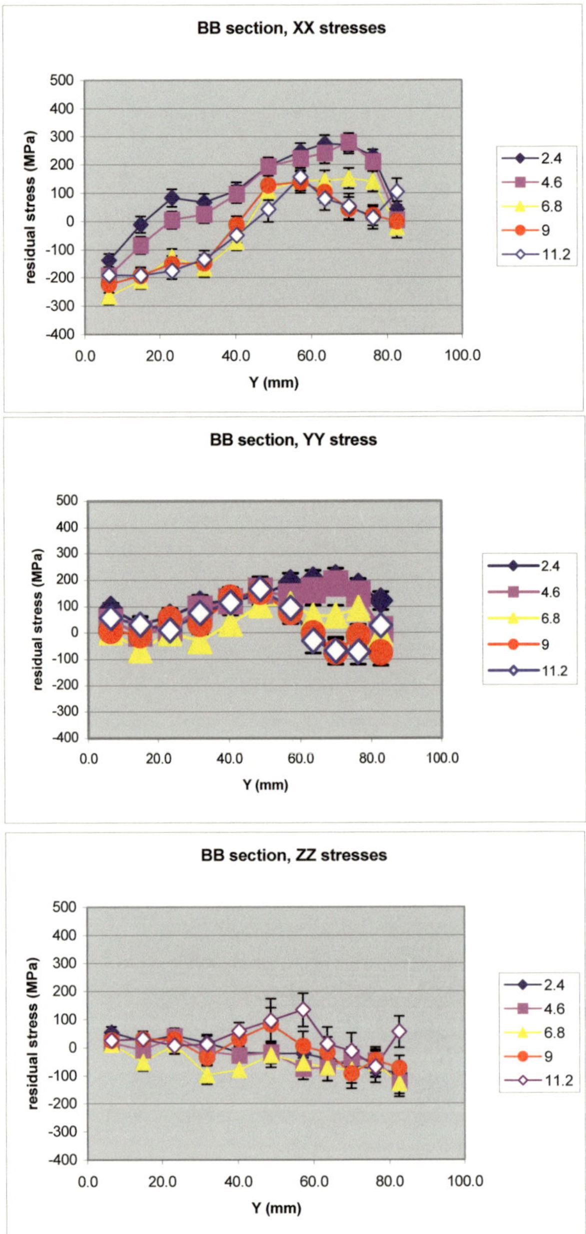

Fig. 18(b). 3-D stress data in B′-B′ plane (base metal) in attachment weld segment (1 ksi=6.895 MPa; beam centerline at 76.2 mm (1 inch=25.4mm))

quantitative correction, it is estimated that X-stresses and Y-stresses of the A-A plane at 2.4 mm should be about equal to the values at 4.6 mm over the scan range, 30mm$\leq$Y$\leq$80mm. The highest residual tensile stresses were near the intersection of attachment flange and attachment web welds. During welding the shrinkage was greater in transverse direction (Y-Y) than in longitudinal direction (X-X).

Residual stresses decayed from outer surface at Z=0, where the attachment was welded, to inner surface. The X-X residual stresses near inner surface (Z=11.2 mm) near the I-beam flange-web welds were much lower than expected. The flange-web welds of the I-beam were expected to produce high tensile residual stresses at the beam flange inner surface, particularly in X-direction. The attachment welds were made on the beam flange after the I-beam was fabricated, and thus a stress re-distribution in the I-beam flange was expected. Residual stresses decayed rapidly away from the weld toe, particularly for stresses in transverse direction (Y-Y).

There was a wrap-around weld at the end of attachment flange. An increase in the longitudinal (X-X) stress was seen near the outer surface (Z=2.4 mm) for both A-A and B′-B′ planes. Similar behavior was not observed for the transverse (Y-Y) stress. In general, higher residual tensile stresses near a weld toe were observed parallel to the weld direction rather than perpendicular to the weld direction.

Figure 19 shows the distribution of X-direction residual stresses (the fatigue stress direction) near the outer surface (Z=2.4mm) and the inner surface (Z=11.2mm) of the beam flange for

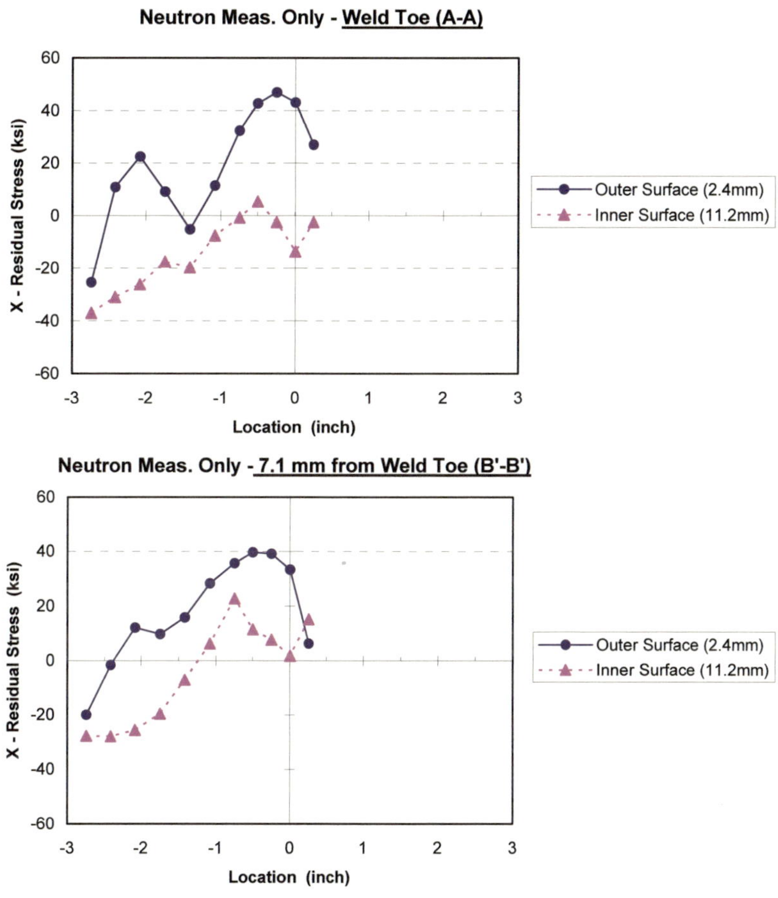

Fig. 19. X-direction stress distribution in attachment weld segment (1 ksi=6.895 MPa; beam centerline at 76.2 mm (1 inch=25.4mm))

A-A (weld toe) and B'-B' (base metal) planes, respectively. These stresses were an average over the 3x3x3 mm cube gage volume measurements. It shows that residual stresses near the outer surface (attachment side) were in tension except near the beam flange tip, and highest near the beam flange-web intersection. The stress was 47 ksi (324 MPa) from neutron diffraction measurement, close to the nominal yield stress of AL-6XN steel. During the fatigue testing program on welded beams with attachments, fatigue cracks were always observed to initiate from the middle of the attachment fillet weld where residual stress was high and from weld start/stop locations. Figure 20 shows a photo of typical fatigue crack that develops under tensile total stress conditions.

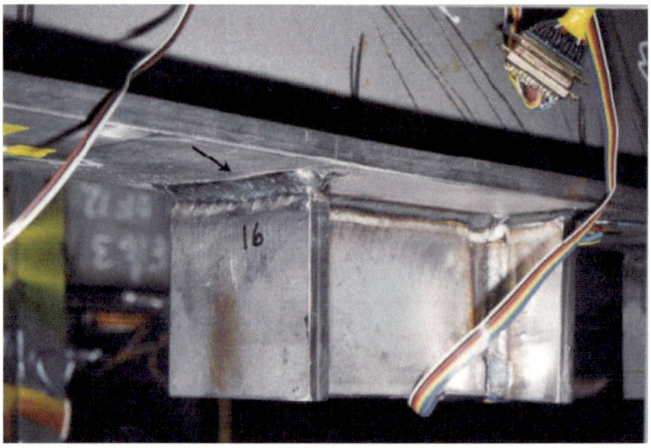

Fig. 20. Typical fatigue crack from attachment fillet weld toe in AL-6XN beam

7. Summary

The objective of this study was to quantify the magnitude and distribution of residual stresses from two types of welded details in AL-6XN steel beams. Residual stresses were measured through destructive saw cutting, and non-destructive neutron diffraction. This chapter showed the measurement results of neutron diffraction. The effect of residual stresses on fatigue strength was discussed. The findings can be summarized as follows.

1. Neutron diffraction method is an effective and powerful tool to measure residual stresses, especially when through-thickness stress distribution and three dimensional stresses are of interest. For large-scale specimens, two measurement methods can be used, such as combination of saw cutting and neutron diffraction as used in this study.
2. Each residual stress measurement technique had advantages and limitations for large scale specimen measurements. The neutron diffraction method provided through-thickness measurements due to its much stronger penetration capacity, and an average measurement within a gage volume normally in ~3x3x3 mm^3 throughout the specimen. It required careful material calibration and it is difficult for the weld metal due to the large grain size. This study showed a good accuracy of neutron diffraction measurements for residual stresses.
3. This study revealed that a large scale welded member has very high tensile residual stresses near weldments, equal to the uniaxial yield stress of the base metal and sometimes even more. Residual stresses decay away from the weld. The attachment-

weld detail exhibited relatively complex distributions along the weld toe and through the flange thickness.

4. Separating a part from a large scale welded assembly provides a partial residual stress release. Such a stress release should not be overlooked when residual stresses are evaluated at a section.

8. Acknowledgment

This study was part of the project on Non-Magnetic Stainless Steel Advanced Double Hall Ships Investigation: Fatigue Resistance of Large Welded Components, supported by the Office of Naval Research (ONR) and directed by Dr. Roshdy G. S. Barsoum, Program Manager. Special thanks are due Dr. Barsoum and his colleauges for their consistent support and help during the entire project. The authors acknowledge the help from the technical staff and fellow workers for the lab work conducted at Fritz Lab and ATLSS Research Center, Lehigh University.

9. Disclaimer

Certain commercial firms and trade names are identified in this report in order to specify aspects of the experimental procedure adequately. Such identification is not intended to imply recommendation or endorsement by the National Institute of Standards and Technology, nor is it intended to imply that the materials or equipment identified are necessarily the best available for the purpose.

10. References

Campus, F. (1954). Effect of Residual Stresses on the Behavior of Structures, *Residual Stress in Metals and Metal Constructions*, William R. Osgood Ed., Prepared for the Ship Structure Committee, Reinhold Publishing Corporation, NY

Cheng, Xiaohua; Fisher, John W.; Prask, Henry J.; Gnaeupel-Herold, Thomas & Luzin, Vladimir (May 2003a). Residual Stress Measurements in AL-6XN Stainless Steel Welded Beams. ATLSS Report No.03-08 (Lehigh University File No. 533612), ONR Grant No. N00014-99-1-0887, Nonmagnetic Stainless Steel for Double Hull Ship Construction, Bethelehem, PA, USA

Cheng, Xiaohua; Fisher, John W.; Prask, Henry J.; Gnaeupel-Herold, Thomas; Yen, Ben T. & Roy, Sougata (2003b). Residual Stress Modification by Post-weld Treatment and Its Beneficial Effect on Fatigue Strength of Welded Structures, *International Journal of Fatigue*, Vol. 25, pp.1219-1269

Danilkin, S.A., Fuess, S., Wieder, T., Hoser, A. (2001). Phonon Dispersion and Elastic Constants in Fe-Cr-Mn-Ni Austenitic Steel, *J. of Materials Science* 36, pp.811-814

Dudt, Philip J. (April 2000). Distortion Control for Stainless Steel Hull Fabrication. NSWCCD-67-TR-2000/07, April, 2000

Fisher, John W.; Yen, Ben T.; Cheng, Xiaohua; Kaufmann, Eric J.; Metrovich, Brian & Ma, Zuozhang (May 2001). Fatigue Resistance of Large Welded AL-6XN Stainless Steel Components with Fillet, Groove, and Attachment Welds, Final Report, Project B, FY 99, ONR Grant No. N00014-99-0887, Nonmagnetic Stainless Steel for Double

Hull Ship Construction, ATLSS Report No.01-04, ATLSS Engineering Research Center, Lehigh University, Bethlehem, PA, USA

HYTEC, Inc. (2001). PRISM System Residual Stress Measurement., www.hytecinc.com

International Association of Classification Society (IACS) (July 1999). No.56 Fatigue Assessment of Ship Structures, *IACS Recom. 56.1*, United Kingdom

The International Nickel Company, Inc. (INCO) (1964). Properties of Some Metals and Alloys, A-297

Lamb, Stephen (Technical Editor) (1999). CASTI Handbook of Stainless Steels & Nickel Alloys, *CASTI Handbook Series*, Vol. 2, CASTI Publishing Inc., Canada

Lu, Le-Wu; Ricles, James M.; Therdphithakvanij, Pholdej; Jang, Seokkwon & Chung, Jin-Oh (Oct. 2002). Compressive Strength of AL-6XN Stainless Steel Plates and Box Columns, Final Report, Project C, FY 99, ONR Grant No. N00014-99-0887, Nonmagnetic Stainless Steel for Double Hull Ship Construction, ATLSS Report No.02-04, ATLSS Research Center, Lehigh University, Bethlehem, PA, USA

NATO ASI Series (1992). Measurement of Residual and Applied Stress Using Neutron Diffraction, Edited by Michael T. Hutchings and Aaron D. Krawitz, *Series E: Applied Sciences*, Vol. 216, (Proceedings of the NATO Advanced Research Workshop on Measurement of Residual and Applied Stress Using Neutron Diffraction, Oxford, UK, March 18-22, 1991,) Kluwer Academic Publishers, Dordrecht, Boston and London

NIST Center for Neutron Research website, www.ncnr.nist.gov, 2011

NIST Center for Neutron Research (2003). The Neutron Diffraction Determination of Residual Stresses in AL-6XN I-Beams, by V. Luzin, T. Gnaeupel-Herold, H. Prask, X. Cheng and J. W. Fisher, NIST internal report

Prask, H. J. & Brand, P. C. (1996). Neutron Diffraction Residual Stress Measurement at NIST, *Material Science Forum*, Vols. 210-213, pp.155-162, Transtec Publications, Switzerland

Prask, Henry J., Gnaupel-Herold, Thomas, Fisher, John W. & Cheng, Xiaohua (June 2001). Residual Stress Modification by Means of Ultrasonic Impact Treatment, *Proc. of Society for Experimental Mechanics Meeting*, Portland, pp. 551-4

Ritchie, D. & Leggatt, R. H. (May 1987). The Measurement of the Distribution of Residual Stresses Through the Thickness of A Welded Joint, *Strain*, Vol. 23, pp. 61-70

Rolled Alloys (April 1997). AL-6XN® Alloy Physical, Mechanical and Corrosion Properties, Technology Department

Sherman, Donald R (April 1969). Residual Stress Measurement in Tubular Members, *Journal of the Structural Division*, Proc. of the ASCE, Vol.95, No.ST4

Society for Experimental Mechanics (SEM), Inc (US) (1996). Handbook of Measurement of Residual Stresses, Edited by Dr. Jian Lu, The Fairmont Press, Inc.

Tebedge, Negussie (1969). Measurement of Residual Stresses – A Study of Methods, MS thesis to Department of Civil Engineering, Lehigh University

Tebedge, N., Alpsten, G. & Tall, L. (Feb 1973). Residual Stress Measurement by Sectioning Method, *Experimental Mechanics*, Vol. 13, No.2, pp. 88-96

3

Modelling Residual Stress and Phase Transformations in Steel Welds

Hui Dai
School of Materials, University of Manchester, Manchester
UK

1. Introduction

Mathematical modelling of welding phenomena is very complex: involving melt pool phenomena, solidification, weldability analysis, microstructure evolution in the heat affected zone, welding heat-flow simulation, electrical-thermal-mechanical simulation etc. Some interactions between these processes are included in Fig. 1. Each topic alone can be intellectually challenging and too hard to be investigated by classical methods. With the increasing power of modern computer systems, numerical modelling and especially finite element analyses make it possible to produce excellent solutions to satisfy engineering demands.

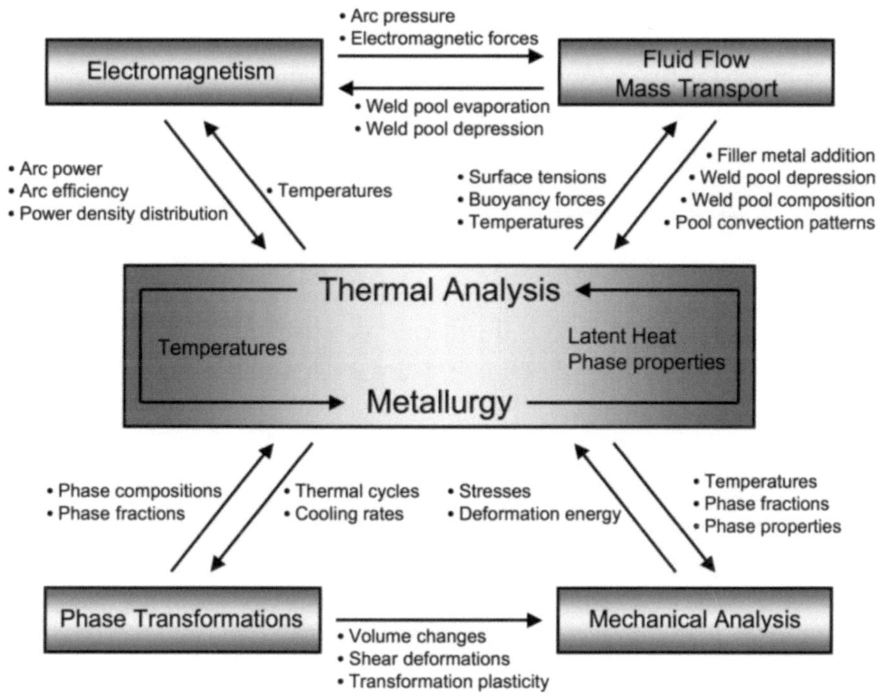

Fig. 1. Interrelated physical phenomena that arise in arc welding of ferritic steels [1, 2]

Finite element analysis offers a powerful technique for elaborating on the factors that affect the formation and distribution of residual stresses in engineering components. For the useful application of this technique for welds in pressure vessel components it is necessary that developments be made to allow analyses of large complex structures and with the successful management of metallurgical effects (e.g. phase transformations and micro-structural evolution).

2. Finite element analysis of welding

This section is concerned with simulations of temperatures, displacements, stresses and strains in welded structure [3] without solid-state phase transformations. Detailed literature review in finite element analysis of welding (2001 and before) can be found in a paper by Lindgren [4].

2.1 Heat conducting equations

Consider a particle with a differential volume $d\Omega$ at the position x at time t. Let u denote its internal energy, KE the kinetic energy, Q_c, the net rate of heat flow by conduction into the particle from its surroundings, Q_s , the rate of heat input due to external sources (such as radiation) and P the rate at which work is done on the particle by body forces and surface forces (i.e., P is the mechanical power input). Then, in the absence of other forms of energy input, the fundamental postulate of conservation of energy states that [5]

$$\frac{d}{dt}(U + KE) = P + Q_c + Q_s \tag{1}$$

And as shown by Lai [5], the energy equation at each point of the continuum reduces to

$$\rho\frac{du}{dt} = tr(\mathbf{TD}) - div\mathbf{q} + \rho\mathbf{q_s} \tag{2}$$

where $\mathbf{q_s}$ is the rate of heat input (known simply as the heat supply or heat flux vector) per unit mass by external sources, $\mathbf{q}$ is a vector whose magnitude gives the rate of heat flow across a unit area by conduction and whose direction gives the direction of heat flow, then by Fourier's law, $\mathbf{q} = -k \cdot \mathbf{grad}T$, k(T) is the thermal conductivity, T is the temperature. The term $tr(\mathbf{TD}) = P_s$ is known as the stress power (per unit volume). It represents the rate at which work is done to change the volume and shape of a particle of unit volume. Finally, $\mathbf{T}$ is the Cauchy stress and $\mathbf{D}$ is the tensor describing the rate of deformation.

Equation (2) represents the energy equation for two-way coupling. The stress power P_s couples the mechanical state to the thermal state; i.e., mechanical work causes heating.

In one way coupling, the stress power term is assumed negligible compared to the heat input terms on the right-hand side of equation (2), and it is also assumed that the internal energy is only a function of the temperature, i.e., independent of strains. Thus, equation (2) becomes:

$$\rho\frac{du}{dt} = -div\mathbf{q} + \rho\mathbf{q_s} \tag{3}$$

Now using Fourier's law $\mathbf{q} = -k \cdot \mathbf{grad}T$, we have

$$\rho \frac{du}{dt} = div(k \cdot \mathbf{grad}T) + \rho \mathbf{q_s} \tag{4}$$

where $\rho(T)$ is the density, and T is the temperature. Introducing the specific heat, C, of the medium define by

$$C = \frac{du}{dt} \tag{5}$$

We can write equation (4) as

$$\rho C \frac{\partial T}{\partial t} = div(k \cdot \mathbf{grad}T) + \rho \mathbf{q_s} \tag{6}$$

$\frac{\partial T}{\partial t}$ is the change in temperature over time, k(T) is the thermal conductivity, the product ρC reflects the capacity of the material to store energy.

For a constant pressure process the change of specific enthalpy is equal to the heat transfer, i.e., specific heat, $C_p = \frac{dH}{dT}$, where H is the specific enthalpy, the subscript 'p' refers to constant pressure. The equation (4) now becomes

$$\rho \frac{dH}{dt} = div(k \cdot \mathbf{grad}T) + \rho \mathbf{q_s} \tag{7}$$

H is a function of temperature.

In gas tungsten arc (GTAW or TIG) welding, the area-specific density of heat flow to the weld pool by the welding arc over a small area of the work-piece is of the order of 5×10^0 to 5×10^2 W/mm² [6]. As a result of this intense local heat flux, there are high temperature gradients in the neighbourhood of the weld pool. It is therefore assumed that the stress power term is small compared to the heat input terms of the energy equation, and that modelling the welding process as one-way coupling, neglecting coupling between mechanical and thermal problems, is reasonable.

2.2 Thermal boundary and initial conditions

Initial and boundary conditions need to be specified to solve equation (7). Various types of conditions are necessary to transform the real physical conditions into mathematical models [5]. Consider an arbitrary fixed volume, Ω, of the medium which is bounded by a closed surface *S*.

a. Temperature conditions

Initial conditions are required only when dealing with transient heat transfer problems in which the temperature field in the material changes with time.

$$T = T(\mathbf{X},0), \mathbf{X} \in \Omega \quad (8)$$

Specified boundary conditions are required in the analysis of all transient or steady-state problems.

$$T = T(\mathbf{X},t), \mathbf{X} \in S \quad (9)$$

b. Surface heat flux

$$\mathbf{q} = \mathbf{q}(\mathbf{X},t), \mathbf{X} \in S \quad (10)$$

c. Volumetric heat flux

$$T = \mathbf{r}(\mathbf{X},t), \mathbf{X} \in \Omega \quad (11)$$

d. Convection

$$\mathbf{q} = h(T - T^0), \mathbf{X} \in S \quad (12)$$

where $h = h(\mathbf{X},t)$ is the film coefficient and $T^0 = T^0(\mathbf{X},t)$ is the sink temperature.

e. Radiation

$$\mathbf{q} = A((T - T^z)^4 - (T^0 - T^z)^4), \mathbf{X} \in S \quad (13)$$

where A is the radiation constant (emissivity multiplied by the Stefan-Boltzmann constant) and T^Z is the temperature corresponding to absolute zero on the scale used. For example T^Z =-273°C or 0 k.

2.3 Moving heat sources

Chemical processes occurring in the weld pool, at elevated temperatures, and the choice of welding consumables affect the weld metal composition [6]. During the heating stage, heat energy has to be supplied to the solidus area. This solidus area includes weld nugget and a portion of the work pieces (weld pool). If the size and shape of a solidus area of the moving molten pool (equivalent heat source) is not determined, then an analytically specified volumetric heat source is used. The parameters of the heat sources are adjusted in a way that the result is approximately the shape of the molten zone. For each welding process a specific type of heat source is most effective. In MIG, TIG, welding residual stresses analysis, most researchers use a simplified 3D double ellipsoid model (see Fig. 2) developed by Goldak [7] for modelling of the heat source [8, 9]. It should be noted that the double ellipsoid heat source uses a Gaussian distribution of heat input. Whatever heat source is used, the mesh must be fine enough to capture the total amount of heat deposited.

In the Goldak model, the fractions of heat deposited in the front and rear of heat source are denoted by f_f and f_r, respectively, and these fractions are specified to satisfy $f_f + f_r = 2$. Let q denote the power density in W/m³ within the ellipsoid, and let a, b, and c denote the semi-axes of the ellipsoid parallel to the x, y, z axes. Then the power distribution inside the front and rear quadrant can be specified by

$$q(x,y,z,t)=\frac{6\cdot\sqrt{3}\cdot f_{f,r}\cdot Q}{a\cdot b\cdot c_{1,2}\cdot\pi\cdot\sqrt{\pi}}\exp^{\frac{-3\cdot x^2}{a^2}}\exp^{\frac{-3\cdot y^2}{b^2}}\exp^{\frac{-3\left(z-v\cdot(\tau-t)\right)^2}{c_{1,2}^2}} \tag{14}$$

In equation (14), Q is the heat available at the source, v is welding speed, τ is a lag time necessary to define the position of the heat source at time $t = 0$. For an electric arc the heat available is

$$Q=\eta VI \tag{15}$$

Where η is the heat source efficiency, $0 \le \eta \le 1$, V is the arc voltage, and I is the arc current. The parameters a, b, c_1, and c_2 are independent, and can take on different values for the front and rear quadrants of the source to properly represent the weld arc.

In the case the moving heat load is applied on the top surface of the model, some researchers [10, 11] employed a modified Gaussian distribution model of the arc heat flux. This states that [11]:

$$q(x,z,t)=\frac{3Q_{f,r}}{\pi a c_{1,2}}\exp^{\frac{-3x^2}{a^2}}\exp^{\frac{-3(z-v(\tau-t))^2}{c_{1,2}^2}} \tag{16}$$

It should be noted that in the case of the modelling of high energy welding process like laser, electron beam, a conical heat source would be more satisfactory [12].

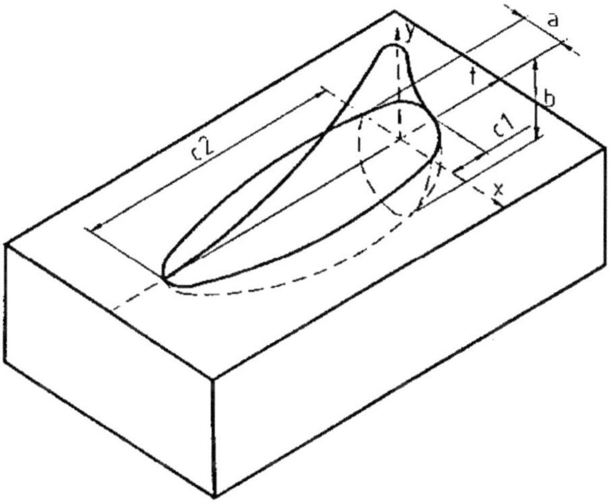

Fig. 2. Goldak double ellipsoid heat source model [6].

2.4 Latent heat effect

Latent heat has to be taken into consideration in case of microstructure transformations and melting solidification. Latent heat influences the formation of the transient temperature field. The metallurgical transformations depend on the thermal history. The thermal properties can be derived from the proportions of phase using mixture rule.

$$k(p_i,T)=\sum_{phases} p_i k_i(T)\ \ \rho(\rho_i,T)=\sum_{phases} p_i\rho_i(T)\ \ H(H_i,T)=\sum_{phases} p_i H_i(T) \tag{17}$$

For a two-phase transformation, we have $H(p_1,p_2,T)=p_1H_1(T)+p_2H_2(T)$ and $p_1+p_2=1$. Substituting them into equation (7), it gives:

$$\rho(p_1\frac{dH_1}{dT}+p_2\frac{dH_2}{dT})\dot{T}=div(k\cdot\mathbf{grad}T)+\rho\mathbf{q_s}-\rho\dot{p}_2(H_2-H_1) \tag{18}$$

where $(p_1\frac{dH_1}{dT}+p_2\frac{dH_2}{dT})$ represent an equivalent specific heat and $\rho\dot{p}_2(H_2-H_1)$ is latent heat of transformation.

The finite element method for thermal computation involves the solution of the system of differential equations as follow [28]:

$$\mathbf{C}\cdot\mathbf{T}+\mathbf{K}\cdot\mathbf{T}=\mathbf{Q} \tag{19}$$

where **T** is a vector of nodal temperature, **C** is the specific heat matrix, **K** is the conductivity matrix and **Q** is the vector of nodal powers equivalent to internal heat sources and boundary conditions.

2.5 Mechanical analysis

One important aspect in weld modelling is melting/re-melting effect. The strains in a weld are annealed at high temperatures. To understand material behaviour at elevated temperature, Dong et al [13] examined a simple one dimension (1D) thermo-plasticity problem. It shows that continuum and structural mechanics based FE codes are not intended to deal with material state (e.g., from solid to liquid) change. As a result, the standard FE computation results in accumulated plastic strains and gives wrong solutions. Therefore, a material point going through melting or re-melting should lose memory (prior plastic strain annihilation) on cooling. It was stated that [13], among all phase change effects, melting/re-melting effects are the most important in residual stress analysis.

As the weld torch moves, weld material is laid down below and behind the torch. Material deposition is another important aspect of finite element modelling of the weld process. Typically the finite element model of the weld joint contains the parent metal plate and all the weld passes in a single mesh. Welding of each pass is simulated in separate steps or sub-analyses. To simulate the first pass of a multi-pass weld the future weld passes are removed using a feature available within the finite element code. Feng et al. [14] use a special user material property subroutine within ABAQUS [15] where elements representing the weld metal are assigned the thermal properties of air during the thermal analysis. The weld bead elements always exist during the thermal analysis, but the thermal conductivity and the heat capacity of these elements are assigned small values to represent air, but then switched to the actual metal properties when the element enters the moving weld pool. For the mechanical analysis, a similar approach is used where the elements to be welded are first assigned a set of artificial, very soft properties. As the elements solidify from the weld pool the actual properties of the metal are reassigned.

SYSWELD [16], a specific welding simulation code use a special process called 'the chewing gum method'. The filler material is declared as an artificial phase which differs from hot but not yet molten material in the property; the modulus of elasticity is a fixed low value, for example 1.0N/mm^2 for solids and 5.0N/mm^2 for shells; the thermal strains are zero; this material - called 'chewing gum' - does not disturb the overall answer of the structure, and it behaves in a purely elastic manner; above the melting temperature, the chewing gum phase transfers to molten material within a small temperature range.

2.6 Case study: bead on plate analysis

This section describes the finite element analyses of an austenitic single bead on plate (BoP) specimen, the geometry and other properties are the same as a paper presented by Dennis et al [17]. A 3D half model, invoking symmetry along the centre of the weld bead is constructed. The ABAQUS finite element mesh consists of 45024 8-node brick elements and 55013 nodes. Based on this mesh, two analyses have been carried out:

Analysis A: Surface flux moving heat source; isotropic hardening assumed; torch moving along the centre line of the weld.

Analysis B: Volumetric flux moving heat source; both mixed isotropic/kinematic hardening and kinematic hardening are investigated; "dynamic fusion boundary method" [17] is adopted and a specific torch path has been derived via trial and error in order to match the observed fusion boundary.

PATRAN and ABAQUS/CAE are used to generate the numerical model; ABAQUS/Standard is used to simulate the welding process. There are two major features in welding simulation with ABAQUS: combined thermal-mechanical analysis procedures in sequentially coupled form; definition of moving torch and weld material deposition via user subroutines (*DFLUX, *GAPCON, *FILM)[15].

A typical sequentially coupled thermal-stress analysis consists of two ABAQUS/Standard runs: a heat transfer analysis and a subsequent stress analysis. To capture the weld deposition, one can use the element birth [14] technique; another approach is to use the GAPCON user subroutine. In the thermal analysis, a thermal contact between the weld and the base plate will be established. The GAPCON subroutine will switch the energy transport across the contact pair from zero to a value representative of welding conditions. The GAPCON subroutine can be written to activate the conductivity across the contact pair. By controlling the torch energy input into the joint using the DFLUX user subroutine, the moving torch can be simulated. The GAPCON subroutine can be used to model continuous deposition along the bead.

The UFILM subroutine should also be considered, the subroutine captures variable convection coefficients. For welding, this subroutine is used to describe the variable convection as the weld metal is laid on the parent material, and to activate convection on the top surface of the weld material as it is deposited.

Once the thermal analysis complete, the structural analysis can be executed. The structural analysis uses the thermal analysis (temperature) results as the loading. Boundary conditions are applied to restrain the system against rigid body motion. The thermal contact between the weld material and base plates is converted to 'tied constraints'. As the weld material cools and becomes stiffer based on temperature dependent material properties, the tied constraint will capture the proper constraint between the weld material and the parent material plate.

In the thermal analysis, it's clear that the penetration at start is much deeper than the rest of the plate, to generate enough penetration, a smaller moving Gaussian ellipse used at start, 1.0 second start dwell assumed in analysis A. It should be noted that the same heat input was applied, regardless of the Gaussian ellipse shape & size used. Figures 3 and 4 provided temperature contour plots showing the 1400°C isotherm along a longitudinal section through the middle of the bead, and a transverse section at weld mid-length, respectively. Figures 5 and 6 show 800°C peak temperature isotherms along the longitudinal section of the plate through the middle of the bead, and a transverse section at weld mid-length, respectively. Figure 7 shows a contour plot of von Mises stresses within the plate.

Fig. 3. 1400°C peak temperature profile along the centre surface (Analysis A).

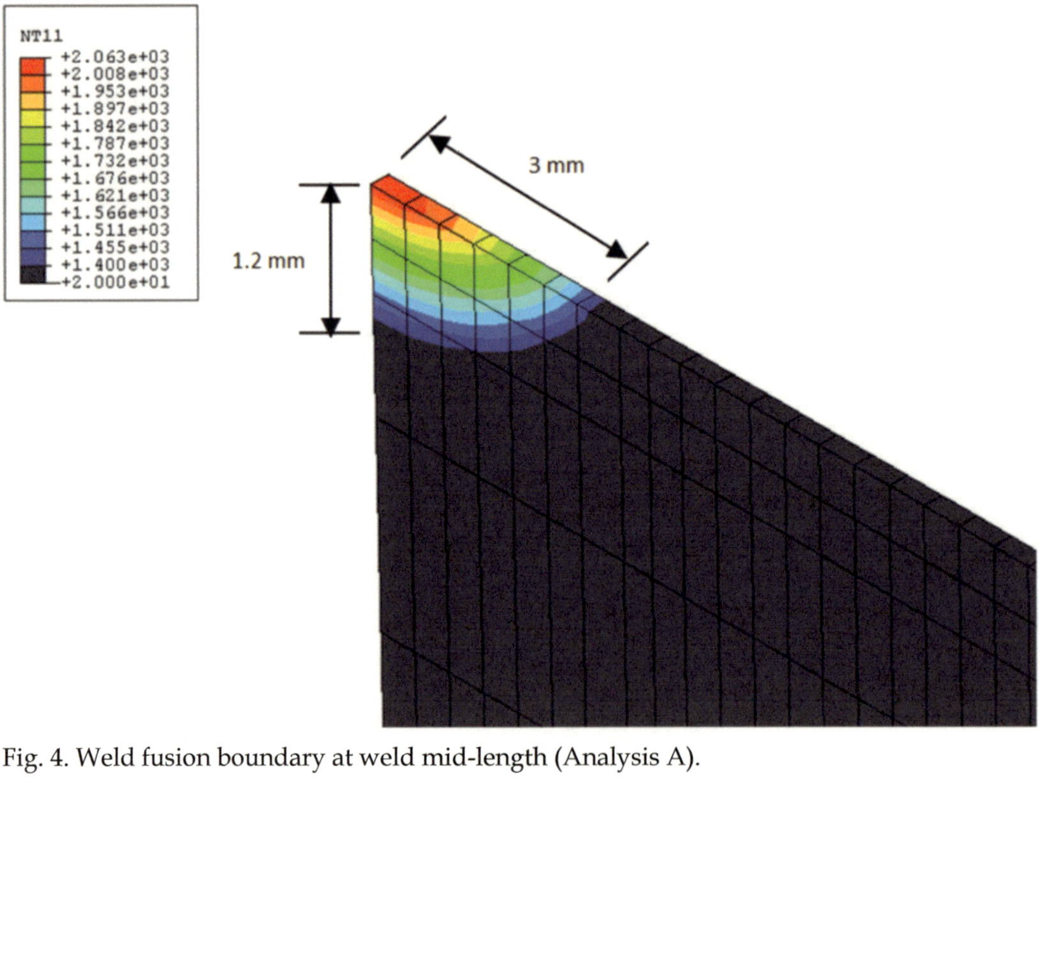

Fig. 4. Weld fusion boundary at weld mid-length (Analysis A).

Fig. 5. 800°C peak temperature isotherms along the centre surface (Analysis A)

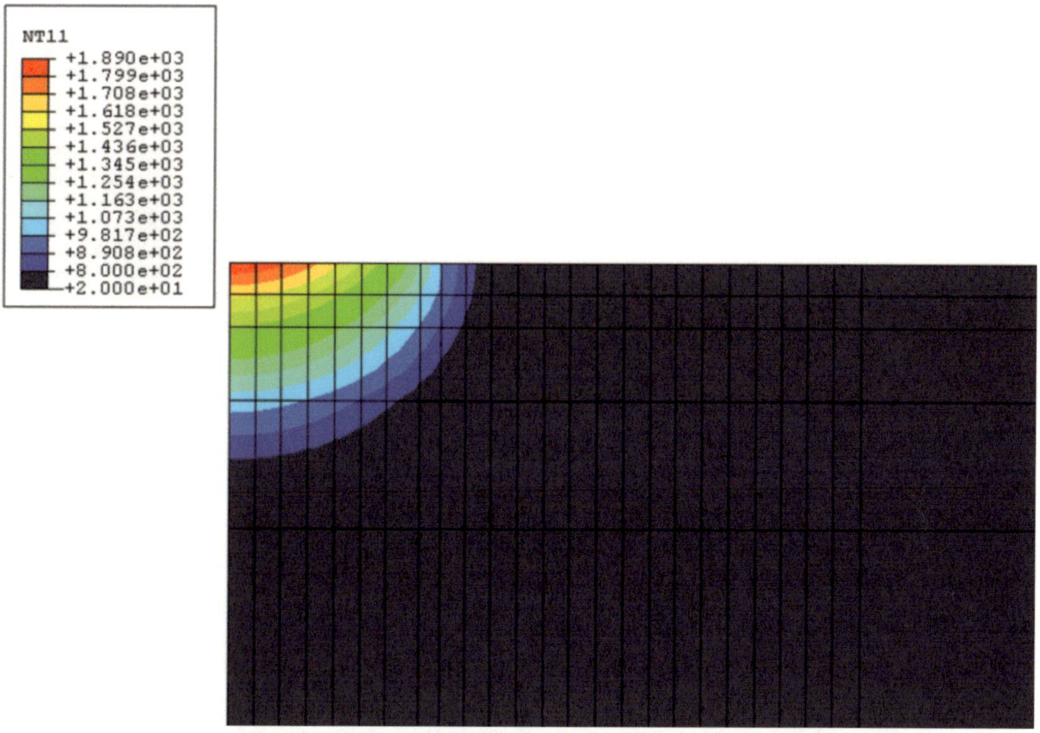

Fig. 6. 800°C peak temperature isotherms at weld mid-length (Analysis A).

Fig. 7. Residual von Mises stress (MPa); isotropic hardening (Analysis A).

In analysis B, because a half model is used, the power density used in equation (14) is multiplied by a coefficient Kq so that only half of the total heat flux deposited on the modelled plate upon integration equals the effective power from the torch, i.e.

$$q(x,y,z,t) = k_q \frac{6\cdot\sqrt{3}\cdot f_{f,r}\cdot Q}{a\cdot b\cdot c_{1,2}\cdot\pi\cdot\sqrt{\pi}}\cdot e^{\frac{-3\cdot x^2}{a^2}}\cdot e^{\frac{-3\cdot y^2}{b^2}}\cdot e^{\frac{-3(z-v\cdot(\tau-t))^2}{c_{1,2}^2}} \tag{20}$$

Figure 8 presents fusion boundary as a longitudinal section along the middle of the bead, showing both start and stop ends of the bead.

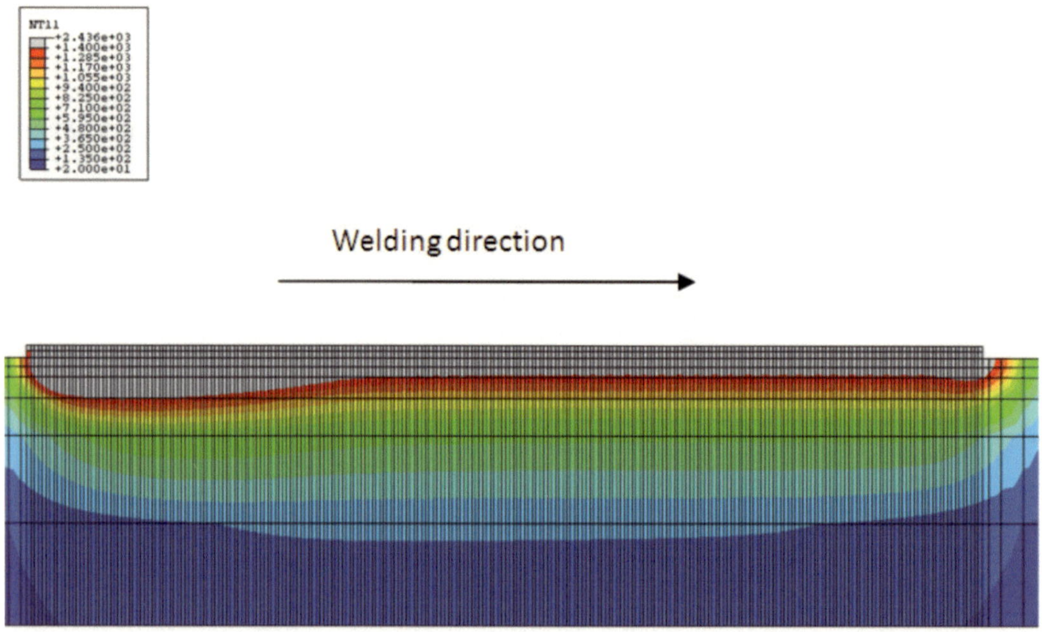

Fig. 8. 1400°C peak temperature profile along the centre surface; torch moved along a specific path, 1.0s start dwell (Analysis B).

While Fig. 9 presents fusion boundary as 5 transverse sections, which shows 'double-lobed' effect at some sections. Figures 10-14 show the predicted 1400°C isotherms compared with the macrographs in 5 transverse sections, respectively. Figure 15 shows the thermocouple measurements of the plate. Figure 16 shows the predicted transient temperature profiles.

The previous part detailed the thermal results from the bead on plate thermal analyses. This part presents the as-welded residual stress results from the subsequent mechanical analyses. Contour plots of the as-welded residual stresses results are similar between the kinematic and mixed hardening model and so contour plots are only presented for the latter in Figs. 17 and 18. These figures present longitudinal and transverse stresses on the bead symmetry plane.

A detailed comparison of the residual stresses has been performed by examination of line plot stresses along a number of sections. Here results are presented for a section on the symmetry plane, parallel to the bead but 2mm below the top surface of the plate. Results are shown in Fig. 19 for longitudinal stresses and Fig. 20 for transverse stresses.

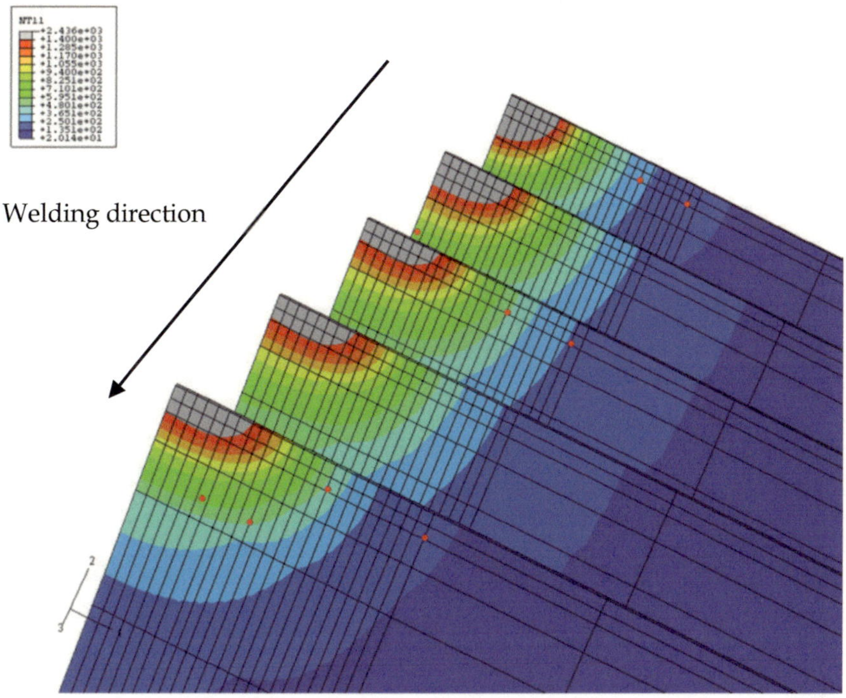

Fig. 9. Fusion boundary at different sections, thermal couples are highlighted (Analysis B).

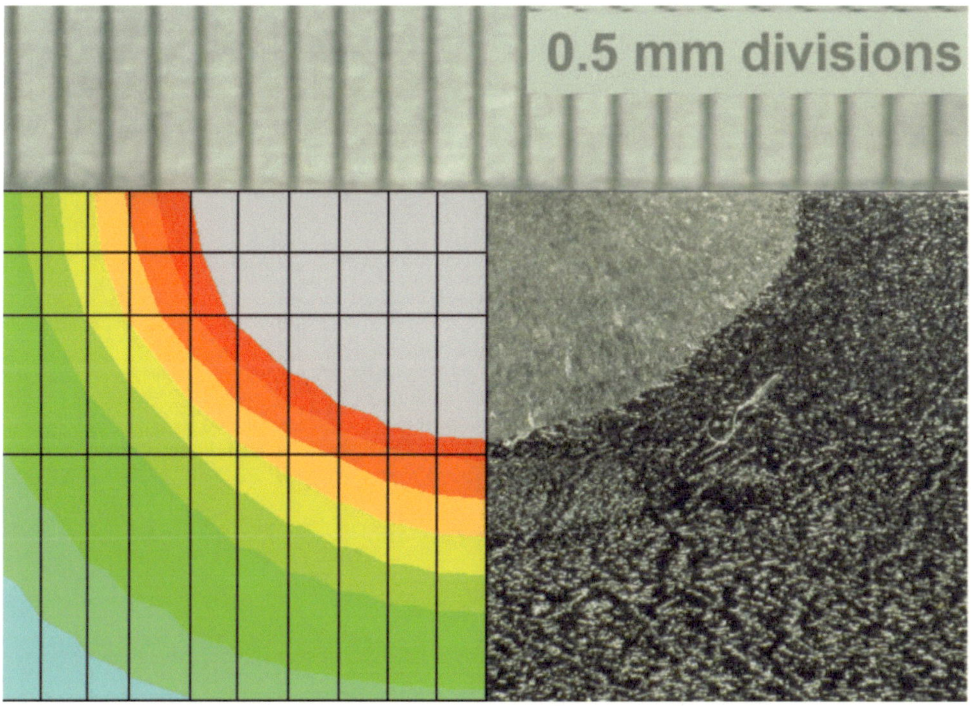

Fig. 10. Transverse section 1 (start-end) through weld bead (Analysis B).

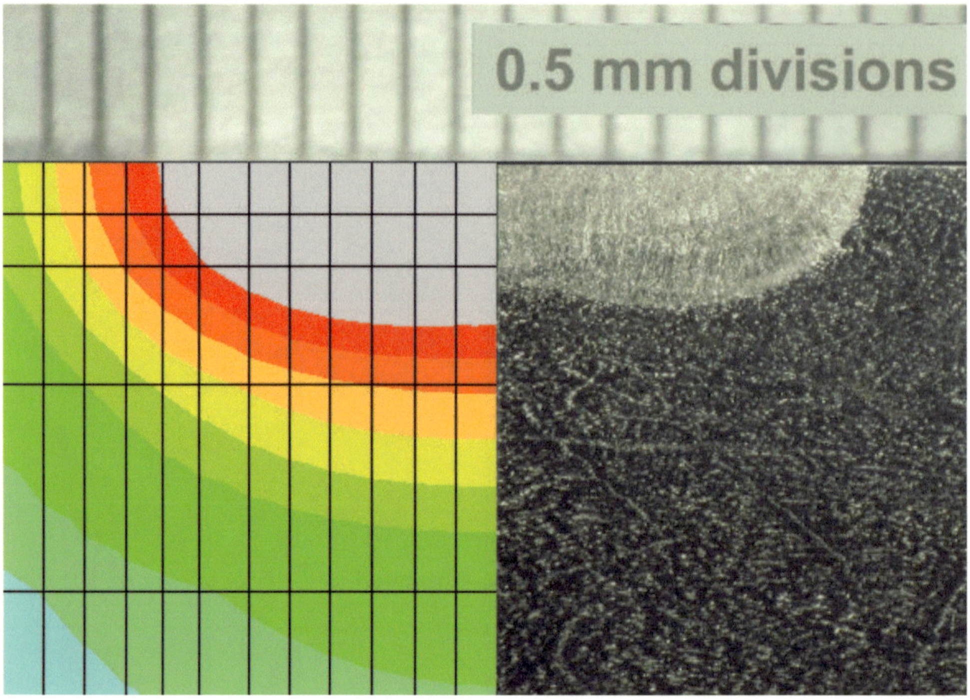

Fig. 11. Transverse section 2 (~25% along bead) through weld bead (Analysis B).

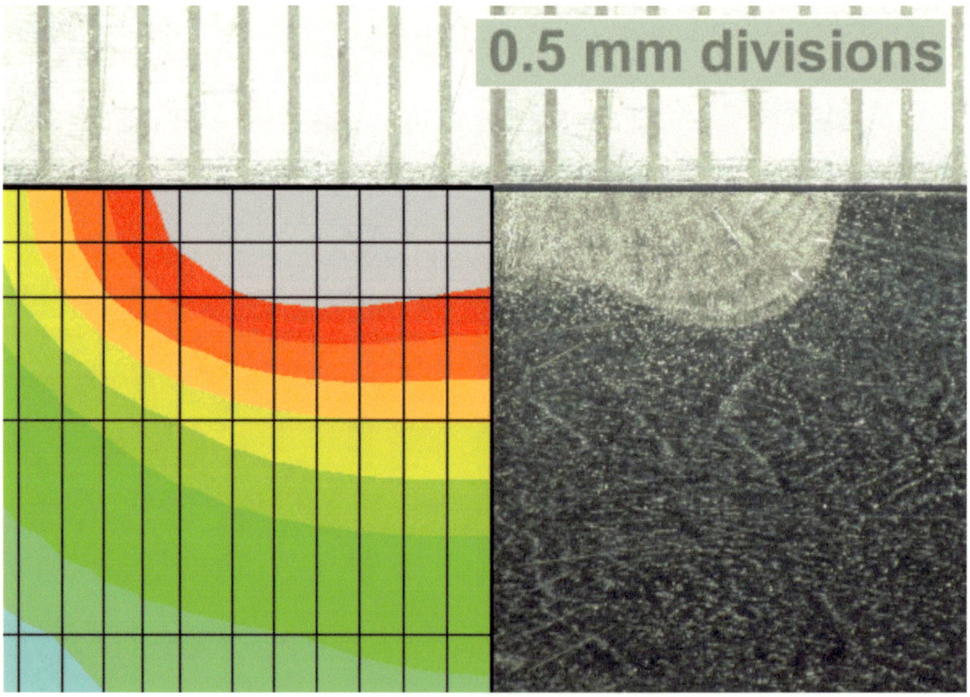

Fig. 12. Transverse section 3 (mid-length) through weld bead (Analysis B).

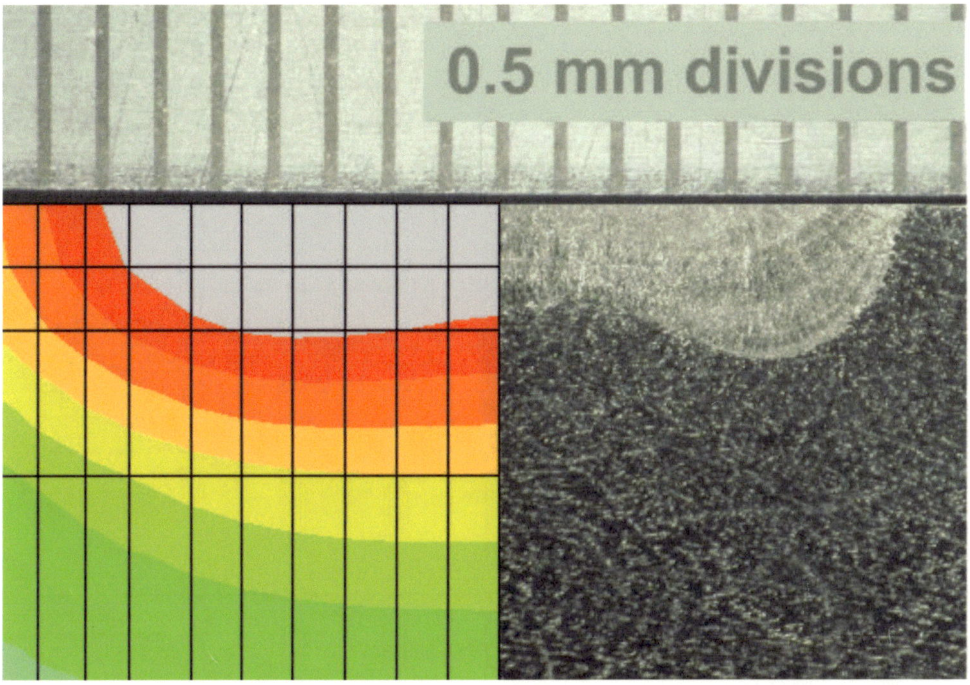

Fig. 13. Transverse section 4 (~75% along bead) through weld bead (Analysis B).

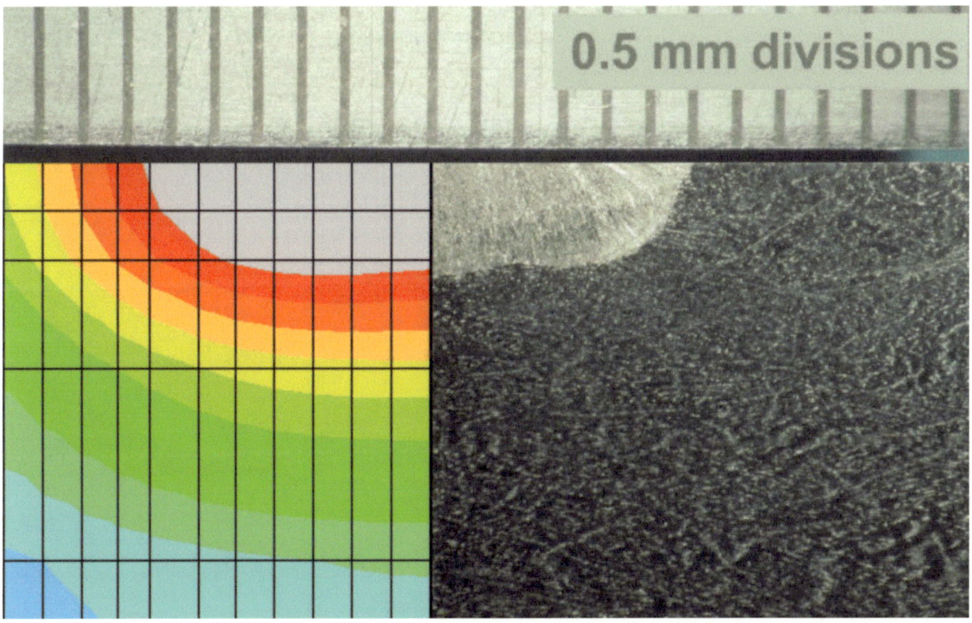

Fig. 14. Transverse section 5 (stop-end) through weld bead (Analysis B).

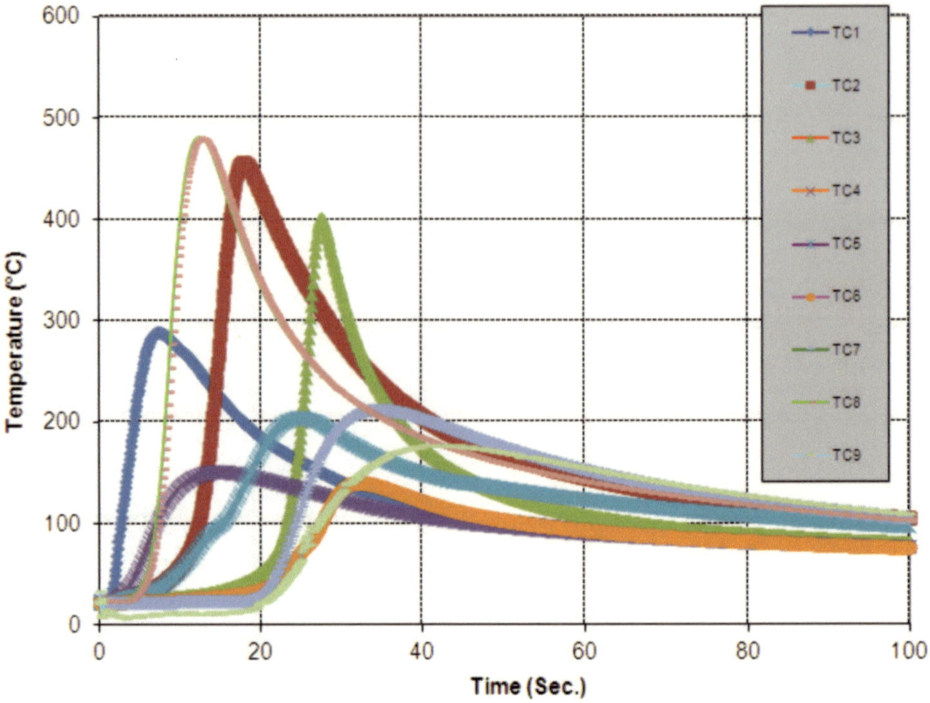

Fig. 15. Thermocouple measurements

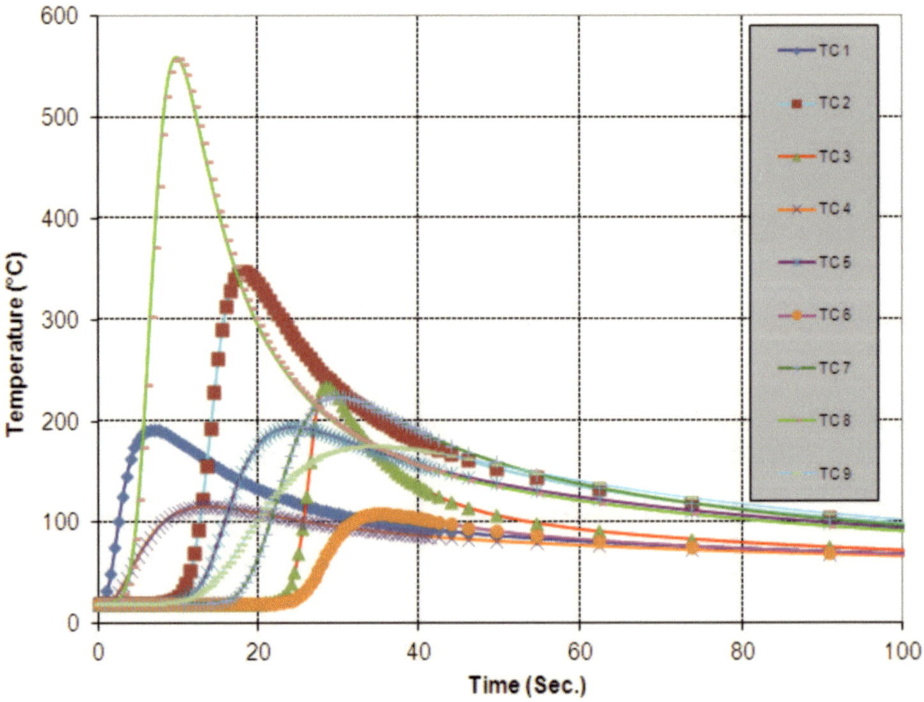

Fig. 16. Temperature predictions (Analysis B).

Fig. 17. Longitudinal stress (MPa); mixed hardening (Analysis B).

Fig. 18. Transverse Stress (MPa); mixed hardening (Analysis B).

Two separated analyses were completed to examine different heat source and hardening models. The first analysis utilize 2D Gaussian ellipse moving heat source with isotropic hardening. The second analysis utilize 3D Gaussian ellipsoidal moving heat source with both linear kinematic hardening and nonlinear isotropic/kinematic hardening, with an annealing temperature of 850°C.

The predictions of transient temperatures and the extent of the melted zone are first compared with thermocouple measurements made during welding, and with the results of destructive metallography. The predicted residual stresses are then presented in order to identify the effects on the predicted residual stresses of the material hardening model, global heat input, mechanical and thermal boundary conditions, and the handling of high temperature inelastic strains.

Due to the absence of an effective heat source fitting tool within ABAQUS, the Gaussian ellipsoidal arc parameters a, b, c_1, and c_2 were calibrated by performing a set of manual iterative transient temperature analyses, which were labour intensive and not very effective.

It appears that the use of isotropic hardening leads to over-conservative predictions of stresses, particularly in the longitudinal direction (refer to Fig. 19). Moving from isotropic to nonlinear isotropic/kinematic or linear kinematic hardening reduces the predicted stresses.

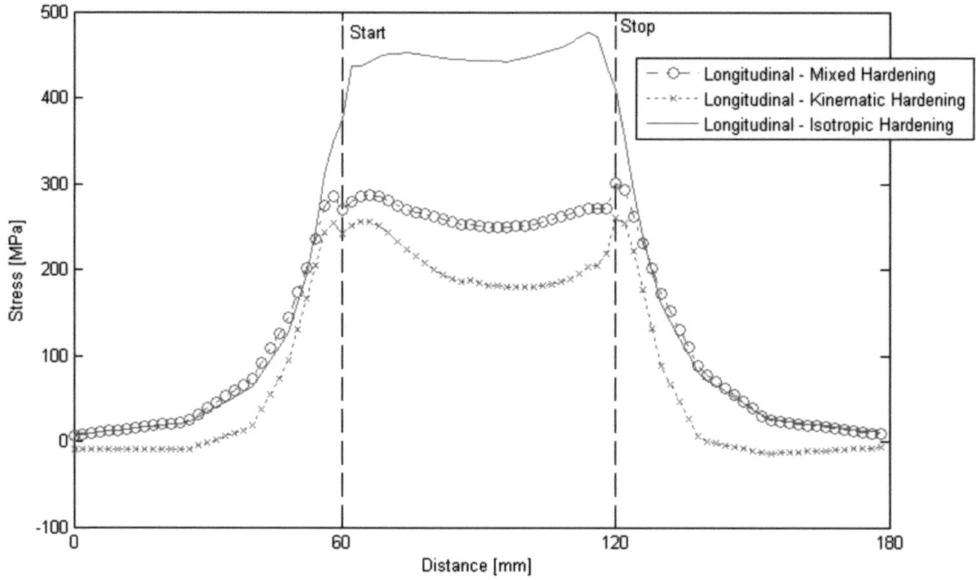

Fig. 19. Longitudinal stresses along section 2mm below top surface of plate (Analyses A and B).

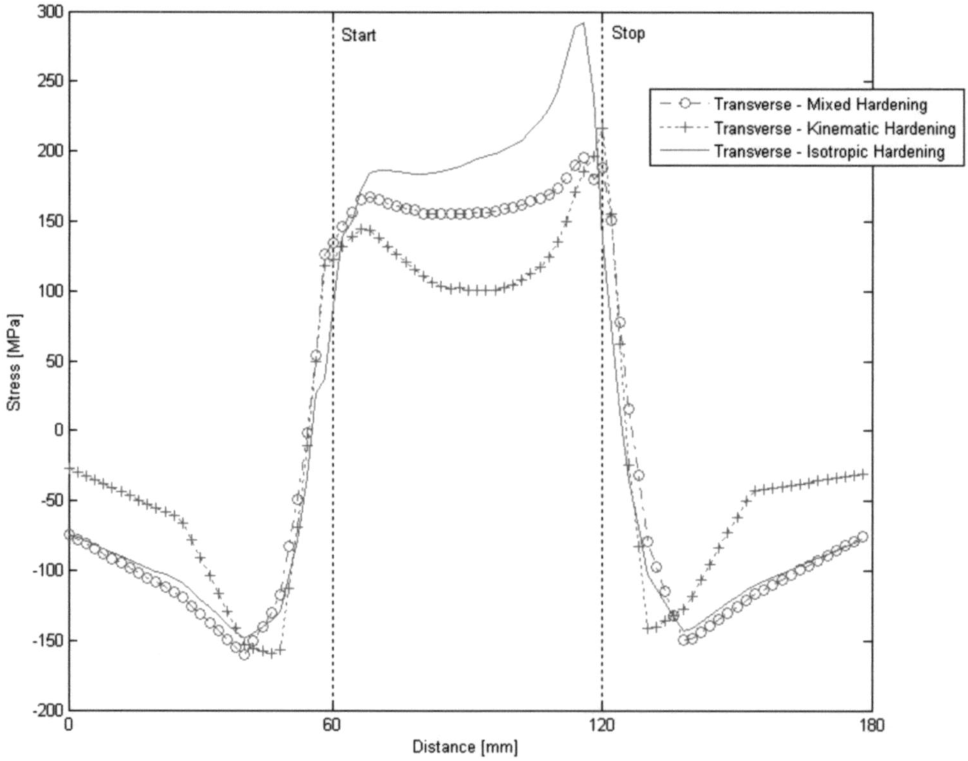

Fig. 20. Transverse stresses along section 2mm below top surface of plate (Analyses A and B).

3. Austenite decomposition kinetics for steels

The estimation of welding residual stresses in thick-section ferritic steel has been less successful than in austenitic stainless steel, largely due to the complexities associated with the solid state phase transformations that occur in multipass welding [1-3, 18-22]. To predict residual stresses in ferritic steels, a phase transformation kinetics model is therefore required for prediction of microstructure evolution. There is more than one model available to address phase transformations in ferritic steels [23-27]. In this section, we focused on one kind of model, namely, Li's model [25]. Li's model is a semi-empirical model for hardenable steels developed by Kirkaldy and Venugopalan [23, 27], and later refined by Li et al. This model is based on a phenomenological approach and on equations of the kinetic-chemical type. The equation is generally written as follow:

The austenite can, during cooling, transform into ferrite, pearlite, bainite and martensite. According to the Li's model, austenite decomposition is characterised as follows:

$$\frac{dX}{dt} = F(G,T,C)X^{0.4(1-X)}(1-X)^{0.4X} \tag{21}$$

Where X is the volume fraction of the transformation product at a given instant of time, G is the ASTM grain size number, C is chemical composition, and and T the absolute temperature.

Austenite is stable at temperatures above the Ae_3 (equilibrium temperature for austenitization end) and is unstable below the Ae_3 line. As the temperature drops below the Ae_3, ferrite begins to form. For the austenite-ferrite decomposition, the function F is expressed by:

$$F(G,T,C) = \frac{2^{0.41G}(Ae_3 - T)^3 \exp(-27500 / RT)}{\exp(-4.25 + 4.12C + 4.36Mn + 0.44Si + 1.71Ni + 3.33Cr + 5.19\sqrt{Mo})} \tag{22}$$

Where, R is the gas constant. For temperatures below eutectoid, and depending upon the cooling rate, the untransformed austenite will tend to decompose to pearlite. For the austenite-pearlite decomposition, the function F is expressed by:

$$F(G,T,C) = \frac{2^{0.32G}(Ae_1 - T)^3 \exp(-27500 / RT)}{\exp(1.0 + 6.31C + 1.78Mn + 0.31Si + 1.12Ni + 2.70Cr + 4.06Mo)} \tag{23}$$

where Ae_1 is quilibrium temperature for austenitization start. When the bainite-start temperature B_s is reached, in this empirical theory, the pearlite is assumed to continue to form bainite at a rate given by:

$$F(G,T,C) = \frac{2^{0.29G}(Bs - T)^2 \exp(-27500 / RT)}{\exp(-10.23 + 10.18C + 0.85Mn + 0.55Ni + 0.90Cr + 0.36Mo)} \tag{24}$$

The martensite transformation is described by the Koistinen-Marburger [28] relationship.

$$X(T) = 1 - \exp(-b(M_s - T) \text{ for } T<M_s \tag{25}$$

There are only two parameters (exponent *b* and martensite start temperature M_s) for this model.

The critical temperatures Ae_3, Ae_1, M_s and B_s are calculated either from thermo-dynamic software, such as MTDATA [29] or from empirical formula [25]. Figures 21 and 22 show two typical phase diagram calculated by MTDATA.

Other researchers have used Kirkaldy based equations to model microstrural evolutions during welding or forming processes, for example Watt et al. [26], Henwood et al. [30] and Akerstrom et al. [31], Lee et al. [32, 33].

Traditionally, the kinetics of transformation is typically described by a standard equation known as the Kolmogorov-Johnson-Mehl-Avrami equation [34-38], named after the individual researchers who derived it. The KJMA equation has been used by many authors to describe phase fraction change [39, 40]. Other methods to model phase transformation include phase-field simulation [41], neural network [42], Monte Carlo simulation [43], etc.

In Li's model a constant grain size is used. Figures 23-25 show continuous cooling transformation (CCT) diagram predicted by using Li's model for SA508 Grade 3 for three different grain sizes [2]. It can be seen that the austenite grain size has significant effect on CCT curves. Further work is in progress in which the author is investigating the phase transformation kinetics incorporating austenite grain growth.

4. Weld residual stress analysis including phase transformation effects

The volume changes that occur in steels as they are heated and cooled can be inferred from dilatometry, where the change in length of an unloaded specimen is measured as a function of temperature. Figure 26 illustrates such an experiment - the upper straight line represents the expansion of the body-centred cubic phase (ferrite, bainite, martensite) and the lower line that of austenite (γ). Data at locations between the upper and lower lines correspond to the co-existence of the parent and product phases. The transformations occurred at different temperatures upon heating and cooling. The transformation temperature is a function of the cooling rate, steel composition and austenite grain size. The measured coefficient of thermal expansion is larger for austenite (~ 23×10^{-6} K^{-1}) than for ferrite (~ 15×10^{-6} K^{-1}). As a consequence, the volume change due to transformation is greater during cooling than during heating. The volume expansion due to the transformation of austenite can partly compensate for thermal contraction strains arising as a welded joint cool. In this section, some finite element modelling cases have been presented. Predictions are compared and rationalised alongside measurements obtained by neutron diffraction.

The thermal analysis is followed by a mechanical analysis in a sequentially coupled model. The total strain rate can be partitioned as follows:

$$\dot{\varepsilon} = \dot{\varepsilon}^{e} + \dot{\varepsilon}^{p} + \dot{\varepsilon}^{th} + \dot{\varepsilon}^{tr} + \dot{\varepsilon}^{tp} \tag{26}$$

where $\dot{\varepsilon}$, $\dot{\varepsilon}^{e}$, $\dot{\varepsilon}^{p}$, $\dot{\varepsilon}^{th}$, $\dot{\varepsilon}^{tr}$, $\dot{\varepsilon}^{tp}$ are total strain rate, elastic strain rate, plastic or viscoplastic strain rate, thermal, transformation (i.e., volume change) and transformation plasticity, strain rates, respectively.

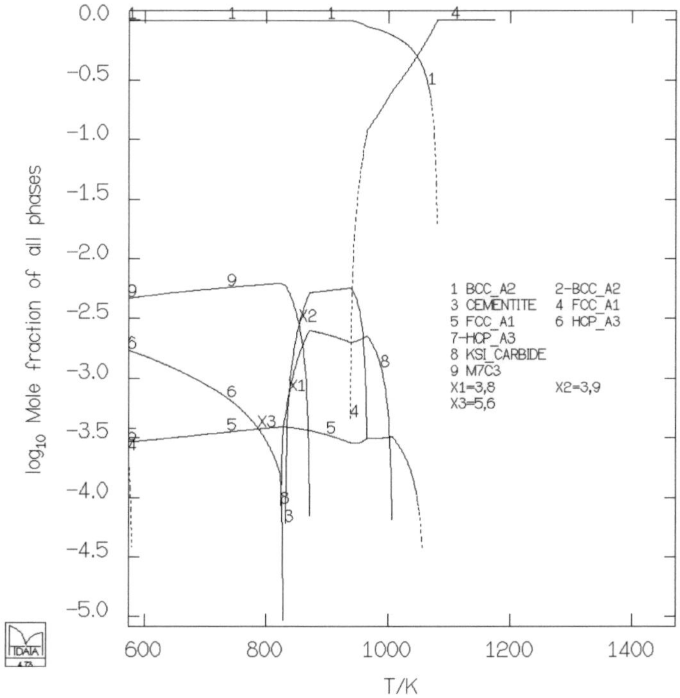

Fig. 21. Equilibrium mole fractions of phases versus temperature for SA508

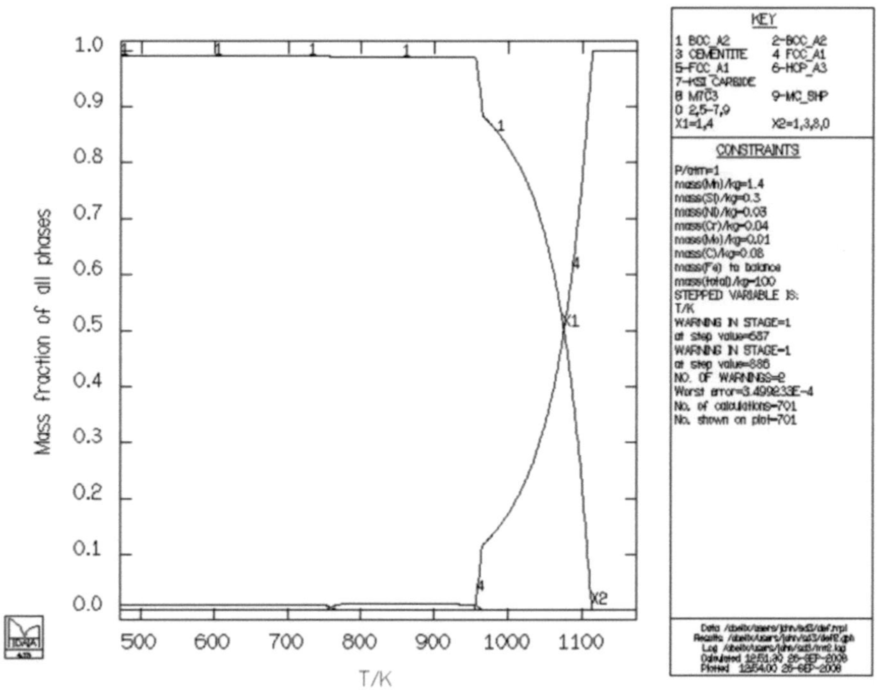

Fig. 22. Equilibrium mass fractions of phases versus temperature for SD3

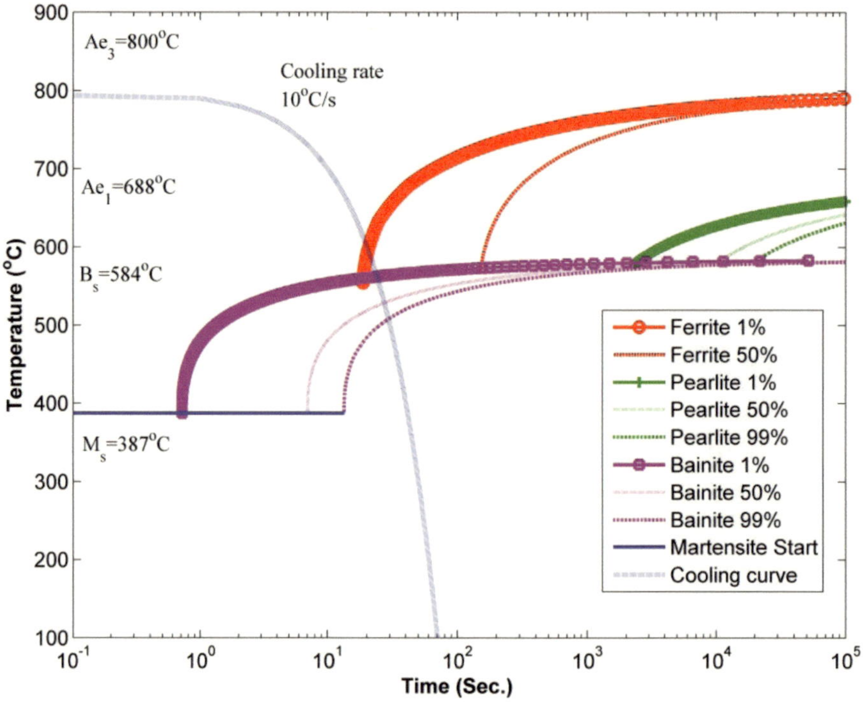

Fig. 23. Predicted CCT curve for SA508, grain size is 10 micrometers

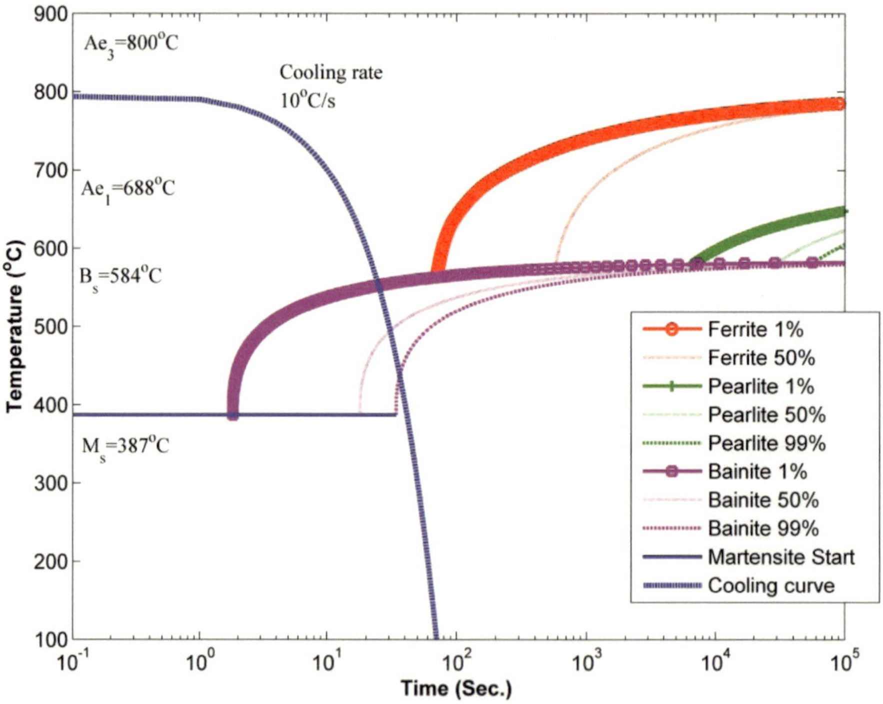

Fig. 24. Predicted CCT curve for SA508, grain size is 50 micrometers

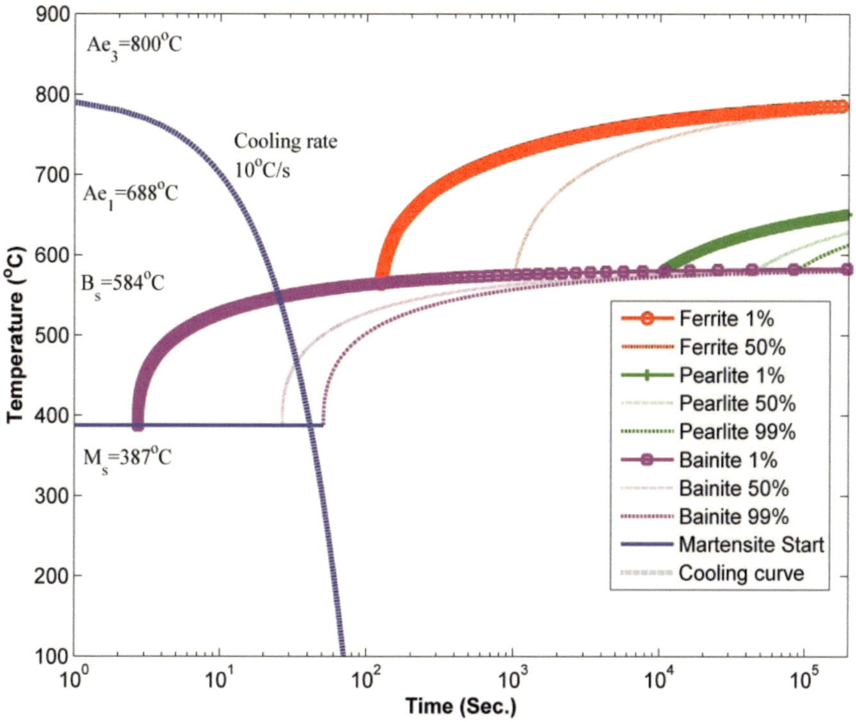

Fig. 25. Predicted CCT curve for SA508, grain size is 100 micrometers

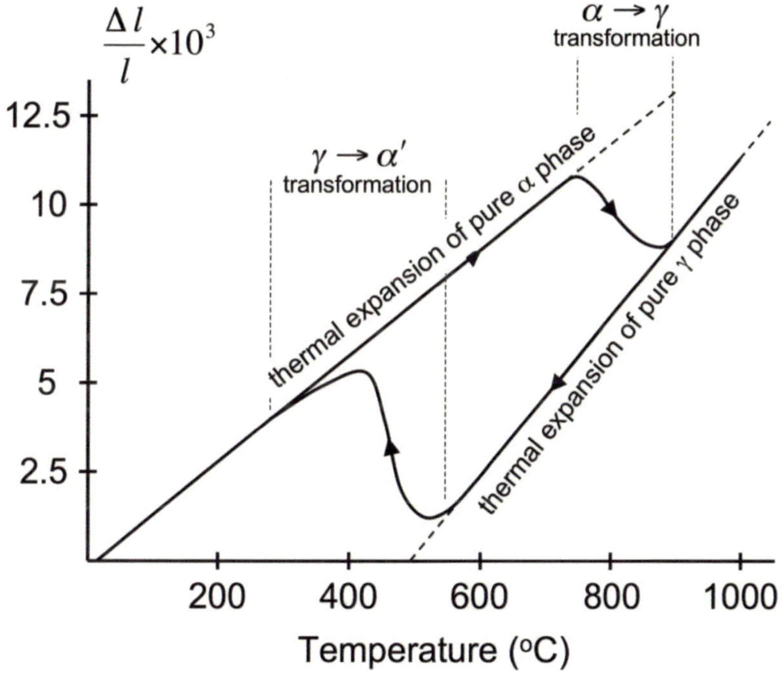

Fig. 26. Dilatometric diagram of SA508 steel heated at 30 K s^{-1} and cooled at 2 K s^{-1} [1, 2].

The thermal strain rate $\dot{\varepsilon}^{th}$ and transformation strain rate $\dot{\varepsilon}^{tr}$ can be expressed as:

$$\dot{\varepsilon}^{th} = (X_A\alpha_A + X_F\alpha_F)\dot{T}\cdot 1 \tag{27}$$

$$\dot{\varepsilon}^{tr} = \varepsilon^{tr^*}\dot{X}_F\cdot 1 \tag{28}$$

Where, X_A, X_F are the volume fractions of austenite and ferrite phases, respectively, α_A, α_F are the thermal expansion coefficients of the corresponding phases, ε^{tr^*} is the dilatation due to austenite decomposition, **1** is the second order identity tensor.

Each phase has its own mechanical properties for plasticity, i.e., yield stress and a strain hardening law. Generally, applying a linear mixture law to the mixture of all phases gives:

$$\sigma^y(T) = \sum_{\text{phases}} X_i\sigma_i^y(T) \tag{29}$$

Leblond et al [44] considered two types of phases, one is a hard phases or alpha phase (mixture of ferrite phases); the other is a soft phase or gamma phase (austenite). A non linear mixture law was used between hard phases and soft phase.

4.1 Case study: single pass weld

This section describes the finite element analyses of two rectilinear single pass welding plates (375×200×12 mm) prepared from the high strength steel Weldox 960 [2, 18]. A 5-mm-deep "V" groove was machined along the center of each plate, with an included angle of 60 deg, into which a single weld bead of either OK75.78 or LTTE was deposited using manual-metal arc welding. The welding was performed in the down hand position with a heat input between 2.2 and 2.5 kJ mm $^{-1}$, while a preheat temperature of 125°C was used and the plates were restrained by clamping during welding. The compositions of the two filler materials and the parent material are given in Table 1.

Material	C	Si	Mn	Cr	Ni	Mo	Cu
Weldox 960	0.20	0.50	1.6	0.7	2.0	0.7	0.3
OK 75.78	0.05	0.19	2.0	0.4	3.1	0.6	—
LTTE	0.07	0.20	1.3	9.1	8.5	—	—

Table 1. Approximate compositions of base plate and filler metals in weight percent

In this work, prior to running SYSWELD models, the transformation temperatures of the parent steel and weld metals were estimated according to software available via MTDATA [29]. Estimates for the elevated-temperature yield stress of each weld metal were obtained by examining the results of the Satoh tests and assuming that once the stress level had reached yield, and prior to the commencement of a transformation, the stress that is recorded in a Satoh test can be assumed to be representative of the yield locus. A room-temperature value for the yield stress of the parent material was obtained from the manufacturer's data sheets. At intermediate temperatures, values for the yield stress were either interpolated or extrapolated. For each weld, a transient 3-dimensional analysis was carried out. A 3D half model, invoking symmetry along the centre of the weld bead is constructed, as shown in Fig. 27. Complex arc and weld pool phenomena were not considered, as is generally the case for the numerical prediction of weld residual stresses. However, a double-ellipsoid heat source was used to represent the welding arc, and this was calibrated using the in-built heat-source fitting tool within SYSWELD, by comparing the predicted geometries for the fusion zones and HAZ's with those measured from macrograph sections through each welded plate. In each case, a 3-dimensional simulation was carried out using 48,240 eight-noded brick elements in both the transient-thermal and mechanical analyses. In order to account for the effects of annealing, the history of any element, including any plastic strain, was erased if the peak temperature exceeded the temperature of fusion. Otherwise, the model was configured to simulate the welding conditions as accurately as possible. Figure 28 presents fusion boundary at mid-length transverse section through weld pass on the plate.

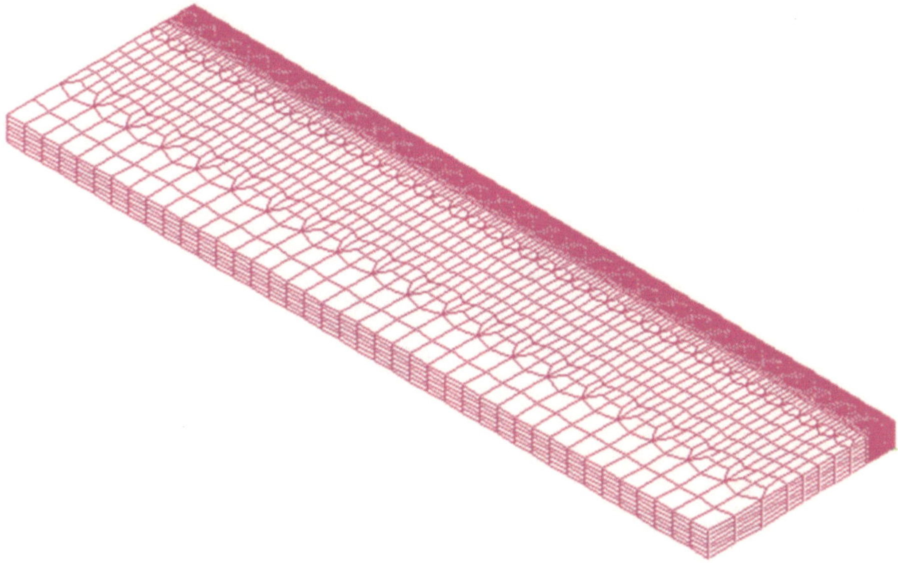

Fig. 27. Mesh of the plate being welded

Fig. 28. Fusion zones produced for both cases.

A comparison of the longitudinal residual stress distributions predicted by SYSWELD with those that were measured by neutron diffraction is given in Fig. 29. The neutron diffraction results (top) show that, when compared to OK75.78, the LTTE weld metal introduces significant longitudinal compressive residual stresses within the fusion zone. As discussed above, this is consistent with LTTE having a lower transformation temperature than OK75.78. Furthermore, the peak tensile residual stresses, which for both welds appear to arise just outside the HAZ, are somewhat lower in the weld made with LTTE, and they arise over a smaller region. In the weld made with the OK75.78 filler material, the phase transformations that have taken place within the HAZ and fusion zone still have significantly reduced the residual stresses to levels that are below the peak tensile stresses found immediately outside the HAZ. Since the transformation temperatures of the weld metal and parent material are similar in this case (Weldox transforms at around 460 °C compared to 440 °C for diluted OK 75.78 [18]), these zones appear to behave similarly, and there is no discernible variation in residual stress with distance down the weld centre-line.

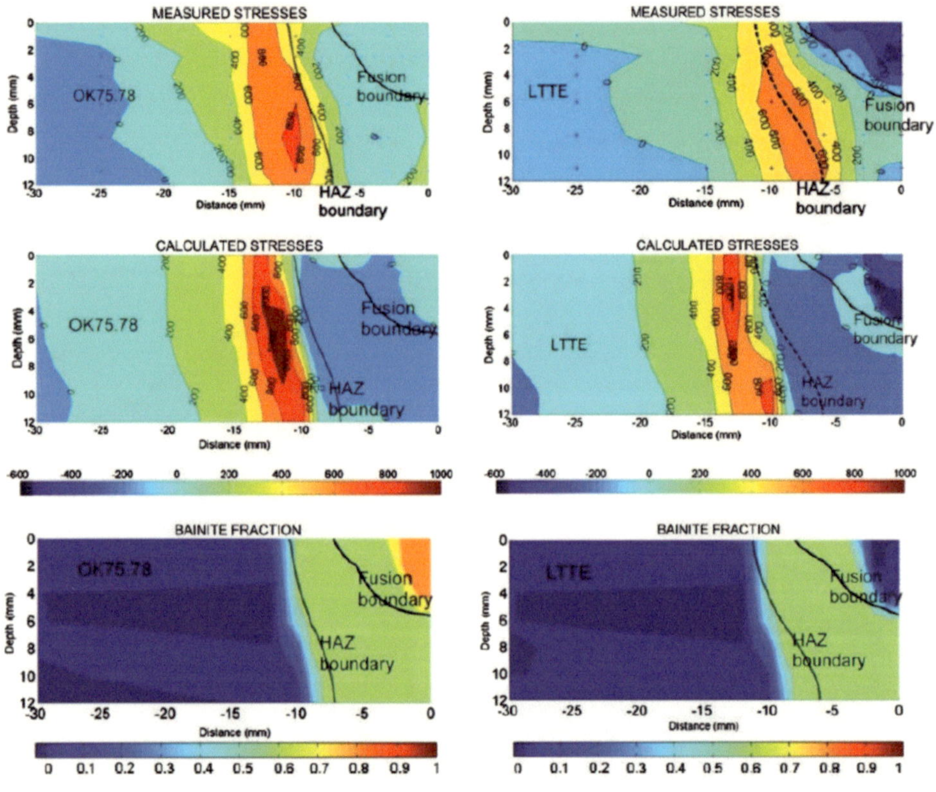

Fig. 29. Comparison of neutron diffraction measurements of longitudinal residual stress (top) in the near weld region as reported in Ref. 10 with the stress distributions predicted by SYSWELD (middle) for welds made using the OK75.78 filler metal (left) and the LTTE filler metal (right).The predicted proportions of bainite formed as a consequence of welding appear for each weld at the bottom of the figure. Note that very little bainite exists in the LTTE weld metal, because this is a martensitic alloy. Beyond the HAZ boundary, the bainite fractions are equal to zero, because no phase transformations took place at these locations during welding. All dimensions are in mm and all stresses are in MPa.

It is evident from Fig. 29 that the major features of the stress distributions have been predicted correctly for both filler metals. For example, it appears that the location of the peak tensile stresses is approximately correct in each case. Furthermore, the models have predicted that the OK75.78 filler metal will lead to relatively low stresses within and directly underneath the weld bead, and that the LTTE filler metal will introduce highly compressive residual stresses to the weld metal region at about the right level (~ -400MPa). The magnitude of the tensile peak stresses, however, appears to be somewhat over-estimated by the SYSWELD models. This may be related to the limitations associated with using data from Satoh tests to estimate the yield locus. In both cases, isotropic hardening was assumed, but the authors also created models that assumed kinematic hardening and it was found that, for the single welding thermal cycle, there did not appear to be any notable sensitivity to the hardening model adopted.

Interestingly, SYSWELD predicts that the transformation temperature of the weld metal does not have a significant effect on the magnitude of the peak stresses beyond the HAZ, although a small reduction in the extent of the peak-stress region does appear to have been captured. It is possible, however, that the effects of the weld metal transformation temperature would have been predicted more accurately if the transformation strains within the fusion zone and HAZ had been represented faithfully. In this respect it should also be noted that variant selection has not been incorporated in the SYSWELD model. This means that only the volume part of the transformation has been accounted for. The anisotropic shear component may be significant, especially for transformations at low temperature, where the stress just prior to transformation will be large and thus may bias variant selection. This would increase the effectiveness of the transformation in reducing the residual stress. This effect will be reported subsequently.

While the overall agreement between the simple SYSWELD model trained only on the basis of Satoh tests and the neutron results is encouraging, the discrepancies between the models and the neutron diffraction measurements highlight the need for validation and improved understanding of the transformation behaviour. In this respect further in-situ synchrotron X-ray diffraction experiments, such as those described in another article [18], offer the potential for the quantification of phase fractions during any simulated welding thermal cycle that may be of interest, and they can also reveal the extent to which stress-induced transformation texture (or variant selection) may contribute to anisotropy in macroscopic transformation strains.

5. Conclusion

Mathematic modelling of austenitic (non-transforming) and ferritic (transforming) steel welds has been investigated by using finite element codes ABAQUS and SYSWELD. This is part of a research programme the author carried out at the University of Manchester, UK. The aim of the work is to understand welding residual stress by numerical modelling and neutron diffraction measurements.

Numerical studies of a single bead on plate austenitic steel weld suggest that the use of isotropic hardening leads to over-conservative predictions of stresses, particularly in the longitudinal direction. Moving from isotropic to nonlinear isotropic/kinematic or linear kinematic hardening reduces the predicted stresses.

Numerical studies of a single pass ferritic steel welds show that phase transformation and transformation temperature have significant effects on residual stress. Neutron diffraction measurements supported numerical results. Utilities were developed within the finite element code ABAQUS to incorporate solid state phase transformations (micro-structural evolutions, phase dependent properties and volume changes) in welding residual stress analysis.

6. Acknowledgements

The author would like to acknowledge support from Rolls-Royce Marine, and the Stress & Damage Characterisation (SDC) unit at the University of Manchester.

7. References

[1] Francis JA, Bhadeshia HKDH, Withers PJ. Welding residual stresses in ferritic power plant steels. Materials Science and Technology 2007;23:1009.

[2] Dai H, Francis JA, Keavey MA, Withers PJ. Modelling of Microstructural Evolution in Thick-Section Ferritic Steel Welds. Manchester: University of Manchester, 2008. p.28.

[3] Dai H, Francis JA, Turski M, Withers PJ. Weld Modelling and Material Models For Steels. Manchester: University of Manchester, 2007. p.42.

[4] Lindgren LE. Finite Element Modeling And Simulation Of Welding Part 1: Increased Complexity. Journal of Thermal Stresses 2001;24:141.

[5] Lai WM, Rubin D, Krempl E. Introduction to Continuum Mechanics (3rd Edition). Elsevier, 1996.

[6] Grong O. Metuallurgical Modelling of Welding. London: The Institute of Materials, 1997.

[7] Goldak J, Breigruine V, Dai N, Hughes E, Zhou J. Thermal Stress Analysis in Solids Near the Liquid Region in Welds. In: Cerjak H, editor. Mathematical Modelling of Weld Phenomena. London: The Institute of Materials, 1997. p.543.

[8] Murugan S, Rai S K, Kumar P V,. Temperature distribution and residual stresses due to multipass welding in type 304 stainless steel and low carbon steel weld pads. International Journal of Pressure Vessels & Piping 2001;78:307.

[9] Parameswaran P, Paul V T, Vijayalakshmi M,. Microstructural evolution in a single pass autogenously welded 2.25Cr-1Mo steel. Trans. Indian Inst. Met. 2004;57:253.

[10] Fanous, Ihab FZ, Younan, Maher YA, Wifi, Abdalla S. 3-D finite element modeling of the welding process using element birth and element movement techniques. New York, NY, ETATS-UNIS: American Society of Mechanical Engineers, 2003.

[11] Wu CS, Yan F. Numerical simulation of transient development and diminution of weld pool in gas tungsten arc welding. Modelling and Simulation in Materials Science and Engineering 2004;12:13.

[12] Goldak J, Bibby M, Moore J, House R, Patel B. Computer modeling of heat flow in welds. Metallurgical and Materials Transactions B 1986;17:587.

[13] Dong P. Residual stresses and distortions in welded structures: a perspective for engineering applications. Science and Technology of Welding & Joining 2005;10:389.

[14] Feng Z, Wang XL, Spooner S, Goodwin GM, Maziasz PJ, Hubbard CR, Zacharia T. A finite element model for residual stress in repair welds. Conference: American Society of Mechanical Engineers (ASME) pressure vessels and piping conference, Montreal (Canada), 21-26 Jul 1996; Other Information: PBD: 28 Mar 1996, 1996. p.Medium: ED; Size: 7 p.

[15] ABAQUS. User's Manual. Hibbit, Karsson and Sorenson, Inc., 2006.

[16] ESI-Group. SYSWELD, Theory Manuals. ESI UK, John Eccles House, The Oxford Science Park, Oxford., 2006.

[17] Dennis RJ, Leggatt NA, Gregg A. Optimisation of Weld Modelling Techniques: Bead-on-Plate Analysis. ASME Conference Proceedings 2006;2006:967.

[18] Dai H, Francis JA, Stone HJ, Bhadeshia HKDH, Withers PJ. Characterizing Phase Transformations and Their Effects on Ferritic Weld Residual Stresses with X-Rays and Neutrons. Metallurgical and Materials Transactions A 2008;39:3070.

[19] Dai H, Keavey MA, Withers PJ. Modelling of Residual Stress in Thick-Section Ferritic Steel Welds Considering Phase Transformation Effects. Manchester: University of Manchester, 2009. p.34.

[20] Dai H, Francis JA, Withers PJ. Prediction of residual stress distributions for single weld beads deposited on to SA508 steel including phase transformation effects. Materials Science and Technology 2010;26:940.

[21] Dai H, Mark AF, Moat R, Shirzadi AA, Bhadeshia HKDH, Karlsson L, Withers PJ. Modelling of residual stress minimization through martensitic transformation in stainless steel welds. In: Cerjak H, Enzinger N, editors. Mathematical Modelling of Weld Phenomena 9. Graz: Technische Universitat Graz, 2010. p.239.

[22] Dai H, Moat R, Mark AF, Withers PJ. Investigation on transformation induced plasticity and residual stress analysis in stainless steel welds. ASME 2010 Pressure Vessels & Piping Division / K-PVP Conference. Bellevue, Washington: ASME, 2010. p.PVP2010.

[23] Kirkaldy JS, Venugopalan D, Marder AR, Goldstein JI. Prediction of microstructure and hardenability in low alloy steels. In: Marder AR, Goldstein JI, editors. Phase Transformations in Ferrous Alloys. New York: AIME, 1984. p.125.

[24] Leblond JB, Devaux J. A new kinetic model for anisothermal metallurgical transformations in steels including effect of austenite grain size. Acta Materialia 1984;32:137.

[25] Li M, Niebuhr D, Meekisho L, Atteridge D. A computational model for the prediction of steel hardenability. Metallurgical and Materials Transactions B 1998;29:661.

[26] Watt D, Coon L, Bibby M, Goldak J, Henwood C. An algorithm for modelling microstructural development in weld heat-affected zones (part a) reaction kinetics. 1988;36:3029.

[27] Kirkaldy JS. Diffusion-controlled phase transformations in steels. Theory and applications. Scandinavian Journal of Metallurgy 1991;20:50.

[28] Koistinen DP, Marburger RE. A general equation prescribing the extent of the austenite-martensite transformations in pure iron-carbon alloys and plain carbon steels. 1959;7:59.

[29] National Physical Laboratory NPL. MTDATA Software. Teddington, UK, 2006.

[30] Henwood C, Bibby M, Goldak J, Watt D. Coupled transient heat transfer-microstructure weld computations (part b). 1988;36:3037.

[31] Akerstrom P, Oldenburg M. Austenite decomposition during press hardening of a boron steel computer simulation and test. Journal of Materials Processing Technology, vol. 174:1-3. Journal of Materials Processing Technology, 2006.
[32] Lee SJ, Lee YK. Finite element simulation of quench distortion in a low-alloy steel incorporating transformation kinetics. Acta Materialia 2008;56:1482.
[33] Lee S-J, Pavlina EJ, Van Tyne CJ. Kinetics modeling of austenite decomposition for an end-quenched 1045 steel. Materials Science and Engineering: A 2010;527:3186.
[34] Avrami M. Kinetics of phase change: 1. General theory. Journal of Chemical Physics 1939;7:1103.
[35] Avrami M. Kinetics of phase change: 2. Transformation-Time Relations for Random Distribution of Nuclei. Journal of Chemical Physics 1940;8:212.
[36] Avrami M. Granulation, Phase Change, and Microstructure Kinetics of Phase Change. III. Journal of Chemical Physics 1941;9:177.
[37] Johnson W, Mehl R. Reaction kinetics in processes of nucleation and growth. Trans. AIME 1939;135.
[38] Kolmogorov A. A statistical theory for the recrystallization of metals. Akad. nauk SSSR, Izv., Ser. Matem 1937;1.
[39] Li JJ, Wang JC, Xu Q, Yang GC. Comparison of Johnson-Mehl-Avrami-Kologoromov (JMAK) kinetics with a phase field simulation for polycrystalline solidification. Acta Materialia 2007;55:825.
[40] Inoue T, Arimoto K. Development and implementation of CAE system hearts for heat treatment simulation based on metallo-thermo-mechanics. Journal of Materials Engineering and Performance 1997;6:51.
[41] Chen LQ. PHASE-FIELD MODELS FOR MICROSTRUCTURE EVOLUTION. Annual Review of Materials Research 2002;32:113.
[42] Dobrzanski L, Trzaska J. Application of neural networks for the prediction of continuous cooling transformation diagrams. Computational Materials Science 2004;30:251.
[43] Maazi N, Penelle R. Introduction of preferential Zener drag effect in Monte Carlo simulation of abnormal Goss grain growth in the Fe-3%Si magnetic alloys. Materials Science and Engineering: A 2009;504:135.
[44] Leblond JB, Mottet G, Devaux JC. A theoretical and numerical approach to the plastic behaviour of steels during phase transformations: -I. Derivation of general relations. 1986;34:395.

4

Data Processing Steps in Neutron Diffraction: From the Raw Data to the Differential Cross Section

J. Dawidowski[1], G. J. Cuello[2] and L. A. Rodríguez Palomino[1]

[1]*Centro Atómico Bariloche and Instituto Balseiro, Avenida Ezequiel Bustillo 9500 (R8402AGP) San Carlos de Bariloche, Río Negro*

[2]*Institut Laue Langevin, 6, rue Jules Horowitz (38042), Grenoble*

[1]*Argentina*

[2]*France*

1. Introduction

Neutron diffraction is a well established tool to investigate the structure of matter in a wide range of disciplines including Physics, Chemistry, Materials Sciences, Life Sciences, Earth Sciences and Engineering. One of its most required applications is the refinement of structures for which a considerable instrumental development has been devoted. In particular, the improvement of the instrumental resolution has been hitherto one of the main concerns in the development of the technique. In other words, most of the efforts in the instrumental development and methods has been devoted to improve the abscissas of the experimental scale (angle or momentum transfer), while on the other hand, the final results in ordinates are normally left in arbitrary units, since most of the applications do not require an absolute normalization.

Nevertheless, there is a growth in the requirements of updated neutron cross section data driven by the need of improved nuclear data libraries by Nuclear Engineers, that currently employ cross sections that sometimes are guessed or extrapolated from very old experiments. Such need could be satisfied by the highly-developed experimental neutron facilities to provide excellent quality data in absolute scales. However, this capacity remains under-exploited, as well as the procedures that are necessary to perform an absolute calibration (in the scale of ordinates), in the sense of transforming the measured number of counts into a physically meaningful scale, and expressing the final result as a cross section. This lack is closely related with the underdevelopment of data processing procedures and methods specific to each experimental configuration in neutron scattering techniques. As an example, neutron diffraction users at big facilities still employ the simple data processing correction procedures developed for X-rays techniques (Blech & Averbach, 1965; Paalman & Pings, 1962) in times when computer resources were limited. However, as shown by many reference works in the literature (Copley et al., 1986; Sears, 1975), the situation in the field of neutron scattering is far more complex, and involves the evaluation of multiple scattering effects that can be tackled efficiently only by numerical simulations, that nowadays can be carried out with the currently available computer power.

To illustrate the consequences of this lack of a developed standard procedure to achieve an absolute normalization, let us consider the long-lasting controversy about the Hydrogen cross section for epithermal neutrons in electron-Volt spectroscopy (eVS), that began when a cross section significantly lower than commonly accepted values was reported in the literature (Chatzidimitriou-Dreismann et al., 1997), and supported by further experimental results [see e.g. (Abdul-Redah et al., 2005)]. The results stirred many criticisms, both on the theoretical likelihood of such phenomenon [see e.g. (Colognesi, 2007)], and also on the general methodology employed in the measurements and data analysis (Blostein et al., 2001) (Blostein et al., 2005). Different experiments contradicting the appearance of anomalies, employing electron-Volt spectroscopy and other techniques were also reported (Blostein et al., 2003; 2009), thus increasing the uncertainty on the matter. As an outcome of the discussion, the idea arose that the data processing methodology employed was not ready to produce a cross section in absolute units in eVs experiments. This thought led to the recent formulation of a whole body of experimental and data processing procedures (Rodríguez Palomino et al., 2011), and its application resulted in Hydrogen cross sections that are in conformity with tabulated values (Blostein et al., 2009).

In the specific case of diffraction, the problem of absolute normalization was also addressed (Rodríguez Palomino et al., 2007) following a similar approach, and the procedures were applied to a set of standard samples, measured at diffractometer D4 (ILL, Grenoble, France)(Fischer et al., 2002), and also recently to a set of light and heavy water with the aim to study the structural characteristics (Dawidowski & Cuello, 2011). The goal of the process (that will be the subject of the present work) is to provide a *modus operandi* that starts from the experimental raw data and ends in the differential scattering cross section. The starting point of the task consists in the description of the measured magnitudes through analytic expressions.

Elementary textbooks state the expressions of the measured magnitude in a diffraction experiment for point-like samples, thus finding a direct relation with the differential cross section. However, in a real experiment where extended samples are used, the formalism that describes the intensity of the scattered beam includes the sample geometry (Sears, 1975), and will be the first point to be treated. The expressions that we will show describe the measured macroscopic magnitude and its relationship with the sought microscopic differential cross section, that is not directly accessible to the experiment due to the multiple scattering, beam attenuation and detector efficiency effects. The evaluation of such effects is made through numerical simulations that follow the general guidelines stated by Copley (Copley et al., 1986). In this work we will make a detailed account of the computer code, as well as the different strategies employed to make use of the experimental data and models of interaction between the neutrons and the systems to feed the simulation program.

2. Preliminary notions

Elementary reference textbooks introduce the cross sections from a microscopic point of view (Lovesey, 1986; Squires, 1978), useful to show the link with quantum-mechanic theories. However, to make a useful comparison with experiments, the correct approach based on macroscopic magnitudes (that we will follow in this work) was formulated by Sears (Sears,

1975). We will see in this section a detail of such expressions, and their relationship with the microscopic cross sections, that is generally the sought magnitude.

2.1 Basic concepts

We will begin by considering the typical diffraction setup shown in Fig. 1, where incident neutrons are monochromatic with a wave vector $\mathbf{k}_0$ (energy E_0) and the scattered neutrons have a wavevector $\mathbf{k}$ (energy E). We will start from the *microscopic double-differential cross section* $\frac{d^2\sigma}{d\Omega dE}$ defined in introductory textbooks as the fraction of incident neutrons scattered into the element of solid angle $d\Omega$, with final energies between E and $E + dE$.

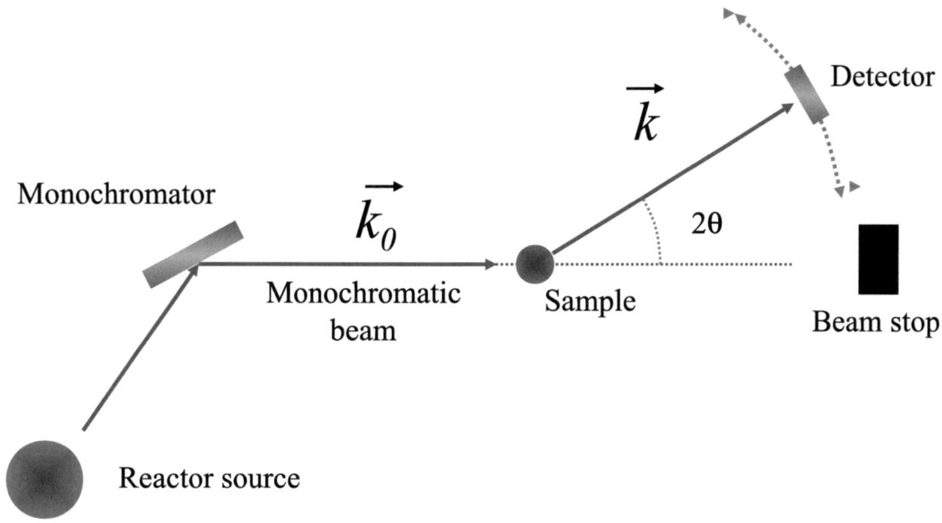

Fig. 1. Diffraction setup, showing the incident and emerging neutrons wavevectors $\mathbf{k}_0$ and $\mathbf{k}$ and the scattering angle 2θ

The microscopic integral magnitudes we will use in this work are the *differential scattering cross section*

$$\left(\frac{d\sigma(E_0)}{d\Omega}\right)_{\text{scatt}} = \int_{\text{all final energies}} \left(\frac{d^2\sigma}{d\Omega dE}\right) dE, \tag{1}$$

and the *total scattering cross section*

$$\sigma_{\text{scatt}}(E_0) = \int_{\text{all directions}} \int_{\text{all final energies}} \left(\frac{d^2\sigma}{d\Omega dE}\right) d\Omega dE. \tag{2}$$

Normally, there are other possible contributions to the total cross section from different nuclear processes in the neutron-nucleus interaction. In this work we will consider only scattering and absorption processes, so the *total cross section* is

$$\sigma_{\text{tot}}(E_0) = \sigma_{\text{scatt}}(E_0) + \sigma_{\text{abs}}(E_0). \tag{3}$$

Both scattering and absorption components are a function of the incident neutron energy E_0. The well-known tabulated free-atom total cross section is the asymptotic value that the total cross section reaches beyond epithermal energies, as can be seen in the total cross section of Polyethylene (Fig. 2), extracted from Granada et al. (1987).

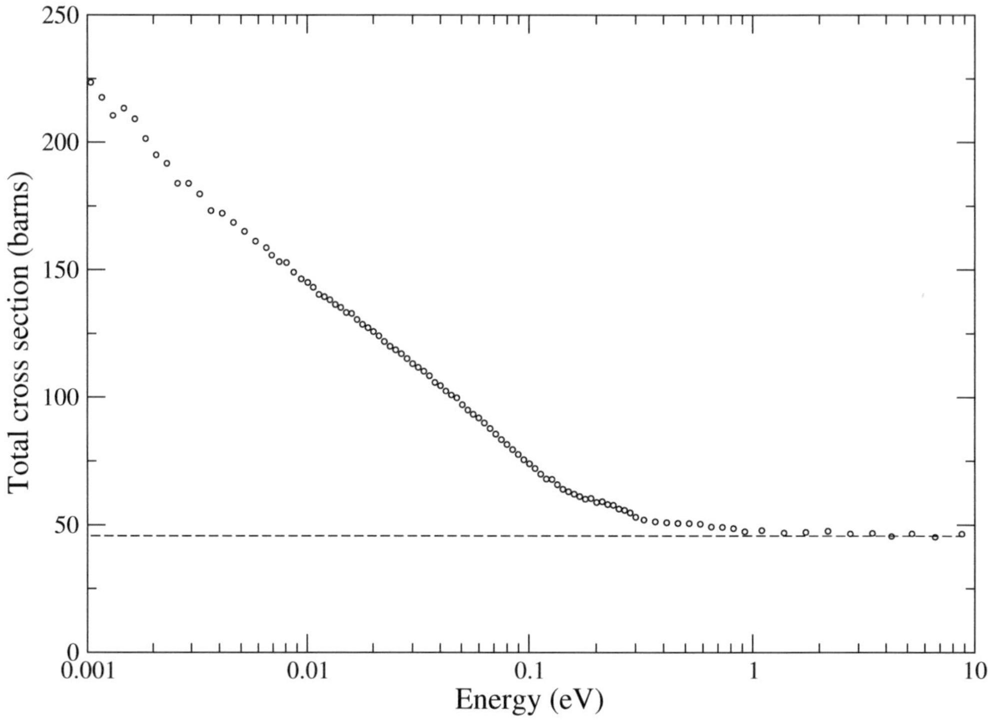

Fig. 2. Total cross section of Polyethylene (from Granada et al. (1987)) (circles). The dotted line indicates the free atom limit $2\sigma_H + \sigma_C$.

The macroscopic total cross section $\Sigma_{tot}(\mathbf{k}_0)$, is defined as

$$\Sigma_{tot}(E_0) = n\sigma_{tot}(E_0), \tag{4}$$

where n is the number density of scattering units in the sample. This magnitude describes the macroscopic aspect of the neutron interactions with the sample. For example, the probability that a neutron will interact after traversing a path x is

$$p(E_0, x) = \Sigma_{tot}(E_0) \exp(-\Sigma_{tot}(E_0)x). \tag{5}$$

Very closely related is the fraction of neutrons of a collimated beam that do not interact with the sample, called *transmission factor*, that in the case of a plane slab of thickness d is

$$t(E_0) = \exp[-\Sigma_{tot}(E_0)d]. \tag{6}$$

2.2 Cross sections for extended samples

Let us treat the case of an extended sample bathed by monochromatic neutrons (flying with velocity v_0), forming a beam of ρ neutrons per unit volume, so the flux is $\Phi = \rho v_0$ (neutrons cm^{-1} sec^{-1}). The distribution function of the incident beam can be represented as

$$f_{inc}(\mathbf{r}, \mathbf{k}) = \rho\delta(\mathbf{k} - \mathbf{k}_0) \tag{7}$$

and the distribution of neutrons arrived at a position $\mathbf{r}$ inside the sample without interaction is (see Eq. (6))

$$f_0(\mathbf{r},\mathbf{k}) = \rho\delta(\mathbf{k}-\mathbf{k}_0)\exp[-\Sigma_{\text{tot}}(\mathbf{k}_0)L(\mathbf{r},\hat{\mathbf{k}}_0)]. \tag{8}$$

In this expression $L(\mathbf{r},\hat{\mathbf{k}}_0)$ is the distance from point $\mathbf{r}$ (inside the sample), to the sample surface in the direction $-\hat{\mathbf{k}}_0$.

To describe the distribution of scattered neutrons effectively observed in an experiment, the double differential-cross section (which is a microscopic concept) must be replaced by the *macroscopic double-differential cross section*, defined as the number of neutrons that *emerge from the sample* with a momentum between $\mathbf{k}$ and $\mathbf{k}+d\mathbf{k}$ of, per unit incident flux.

$$\frac{\mathrm{d}^2\Sigma}{\mathrm{d}\Omega\mathrm{d}E} = \left[\frac{mk^2}{\rho\hbar^2 k_0}\int_{S(\hat{\mathbf{k}})}\mathrm{d}S\,\hat{\mathbf{e}}\cdot\hat{\mathbf{k}}f(\mathbf{r},\mathbf{k})\right]. \tag{9}$$

where m is the neutron mass and the integral is performed on $S(\hat{\mathbf{k}}$, the surface of the sample that is visible from direction $\hat{\mathbf{k}}$ and $\hat{\mathbf{e}}$ is its normal unit vector. From this equation, we can derive that the expression for $\mathrm{d}^2\Sigma/\mathrm{d}\Omega\mathrm{d}E$ per unit cross sectional area exposed at the incident beam $A(\hat{\mathbf{k}}_0)$, can be decomposed in a transmitted beam plus a distribution of scattered neutrons

$$\frac{1}{A(\hat{\mathbf{k}}_0)}\frac{d^2\Sigma}{d\Omega dE} = \mathcal{T}(k_0) + \mathcal{S}(\mathbf{k}_0,\mathbf{k}), \tag{10}$$

with

$$\mathcal{T}(k_0) = \frac{mk_0}{\hbar^2}t(k_0)\delta(\mathbf{k}-\mathbf{k}_0) \tag{11}$$

and

$$\mathcal{S}(\mathbf{k}_0,\mathbf{k}) = \frac{V}{A(\hat{\mathbf{k}}_0)}\frac{n\sigma_s}{4\pi}\frac{k}{k_0}s(\mathbf{k}_0,\mathbf{k}). \tag{12}$$

In Eq. (12) σ_s is the bound-atom scattering cross section of the sample and V its volume.

The transmission factor, introduced for plane slabs in Eq. (6), for general geometries can be expressed alternatively either as a volume or a surface integral, as

$$\begin{aligned} t(k_0) =& \frac{1}{A(\hat{\mathbf{k}}_0)}\int_{S(\hat{\mathbf{k}})}\mathrm{d}S\;\hat{\mathbf{e}}\cdot\hat{\mathbf{k}}_0\;\exp[-\Sigma_{\text{tot}}(\mathbf{k}_0)L(\mathbf{r},\hat{\mathbf{k}}_0)] \\ =& 1-\frac{\Sigma_{\text{tot}}(\mathbf{k}_0)}{A(\hat{\mathbf{k}}_0)}\int_V \mathrm{d}\mathbf{r}\;\exp[-\Sigma_{\text{tot}}(\mathbf{k}_0)L(\mathbf{r},\hat{\mathbf{k}}_0)]. \end{aligned} \tag{13}$$

The second term in Eq. (10) contains the effective scattering function $s(\mathbf{k}_0,\mathbf{k})$, that includes a component due to singly scattered neutrons in the sample $s_1(\mathbf{k}_0,\mathbf{k})$, another due to singly scattered neutrons in the container $s_C(\mathbf{k}_0,\mathbf{k})$, and a third due to multiply scattered neutrons $s_M(\mathbf{k}_0,\mathbf{k})$

$$s(\mathbf{k}_0,\mathbf{k}) = s_1(\mathbf{k}_0,\mathbf{k}) + s_M(\mathbf{k}_0,\mathbf{k}) + s_C(\mathbf{k}_0,\mathbf{k}). \tag{14}$$

Eq. (10) is the analytical expression of the number of neutrons that interact with a sample in a scattering experiment, and is the basis on which we will develop the expressions that must be compared with experiments.

3. Numerical simulations basic theory

3.1 General expressions

The task of developing analytical expressions for $s(\mathbf{k}_0, \mathbf{k})$ is involved, and does not lead to results of practical application. Far more profitable are the numerical simulations, since they allow a direct comparison with the experiment. Therefore, rather than developing the expressions for $s(\mathbf{k}_0, \mathbf{k})$ (than can be found in Sears (1975)), we will will examine the expressions that link the magnitudes calculated in simulations with the experimental count rates.

Neutrons detected in a scattering experiment generally undergo a variable number of collisions until they emerge from the sample. We will develop the expressions for the probabilities of detection of such neutrons after n collisions. Since the experiment does not discriminate the number of collisions of the detected neutrons, our task will be to assess the contributions of multiply scattered neutrons, to subtract them to the measured spectra, in order to keep the singly scattered component, that is directly related with the microscopic cross sections.

The basic interactions that will be described in this work are scattering and absorption. In the scattering interaction, the probability that a neutron changes from wavevector $\mathbf{k}_0$ to $\mathbf{k}$ is

$$P(\mathbf{k}_0, \mathbf{k}) = \frac{1}{\sigma_{\text{scatt}}(E_0)} \frac{\mathrm{d}^2\sigma}{\mathrm{d}\Omega \mathrm{d}E}. \tag{15}$$

The starting point of our numerical calculations is the probability that an incident neutron arrives to a point $\mathbf{r}$ and from that position is scattered with a final wave vector $\mathbf{k}$, based on Eqs. (5) and (15)

$$z_1(\mathbf{r}, \mathbf{k}_0, \mathbf{k}) = \underbrace{\frac{\Sigma_{\text{scatt}}(\mathbf{k}_0)}{A(\hat{\mathbf{k}}_0)} \exp[-\Sigma_{\text{tot}}(\mathbf{k}_0) L(\mathbf{r}, -\hat{\mathbf{k}}_0)]}_{\substack{\text{Prob. per unit area that the neutron} \\ \text{arrives at } \mathbf{r} \text{ and is scattered}}} \underbrace{P(\mathbf{k}_0, \mathbf{k})}_{\substack{\text{Prob. scatt.} \\ \text{distribution}}} . \tag{16}$$

Integrating (16) over all the sample volume, we get the *distribution of neutrons after the first scattering*. In the integral we employ the result of Eq. (13), thus obtaining

$$z_1(\mathbf{k}_0, \mathbf{k}) = \frac{\Sigma_{\text{scatt}}(\mathbf{k}_0)}{\Sigma_{\text{tot}}(\mathbf{k}_0)} (1 - t(E_0)) P(\mathbf{k}_0, \mathbf{k}). \tag{17}$$

We observe that $z_1(\mathbf{k}_0, \mathbf{k})$ is directly composed by a factor related with macroscopic properties, times another related with the sought microscopic scattering cross section. This magnitude, however, cannot be directly observed in the experiment since in general singly-scattered neutrons will undergo more collisions. So the *distribution of neutrons detected after the first scattering*, will be one of the components of the total detected spectrum, and its expression is based on Eq. (17), times the attenuation undergone by the neutron beam in the outgoing path in the sample towards the detector, times the probability that the neutron is detected (detector efficiency). The contribution from point $\mathbf{r}$ is

$$\tilde{z}_1(\mathbf{r}, \mathbf{k}_0, \mathbf{k}) = z_1(\mathbf{r}, \mathbf{k}_0, \mathbf{k}) \underbrace{\mathrm{e}^{-\Sigma_{\text{tot}}(\mathbf{k}) L(\mathbf{r}, \hat{\mathbf{k}})}}_{\substack{\text{Attenuation} \\ \text{undergone by n in} \\ \text{the outgoing path}}} \underbrace{\varepsilon(k)}_{\substack{\text{Detector} \\ \text{efficiency}}} , \tag{18}$$

and the total contribution of the sample is the integral over all the volume

$$\tilde{z}_1(\mathbf{k}_0,\mathbf{k}) = \frac{1}{A(\hat{\mathbf{k}}_0)}\Sigma_{\text{scatt}}(\mathbf{k}_0)P(\mathbf{k}_0,\mathbf{k})\varepsilon(k)\underbrace{\int_V \mathrm{d}\mathbf{r}\,\mathrm{e}^{-\Sigma_{\text{tot}}(\mathbf{k}_0)L(\mathbf{r},-\hat{\mathbf{k}}_0)}\mathrm{e}^{-\Sigma_{\text{tot}}(\mathbf{k})L(\mathbf{r},\hat{\mathbf{k}})}}_{VH_1(\mathbf{k}_0,\mathbf{k})}, \tag{19}$$

where $H_1(\mathbf{k}_0,\mathbf{k})$ is the primary attenuation factor defined by Sears (1975).

This magnitude is directly related with $s_1(\mathbf{k}_0,\mathbf{k})$ of Eq. (14) through

$$\tilde{z}_1(\mathbf{k}_0,\mathbf{k}) = \frac{N\sigma_{\text{bound}}}{4\pi A(\hat{\mathbf{k}}_0)}\frac{k}{k_0}\varepsilon(k)s_1(\mathbf{k}_0,\mathbf{k}), \tag{20}$$

The expression of the *distribution of neutrons detected after the n-th scattering*, $\tilde{z}_n(\mathbf{k}_0,\mathbf{k})$ is mathematically involved and will be omitted. Its calculation will be done with numerical simulations.

3.2 Application for diffraction experiments

In the case of diffraction experiments, the observed angular distributions result from the integrals in final energies of the former expressions. From Eq. (12)

$$\frac{1}{A(\hat{\mathbf{k}}_0)}\left(\frac{\mathrm{d}\Sigma}{\mathrm{d}\Omega}\right)_{\text{scatt}} = \frac{V}{A(\hat{\mathbf{k}}_0)}\frac{n\sigma_s}{4\pi}\int \mathrm{d}E\frac{k}{k_0}s(\mathbf{k}_0,\mathbf{k}). \tag{21}$$

The equivalent angular magnitudes (17) and (19) are

$$z_1(\mathbf{k}_0,\theta) = \frac{\Sigma_{\text{scatt}}(\mathbf{k}_0)}{\Sigma_{\text{tot}}(\mathbf{k}_0)}(1-t(E_0))\frac{1}{\sigma_{\text{scatt}}(E_0)}\frac{\mathrm{d}\sigma}{\mathrm{d}\Omega}(E_0,\theta) \tag{22}$$

$$\tilde{z}_1(\mathbf{k}_0,\theta) = \frac{V}{A(\hat{\mathbf{k}}_0)}\Sigma_{\text{scatt}}(\mathbf{k}_0)\int \mathrm{d}E\,\frac{1}{\sigma_{\text{scatt}}(E_0)}\sigma(E_0,E,\theta)\,\varepsilon(E)H_1(\mathbf{k}_0,\mathbf{k}), \tag{23}$$

where Eq. (23) is the single scattering component of Eq. (21).

4. Experimental procedure

The neutron diffraction experimental procedure was treated in detail by Cuello (2008). Since we will refer in this work to experiments performed at instrument D4 (Fischer et al. (2002)) of Institut Laue Langevin, we will draw upon its customary *modus operandi* to describe the steps of a typical diffraction experiment in a steady source.

The diffractometer D4 is essentially a two-axis diffractometer as those commonly used in powder diffraction. However two main characteristics distinguish this instrument from other powder diffractometers. First, the use of hot neutrons of energy of some hundreds of meV, which allows to reach higher momentum transfer than in a conventional thermal neutron diffractometer. Second, the collimation and evacuated tubes all along the flight path from the monochromator up to the detectors, which allows to reduce the background to extremely low level.

The epithermal neutron spectrum produced in the fission reactions is thermalised by a piece of graphite near the reactor core. This hot source, in thermal equilibrium at 2400 K, serves to shift the Maxwellian distribution of velocities to higher energies. The extraction tube is at no more that 1 mm of the hot source to avoid the thermalisation in the heavy water. At the end of this extraction tube is located a double-focusing copper monochromator that allows to choose the incident energy or wavelength; the usual wavelengths at D4 are 0.7 and 0.5 Å. This monochromatic beam passes first through a transmission monitor (an almost transparent detector) and then is collimated by means of a series of diaphragms and silts, defining the size of the beam at the sample position. Between the monochromator and the monitor there is an evacuated tube, and the sample is located in the center of a cylindrical evacuated bell-jar. Cylindrical sample shapes are preferred, in accordance with the detector array geometry. In this case the incident beam is perpendicular to the cylinder axis. When dealing with non self-supporting samples, a vanadium cell is used as container.

The detection ensemble is composed of 9 banks of 64 ^{3}He detection cells. In front of each bank an evacuated collimation tube reduces the background produced by the bell-jar wall. The flat detection surfaces are arranged in an arc of circumference in whose center the sample is placed. The detector ensemble is moved around the sample in order to register a complete diffractogram, covering a scattering angle ranging 1 to 140°.

Besides the measurement of the sample itself, a series of ancillary measurements must be performed. The most important one is the empty (vanadium) container. The empty instrument, *i.e.*, the empty bell-jar for experiments at ambient conditions or the empty sample environment for other experiments, is the background contribution. Finally, the customary procedure to normalise the measured intensity to an absolute scale, in barns per steradian, is carried out through the measurement of a vanadium solid sample. This measurement is also used to take into account the instrumental resolution.

There are two kinds of corrections to be performed on the experimental data: one set coming from the experimental conditions and the other one from the theoretical assumptions made to derive the structure factors. There are more or less standard programs available to perform these corrections like, the code CORRECT (Howe et al. (1996)).

This program performs the main experimental corrections, such as the container (empty cell) and background subtraction (empty instrument). The knowledge of dimensions and materials of each component in the beam allows the calculation of the absorption coefficients and the extraction of the sample diffractogram. In the case of cylindrical geometry, these absorption coefficients are calculated using the Paalman and Pings corrections Paalman & Pings (1962), and the multiple scattering, is evaluated using the Blech and Averbach correction Blech & Averbach (1965), is subtracted from the experimental data.

Another instrumental parameter to take into account is the detector efficiency as a function of the neutron energy, that depends on the geometry and the filling gas pressure of the detectors. In the case of D4, it is well described by (Fischer et al. (2002))

$$\varepsilon(E) = 1 - \exp\left(-\frac{0.9599}{\sqrt{E}}\right) \tag{24}$$

The last experimental correction to perform is the correction for the instrumental resolution. Knowing the instrumental resolution, one can attempt to extract the structure factor by performing a deconvolution process, but this is a difficult task. Instead, one can measure a standard vanadium sample (an almost incoherent scatterer) which should give a flat diffractogram. In fact this diffractogram is not flat because (mainly) the resolution of the instrument and the resolution-corrected data are obtained by taking the ratio of the sample and vanadium diffractograms. In doing this, and because the cross section of vanadium is well known, the data are also normalized to an absolute scale.

5. Numerical procedure

The probability functions shown in Section 3.1 are the expressions of the distributions followed by numerical the simulations of neutron scattering experiments we will refer to. In this section we will describe the numerical procedure employed in the simulation and the data correction analysis.

The scheme we will develop involves

- A numeric simulation program
- An iterative correction scheme
- A method to perform absolute normalizations from experimental data

We will firstly refer to the numeric simulations developed to describe diffraction experiments.

5.1 Numeric simulations

The goal of the Monte Carlo simulations is to describe the real experiment as closely as possible, to provide a means to assess the effects of

- multiple scattering
- beam attenuation
- detector efficiency
- scattering from the container

The simulations we will describe involve the incident beam and the sample-container set. We do not simulate the sample environment (collimators, bell-jar container, etc.). This background contribution is assumed to be heuristically subtracted from the experimental data. A closer simulation of the sample environment falls in the domain of instrument design, and is out of the scope of the present work.

The Monte Carlo program, is based on the method proposed by Bischoff (Bischoff (1969)) and Copley (Copley et al. (1986)). In this method, discrete neutron histories are tracked and averaged in a random walk directed by interaction probabilities obtained either by measurements or by models. In this section we will describe how neutron histories are generated, and then how they are scored and recorded, following closely the formalism developed in Sect. 3.1.

5.1.1 Neutron histories

A history consists in a trajectory of the neutron inside the sample composed by a series of scattering steps. Neutron histories are originated in a random point of the sample surface that faces the incoming beam, and consists of a series of tracks governed by the path-length estimator followed by collisions governed by a suitable angle and energy-transfer probability. A variance reduction technique is employed (following Bischoff (1969); Spanier & Gelbard (1969)) to make more efficient the process. To this end the probabilities are altered, so that the neutron never leaves the sample and never is absorbed. To compensate the biased probabilities a weight is assigned to the event. Thus a neutron history has an initial weight 1, that decreases as the neutron progresses until it drops below a significant value. In such case the history is finished.

5.1.1.1 Path lengths

The step lengths of the trajectories are governed by the macroscopic cross sections of the traversed materials. The distance traveled by the neutron between two scattering steps is randomly drawn from the distribution

$$p(E,x) = \frac{\Sigma_{tot}(E,x)t(E,x)}{1 - t(E,d)}, \tag{25}$$

that is the track-length distribution (5) altered , in order that the neutron never gets out of the sample. In Eq. (25) $\Sigma_{tot}(E,x)$ is the macroscopic total cross section of the sample-container set, a distance x away from the neutron previous scattering position, taken in the current flight direction, $t(E,x)$ is the fraction of transmitted neutrons in that direction after traversing a distance x, and d is the distance to the sample surface in that direction.

As mentioned above, the neutron is also forced to scatter (since absorption is forbidden in this altered scheme). To compensate the biases in the probability, a weight is assigned to each neutron that decreases according to the transmitted fraction in the traversed path, being 1 the initial value. Given the weight at step $i-1$ the weight at step i is calculated as in Spanier & Gelbard (1969)

$$w_i = w_{i-1}(1 - t(E,d))\frac{\Sigma_{scatt}(E,0)}{\Sigma_{tot}(E,0)}, \tag{26}$$

where the ratio $\Sigma_{scatt}(E,0)/\Sigma_{tot}(E,0)$ of the macroscopic scattering and total cross sections, at position $i-1$ indicates the probability that the neutron will not be absorbed, and $1-t(E,d)$ the probability that it will interact in the considered path. A history is finished when the weight drops under a predetermined cut-off value, so the number of scattering events is not predetermined.

To evaluate (25) and (26) the program requires as input the tabulated values of the macroscopic total cross sections of the sample and container materials, as well as their absorption probabilities as a function of energy. In Fig. 3 we show the input values employed for the case of a D_2O sample, with a vanadium cell as container. We observe the macroscopic cross section of D_2O (taken from Kropff et al. (1984)) and V (from Schmunk et al. (1960)), and the typical "$1/v$" absorption cross sections taken from Mughabghab et al. (1984).

5.1.1.2 Direction and energy after a collision

After a collision, a new energy and flight direction must be assigned for the next step in the neutron history. If E_i is the energy before the collision, the final energy E_{i+1} and the collision angle θ, must be drawn from the two-variable probability density (15), that can be rewritten as

$$P(E_i, E_{i+1}, \theta) = \frac{1}{\sigma_{scatt}(E_i)}\sigma(E_i, E_{i+1}, \Omega). \tag{27}$$

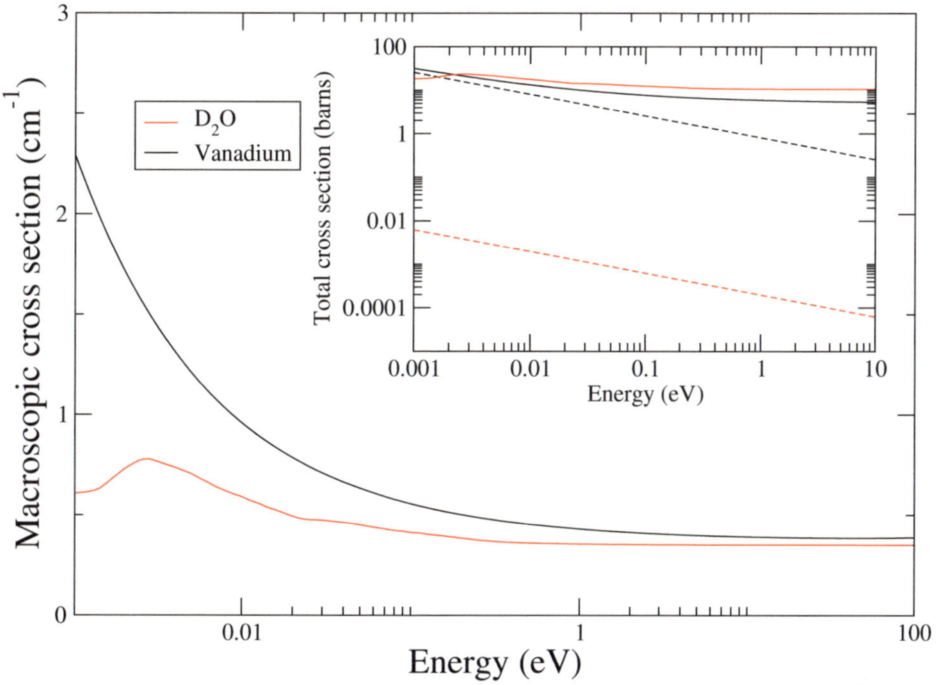

Fig. 3. Input parameters employed in the simulation of a D_2O sample, in a vanadium container. In the main frame, the macroscopic total cross sections of D_2O (after Kropff et al. (1984)) and the vanadium cross section (after Schmunk et al. (1960)). Inset: The microscopic total and absorption cross sections (Mughabghab et al. (1984)).

The process must be done in two steps (Spanier & Gelbard (1969)). First we define the marginal density as the angular integral

$$P_1(E_i, E) = \frac{1}{\sigma_{\text{scatt}}(E_i)} \int \sigma(E_i, E, \Omega) d\Omega \tag{28}$$

that for the special case of systems with azimuthal symmetry (such as liquids) can be written as

$$P_1(E_i, E) = \frac{1}{\sigma_{\text{scatt}}(E_i)} \int_0^{2\pi} 2\pi \sin\theta \sigma(E_i, E, \theta) d\theta. \tag{29}$$

The integral in angles is known as the *energy-transfer kernel* $\sigma(E_i, E)$ (Beckurts & Wirtz (1964)), so

$$P_1(E_i, E) = \frac{1}{\sigma_{\text{scatt}}(E_i)} \sigma(E_i, E). \tag{30}$$

The energy E_{i+1} is drawn solving the equation

$$\rho_1 = \int_{-\infty}^{E_{i+1}} P_1(E_i, E) dE, \tag{31}$$

where ρ_1 is a random number uniformly distributed between 0 and 1. In the second step, the angle is defined with the probability (27), with E_i and E_{i+1} as fixed values and a second random number

$$\rho_2 = \frac{1}{\sigma_{\text{scatt}}(E_i)} \int_0^{\theta_1} 2\pi \sin\theta \, \sigma(E_i, E_{i+1}, \theta) d\theta. \tag{32}$$

Since diffraction experiments do not provide a measurement of energy transfers we must rely on models to evaluate $P(E_i, E_{i+1}, \theta)$ for the different materials composing the sample. For molecular systems the Synthetic Model formulated by Granada (Granada (1985)) proved to be an adequate description for the energy transfer kernels and total cross sections. The model requires a minimum input dataset to describe the molecular dynamics, and is fast and amenable to be included in a calculation code. Since the model describes only the incoherent cross section, it is not adequate to describe the coherence manifested in the angular distributions, but is still a good description of the energy transfers, as tested in the calculation of spectra from moderators. Thus, the energies drawn through Eq. (31) can be computed with this model. In practice, the lower and upper integration limits are placed at finite values, beyond which the contribution to the integral is negligible. In Fig. 4 we show the energy transfer kernel (multiplied by E, for practical purposes) for D_2O from the Synthetic model. In yellow dashed curves we show the integration limits as a function of the incident energy employed in the calculations. The curves are calculated previously and used also as input for the simulation program, together with the input parameters of the Synthetic Model. A

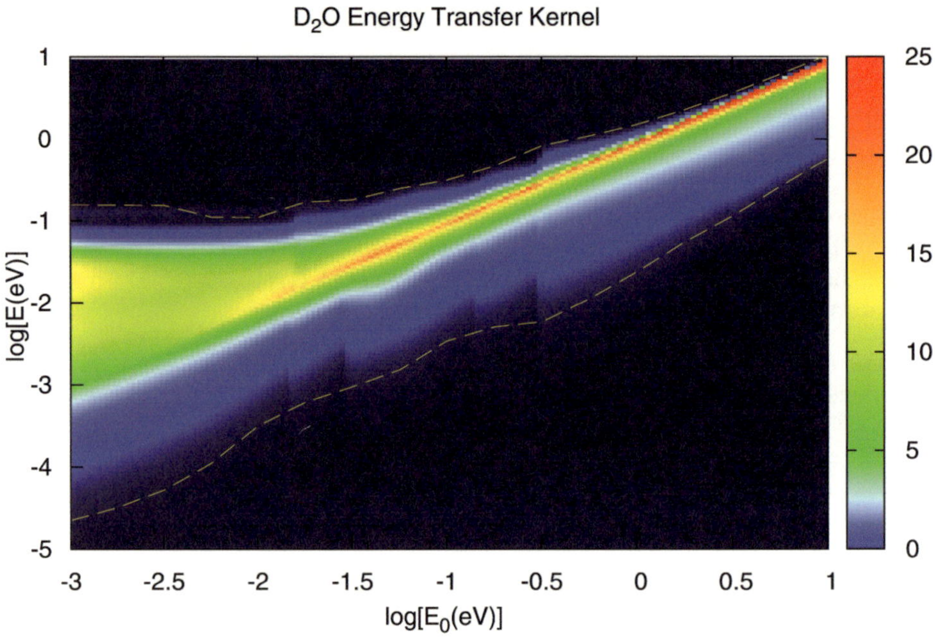

Fig. 4. Energy-transfer kernel $E\sigma(E_i, E)$ for D_2O calculated with Granada's Synthetic Model. Yellow dashed curves show the lower and upper integration limits employed in Eq. (31). The colored scale represents the intensity in barns.

basic input data table of the Synthetic Model for D_2O is shown in Table 1. The data include the energies of the oscillators $\hbar\omega_i$ and their widths $\hbar\sigma_i$ describing the molecular vibrational modes, the vibrational masses, the bound and absorption cross sections of the atoms, their masses, and the chemical composition of the molecule.

To draw angles from Eq. (32) we can follow two alternative procedures. The first is to employ also the Synthetic Model, that describes only incoherent effects as mentioned above, so it will have a narrow range of applicability, as was described by Rodríguez Palomino et al. (2007).

atom	$\hbar\omega_1$	$\hbar\omega_2$	$\hbar\omega_3$	$\hbar\omega_4$	M_1	M_2	M_3	M_4
D	14	50	150	306	40	5.2	12.1	5.87
O					40	214	290	169
atom	$\hbar\sigma_1$	$\hbar\sigma_2$	$\hbar\sigma_3$	$\hbar\sigma_4$	σ_{bound}	σ_{abs}	M atom	Numb.
D	1	21	18	19	7.63	8.2×10^{-5}	2	2
O					4.234	3.0×10^{-5}	16	1

Table 1. Values of the parameters of the Synthetic Model for D_2O used in the calculations (Granada (1985)). Energies ($\hbar\omega_i$ and $\hbar\sigma_i$) are given in meV, masses (M_i,M atom) in neutron mass units and cross sections (σ_{bound} and σ_{abs}) in barns. "Numb." indicates the number of atoms per molecule.

The second alternative is to employ the angular distributions determined by the experimental data, as will be described in this work.

The experimental data $\mathcal{E}^{(0)}(\theta)$ are usually tabulated in the elastic-Q scale,

$$Q_{el} = 2k_0 \sin(\theta/2), \tag{33}$$

where θ is the angle between the incident beam and the sample-detector path. However a Q_{el}-bin collects also all the inelastic contributions, from different initial and final energies. A practical approach in the application of Eq. (32), is to replace $\sigma(E_i, E_{i+1}, \theta)$ by a function based on the experimental data $f(\theta)$

$$f(\theta) = N\,\mathcal{E}^{(0)}(Q_i(\theta)), \tag{34}$$

where

$$Q_i(\theta) = 2k_i \sin(\theta/2), \tag{35}$$

and N is a normalization constant. This approach involves the idea that the angular distribution for a specific inelastic interaction is the same as the total measured distribution for elastic and inelastic processes. This question is further discussed in Sect. 7. In Fig. 5 we show the experimental data from a diffraction experiment on D_2O at D4. The raw data and and background measured with the empty bell-jar commented in Sect. 4 are shown, together with the data after background subtraction, in Q_{el} scale. The main experimental parameters are summarized in table 2.

Parameter	Value
Incident energy	324.3 meV
Beam height	5.0 cm
Beam width	0.6 cm
Cell external diameter	0.6 cm
Cell internal diameter	0.58 cm

Table 2. Experimental parameters related to the beam and sample size of the diffraction experiment in D_2O taken as example.

The justification to employ this hybrid model is that diffraction data provide an excellent approximation to the real angular distributions, while the Synthetic Model proved to be an adequate description for the energy transfer kernels for a wide variety of samples. Even in the case of coherent scatterers, the model was able to reproduce moderator spectra and the inelastic pedestal in diffraction experiments Granada (1985).

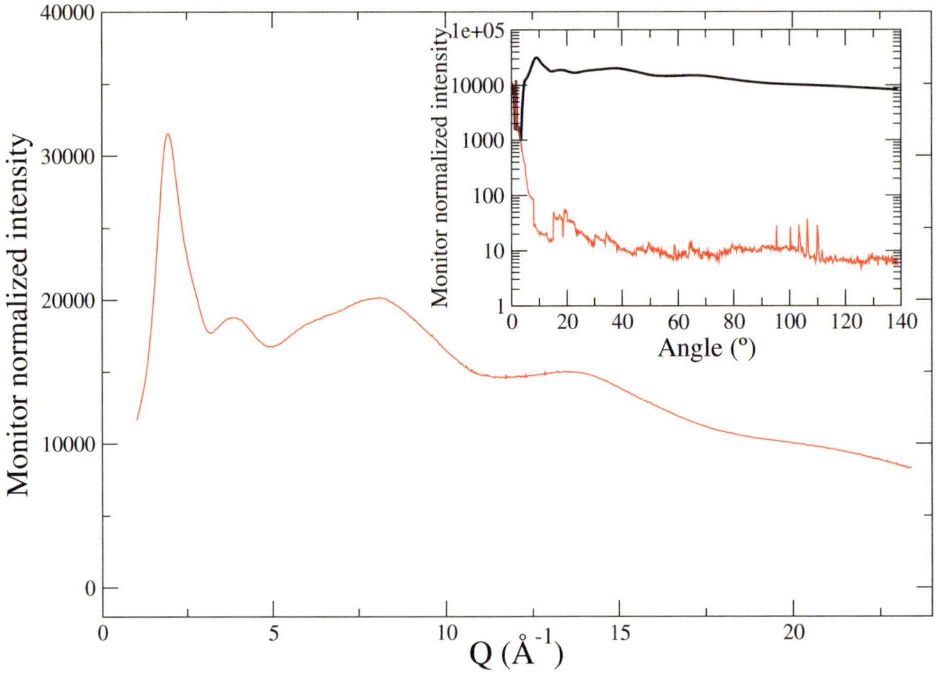

Fig. 5. Experimental data from a neutron diffraction experiment on D_2O in a vanadium container. Inset: Raw data and background measured with the empty bell-jar. Main frame: Experimental data after background subtraction, in $Q_{\rm el}$ scale.

Finally, if the scattering process takes place in the container, a gas model is assumed. The assumption is a good approach, since a heavy gas describes mostly elastic processes, that is a good description of the actual interactions taking place.

5.1.2 Scoring

At each step, the contribution of the current history to the final spectrum is scored for the set of detectors that compose the bank. For a detector placed at an angle θ with respect to the incident beam (taken as the z axis), with a direction characterized by the unit vector $\hat{\mathbf{d}} = (0, \cos\theta, \sin\theta)$, and if the neutron is flying inside the sample in the direction $\hat{\mathbf{k}} = (k_x, k_y, k_z)$, the scattering angle ϕ of the neutron to the detector position is calculated trough

$$\cos\phi = k_y \cos\theta + k_z \sin\theta \tag{36}$$

The scored magnitudes are summarized in table 3

Estimator	Description
$z_1(\mathbf{k}_0, \theta)$	Singly scattered neutrons
$\tilde{z}_1(\mathbf{k}_0, \theta)$	Detected singly scattered neutrons
$\tilde{z}_C(\mathbf{k}_0, \theta)$	Detected singly scattered neutrons from cell
$\tilde{z}_M(\mathbf{k}_0, \theta)$	Detected multiply scattered neutrons

Table 3. Estimators calculated in the numerical simulations

The contribution to the *distribution of neutrons after the first scattering* (Eq. (17)), can be obtained by scoring the estimator composed by the weight w_1 (Eq.(26))

$$z_1(\mathbf{r}_1, \mathbf{k}_0, \theta) = w_1 \mathcal{E}^{(0)}(Q), \tag{37}$$

where the normalized experimental angular distribution is employed as angular probability expressed in the elastic-Q scale. This scale has to be recalculated at every new neutron energy and renormalized in the accessible Q-range. It can be shown that the average over a large number of histories of (37) leads to the desired result (22). Similarly, the *distribution of detected neutrons after the i-th scattering* results from the average of the estimator

$$\tilde{z}_i(\mathbf{r}_\mathrm{i}, \mathbf{k}_{\mathrm{i}-1}, \theta) = w_i \mathcal{E}^{(0)}(Q) t(E_{\mathrm{i}-1}, \mathbf{r}_i, -\hat{\mathbf{k}}) \varepsilon(E), \tag{38}$$

where i is the step of the history, w_i is the weight of the neutron, and $t(E_i, \mathbf{r}_i, -\hat{\mathbf{k}})$ is the transmission coefficient from the point $\mathbf{r}_i$ inside the sample to the sample surface in the direction $\hat{\mathbf{k}}$ to the detector position. The average of (38) over a large number of histories converges to (23).

As we will show in the next section the experimental $\mathcal{E}^{(0)}(Q)$ is corrected in successive runs, resulting the distribution $\mathcal{E}^{(j-1)}(Q)$ that is employed in Eqs. (37) and (38) as the experimental distributions corrected in the preceding run.

The Monte Carlo process also records the sum of multiple scattering contributions occurring either in the sample or the container

$$\tilde{z}_\mathrm{M}(\mathbf{r}_i, \mathbf{k}_0, \theta) = \sum_{i=2} \tilde{z}_i(\mathbf{r}_i, \mathbf{k}_0, \theta) \tag{39}$$

and the contribution of single-scattering events in the container $\tilde{z}_\mathrm{C}(\mathbf{r}_1, \mathbf{k}_0, \theta)$. The average after a large number of histories of these magnitudes are $\tilde{z}_\mathrm{M}(\mathbf{k}_0, \theta)$ and $\tilde{z}_\mathrm{C}(\mathbf{k}_0, \theta)$, *i.e.* the contribution of the multiple-scattered and container-scattered neutrons to the detected spectrum.

5.2 Iterative method

The magnitudes $\tilde{z}_1(\mathbf{k}_0, \theta)$ and $z_1(\mathbf{k}_0, \theta)$ defined in Eqs. (22) and (23) respectively can be determined with the Monte Carlo procedure, as well as $\tilde{z}_\mathrm{M}(\mathbf{k}_0, \theta)$ and $\tilde{z}_\mathrm{C}(\mathbf{k}_0, \theta)$. After the *j-th* Monte Carlo run is completed we can define the *multiple scattering factor*

$$f_\mathrm{MS}^{(j)}(\mathbf{k}_0, \theta) = \frac{\tilde{z}_1^{(j)}(\mathbf{k}_0, \theta)}{\tilde{z}_1^{(j)}(\mathbf{k}_0, \theta) + \tilde{z}_\mathrm{C}^{(j)}(\mathbf{k}_0, \theta) + \tilde{z}_\mathrm{M}^{(j)}(\mathbf{k}_0, \theta)}, \tag{40}$$

and the ratio of detected singly-scattered neutrons to the total singly-scattered neutrons, which we will call *generalized attenuation factor*

$$\mathcal{A}^{(j)}(\mathbf{k}_0, \theta) = \frac{\tilde{z}_1^{(j)}(\mathbf{k}_0, \theta)}{z_1^{(j)}(\mathbf{k}_0, \theta)} = \frac{V}{A(\hat{\mathbf{k}}_0)} \frac{\Sigma_\mathrm{tot}(\mathbf{k}_0)}{(1 - t(E_0))} \int dE \frac{\sigma(E_0, E, \theta)}{\frac{d\sigma}{d\Omega}(E_0, \theta)} \varepsilon(E) H_1(\mathbf{k}_0, \mathbf{k}). \tag{41}$$

It is illustrative to point out that in the case of purely elastic scattering this magnitude is simply related with the first order attenuation factor defined in Sect. 3.1 as

$$\mathcal{A}_{\text{elast}} = \frac{V}{A(\hat{\mathbf{k}}_0)} \frac{\Sigma_{\text{tot}}(\mathbf{k}_0)}{(1-t(E_0))} \varepsilon(E_0) H_1(\mathbf{k}_0, \mathbf{k}_0). \tag{42}$$

The iterative correction process consists in applying to the experimental angular distribution $\mathcal{E}^{(0)}(Q)$, the correction factors defined in Eqs. (40) and (41) after a Monte Carlo run. The angular distribution thus corrected serves as input for the next run. Thus, in the first run, the raw experimental angular distribution (background subtracted) $\mathcal{E}^{(0)}$ is employed as input, and in the run $j+1$, we employ the distribution originated in run j as

$$\mathcal{E}^{(j+1)}(Q) = \frac{f_{\text{MS}}^{(j)}(\mathbf{k}_0, \theta)\ \mathcal{E}^{(0)}(Q)}{\mathcal{A}^{(j)}(\mathbf{k}_0, \theta)}. \tag{43}$$

The process finishes when no appreciable changes in the angular distribution are observed.

5.3 Summary of input data

Table 4 summarizes the input data required to perform a simulation of a diffraction experiment, with references to the specific example presented in this work. This set of data constitutes the minimum knowledge that must be gathered to perform the experimental program proposed in this work.

Data	Reference
Measured angular distribution, Background subtracted	Figure 5
Total cross section of the sample and the container	Figure 3
Absorption cross section of the sample and the container	Figure 3
Detector bank efficiency as a function of energy	Eq. (24)
Input parameters for the model that describes inelastic interactions	Table 1, Figure 4
Geometry parameters for the proposed experimental setup and sample environment	Table 2

Table 4. Summary of input data required in the numeric simulation of diffraction experiments. The reference indicates the specific data presented in this work.

6. Examples and applications

In this section we will show selected results of the collected experience in the application of this method (Dawidowski & Cuello (2011); Rodríguez Palomino et al. (2007)).

We begin showing diffraction experiments carried out on a series of D_2O and H_2O mixtures at room temperature, at the diffractometer D4 already mentioned in Sect. 4 (Institute Laue Langevin). We will mainly concentrate on the results for pure H_2O and D_2O. The incident neutron beam wavelength was $\lambda_0 = 0.5$ Å (energy 0.324 eV). The sample holder was a

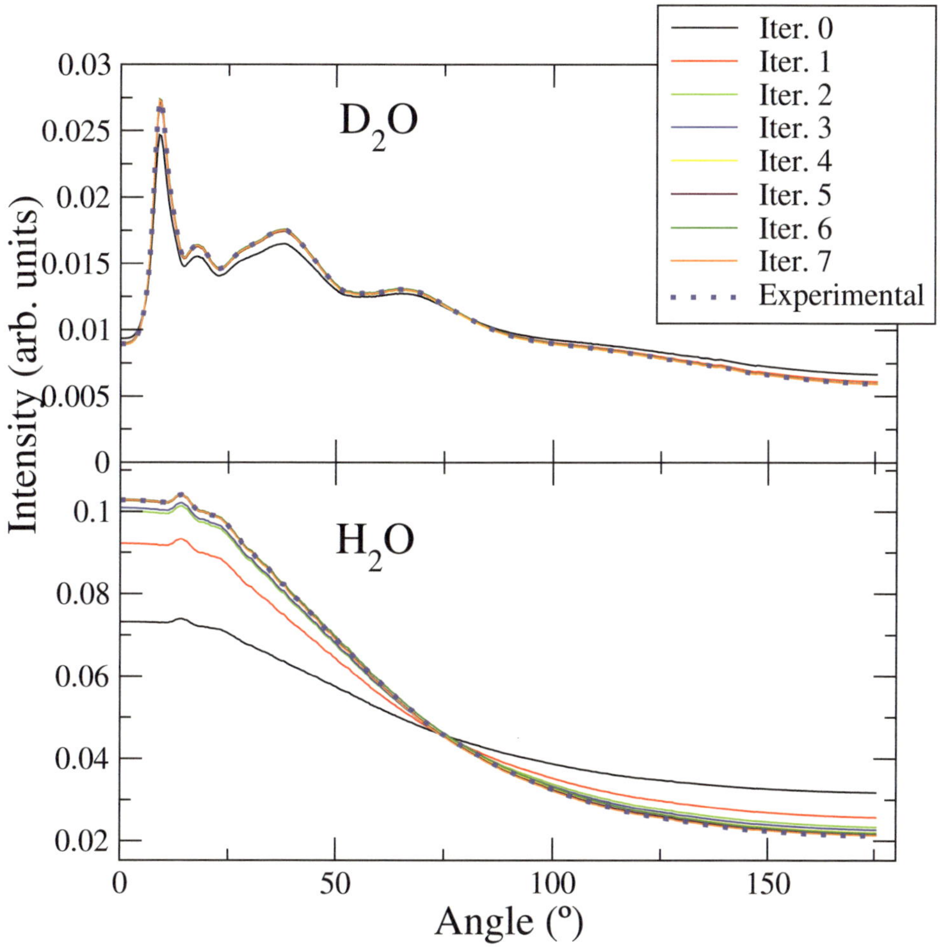

Fig. 6. Total scattering distribution calculated in successive iterations, compared with the experimental data, for D_2O and H_2O.

thin-walled cylindric vanadium can, 6 mm inner diameter and 60 mm height, situated at the centre of an evacuated bell jar.

To illustrate the iterative process we compare results from pure D_2O and H_2O in Figs. 6 and 7. Fig. 6 shows the total scattering calculated in 8 iterations. In H_2O where the level of multiple scattering is higher, the total scattering calculated in the first two iterations differ substantially from the original experimental data. The reason is that the experimental data firstly employed as the input for the angular distribution, also includes the components of multiple scattering and the attenuation effects that have to be corrected. As the iterative process progresses, the corrected experimental data approach closer to the distribution of singly scattered neutrons. We observe that the convergence is achieved starting from iteration 4 in H_2O and from iteration 2 in D_2O.

The effect of the iterative process is also illustrated in Fig. 7, where we show the results on both samples in the first and eighth iterations. In the Figure, we show the distribution

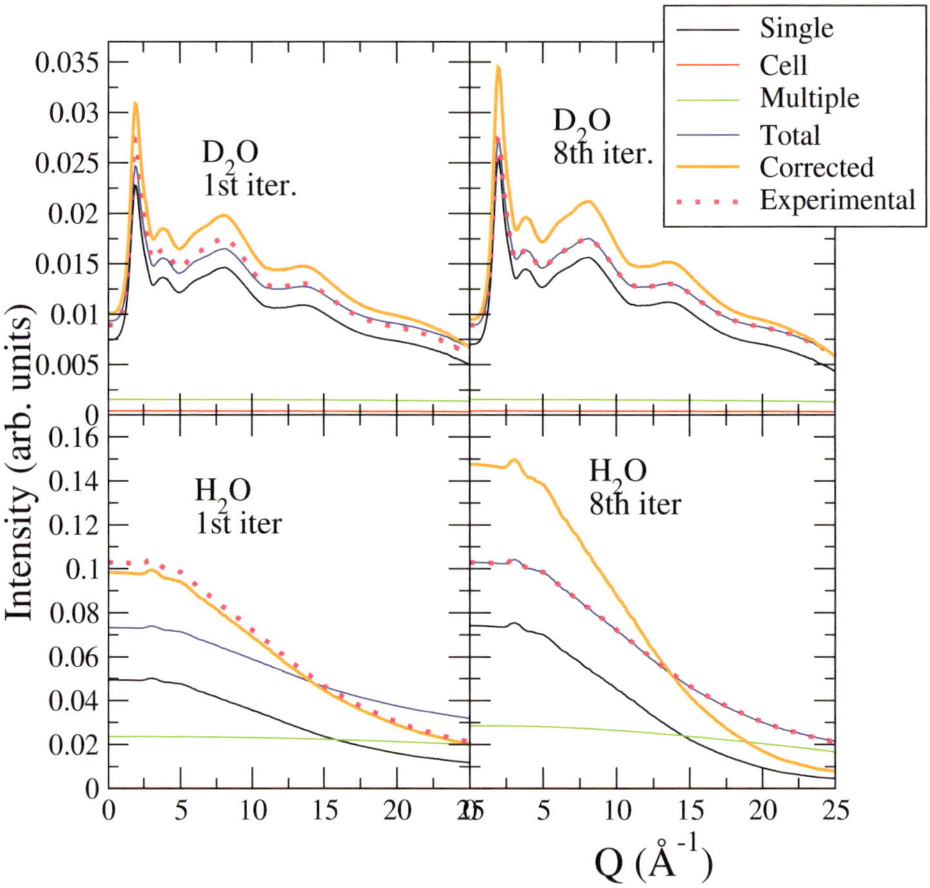

Fig. 7. Calculated components of a diffractogram compared with the experimental data for D_2O and H_2O. Also shown the distribution of singly scattered neutrons $z_1(\mathbf{k}_0, \theta)$ (orange line).

of singly scattered detected neutrons $z_1(\mathbf{k}_0, \theta)$, and the contributions from the cell $z_c(\mathbf{k}_0, \theta)$ and multiple scattering $z_M(\mathbf{k}_0, \theta)$. Also shown in the same figure are the distributions of singly scattered neutrons $z_1(\mathbf{k}_0, \theta)$, that corresponds also to the corrected experimental data in the current step of iteration. As the total scattering resulting from the calculation process converges to the experimental data, the rest of the components also converge to stable values. In Fig. 8, we show the attenuation ($\mathcal{A}^{(8)}(\mathbf{k}_0, \theta)$) and multiple scattering ($f_{\mathrm{MS}}^{(8)}(\mathbf{k}_0, \theta)$) factors after the convergence of the process.

The conclusion of the correction process described in this work is portrayed in Fig. 9, where we show the experimental data and their corresponding corrected data.

6.1 Absolute normalization

The procedure shown in the previous sections achieves the goal to correct the experimental data for attenuation, detector efficiency and multiple scattering effects. The corrected diffractogram is thus proportional to the desired *angular distribution of singly scattered neutrons*

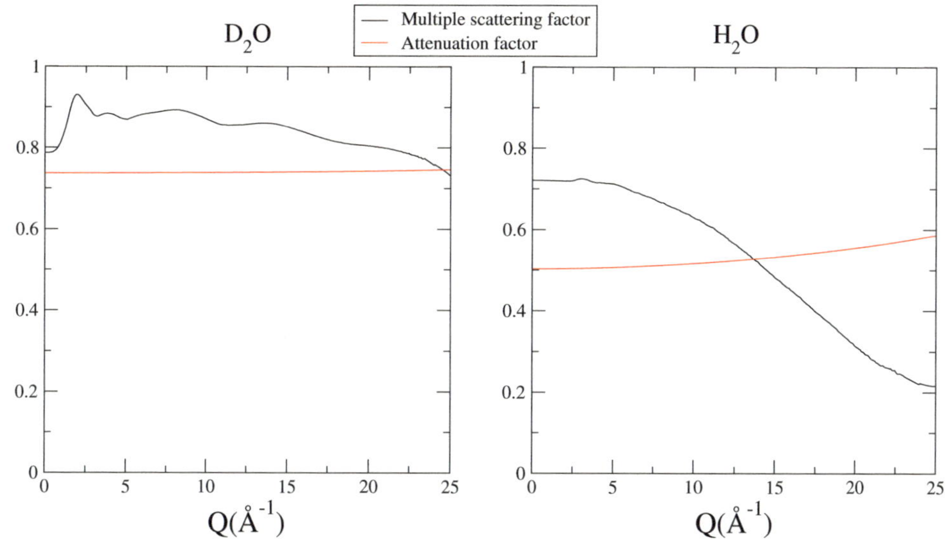

Fig. 8. Attenuation factor (Eq.(41)) and multiple scattering factor (Eq. (40)) for D_2O and H_2O.

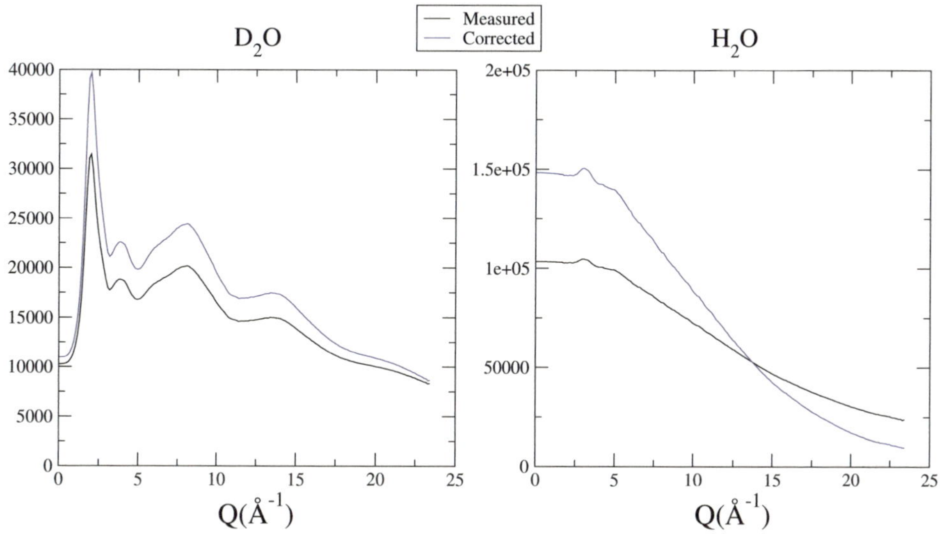

Fig. 9. Comparison between the measured and the corrected diffractograms for D_2O and H_2O after the correction procedure described in this work.

developed in Sect. 3.1. As we showed, Equation (22) is the analytic expression of this magnitude that is directly related with $d\sigma/d\Omega$, the sought result in diffraction experiments. Then,

$$Z_1^{\exp}(\mathbf{k}_0, \theta) = K z_1(\mathbf{k}_0, \theta). \tag{44}$$

The constant K encompasses experimental magnitudes that are independent of the sample, such as the intensity of the incident flux, the efficiency of the detector system, and the monitor normalization.

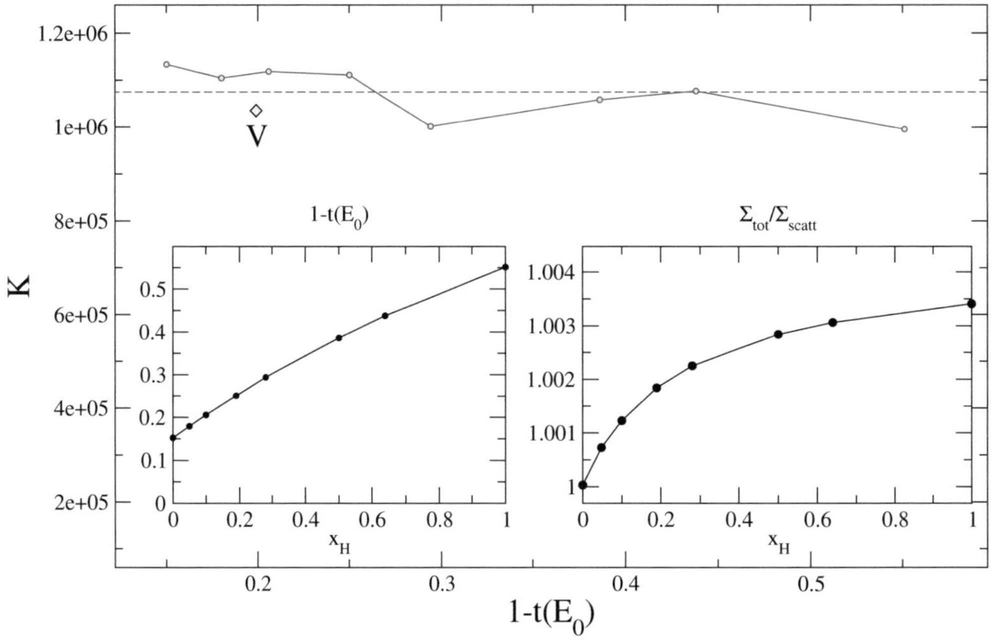

Fig. 10. Main frame: Values of K for the different D_2O-H_2O mixtures as a function of the scattering power of the samples. The mean K value is indicated as an horizontal dotted line. Also shown is K calculated from a vanadium rod 6 mm diameter, 6 cm height. Left inset: scattering power of the samples as a function of hydrogen concentration. Right inset: the ratio $\Sigma_{tot}(\mathbf{k}_0)/\Sigma_{scatt}(\mathbf{k}_0)$ at the incident neutron energy (324 meV) as a function of hydrogen concentration.

The value of K is obtained by integration in angles or its equivalent in Q in Eqs. (22) and (44)

$$
\begin{aligned}
K &= \frac{\Sigma_{tot}(\mathbf{k}_0)}{\Sigma_{scatt}(\mathbf{k}_0)} \frac{1}{1-t(E_0)} \int_0^{2\pi} Z_1^{exp}(\mathbf{k}_0,\theta) 2\pi \sin\theta d\theta \\
&= \frac{\Sigma_{tot}(\mathbf{k}_0)}{\Sigma_{scatt}(\mathbf{k}_0)} \frac{1}{1-t(E_0)} \int_0^{2k_0} Z_1^{exp}(\mathbf{k}_0,Q) \frac{2\pi}{k_0^2} Q dQ.
\end{aligned}
\tag{45}
$$

To perform the integral, the data must be extrapolated up to $\theta = 180^0$.

In Fig. 10 we show the values of K obtained for the the different D_2O-H_2O mixtures as a function of the scattering power of the samples. Also shown is K calculated from the diffractogram of a vanadium rod 6 mm diameter, 6 cm height (typically employed as calibrator sample), subjected to the same corrections. We observe a variation about 5% around the mean value. The reason for it will be analyzed in the next section.

The result of the calibration process is shown in Fig. 11, where the differential cross sections of D_2O and H_2O are shown expressed in barns per steradian. The data are thus properly normalized, and their integral in Q is the total cross sections an the incident neutron energy (324 meV).

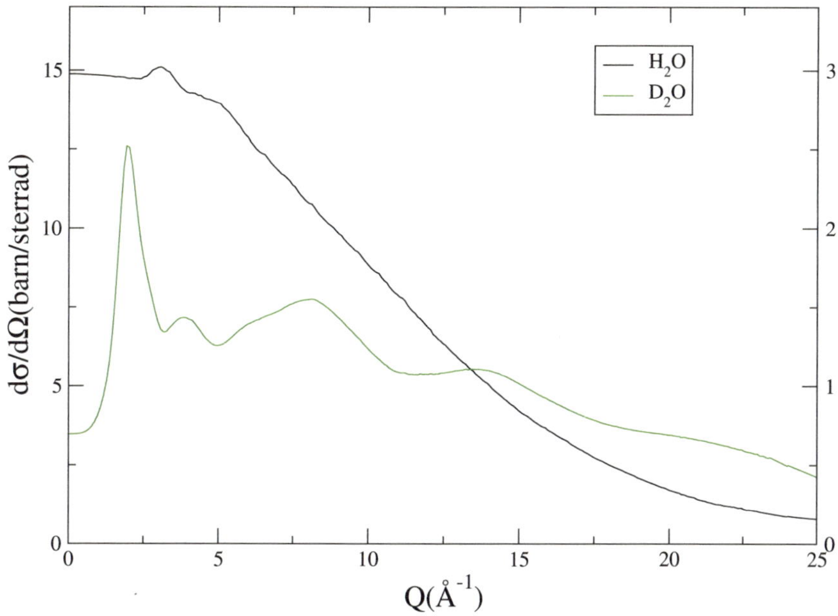

Fig. 11. Differential cross sections of D_2O and H_2O normalized with the procedure of this paper. The left y-axis corresponds to H_2O and the right y-axis to D_2O.

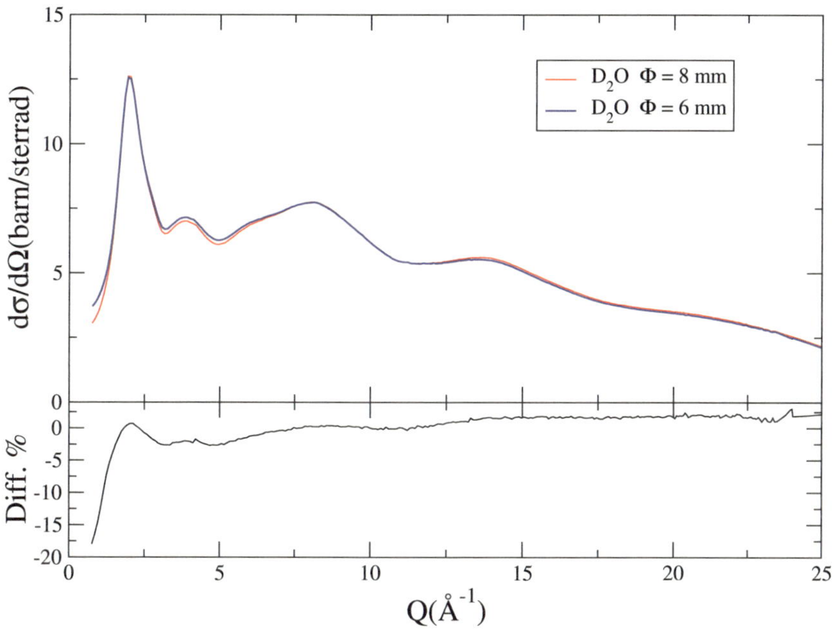

Fig. 12. Upper frame: Assessment of the data correction procedure applied to D_2O samples of different diameters. Lower frame the percentage difference.

7. Discussion

It is worth to comment about the coefficient K, the constant that links the scattering power of the samples with the number of registered counts (Eq. (44)). If the correction procedure is accurate in its different aspects, this coefficient should be an instrumental constant independent of the sample. The observed decreasing trend with the higher Hydrogen-containing samples, has been already noticed in Ref. Rodríguez Palomino et al. (2007). The causes there analyzed were possible systematic errors in the description of the detector bank efficiency, and the presence of a sample-dependent background, that increases with the higher Hydrogen-containing samples. Regarding the detector efficiency, it was exhaustively checked during the process of calibration of D4C Fischer et al. (2002). However, the sample-dependent background is a subject that remains to be treated, both from the numerical and the experimental points of view, and is not only a problem in diffraction experiments but a general issue in neutron scattering experiments.

It is interesting also to explore the limits of the present prescription by comparing the corrected data of D_2O from samples of different sizes, shown in Fig. 12. There is a general good agreement, except for the largest discrepancies shown at lower Q values, where the relative importance of the multiple scattering component is more significant. Improvements on the model should be explored regarding the angular distributions we employed, that were based on the experimental data. The present model involves the assumption that the angular distribution for inelastic processes is the same as for elastic processes. A more accurate procedure would involve a detailed knowledge of the double differential cross section of the studied material, that is normally out of the possibilities of the experimenter that performs a diffraction experiment. The present prescription is intended to keep as reasonable as possible the number of required parameters that are external to the experiment itself. A detailed assessment of this approach with a combination of diffraction and inelastic experiments, remains yet to be done, and should be left for special cases where the inelasticity plays a primary role. The convergence of the iterative process proves the self-consistency of the method. However we must be cautious, because the accuracy of its results will still depend on the goodness of the model employed to describe the system.

8. Conclusions

In this paper we showed a procedure to obtain the link between the arbitrary experimental scale in neutron diffraction experiments (number of recorded counts per monitor counts) with the corresponding cross sections that collects updated knowledge about simulations and cross sections, to describe the processes as realistically as possible. The process involves multiple scattering, attenuation and detector efficiency corrections, that are calculated by a Monte Carlo method. We also developed the mathematical expressions of the magnitudes calculated by Monte Carlo and their experimental counterparts. Those expressions link the measured macroscopic magnitudes with the sought microscopic one.

It is also important to stress that the present prescription is a step ahead over the Paalman-Pings and Blech-Averbach corrections customarily employed, that assume a model of elastic and isotropic scattering. The shortcomings of those simplified approaches compared with the present prescription was analyzed in Rodríguez Palomino et al. (2007), where differences up to 20% were found in hydrogenated samples in which the elastic isotropic model is completely inaccurate.

An important issue, yet to be tackled, is the role of vanadium as a normalizer. The usual measurement of a vanadium sample in D4 does not yield, in general, angular distributions according to theoretical differential cross sections (Cuello & Granada (1997)), because of the instrumental resolution effects. A complete simulation of the instruments with the available simulation software (e.g. McStas), together with the sample simulation must be the next step to solve the issue. The approach we employed in our normalization procedure based on an integral of the diffractograms, avoids the resolution problems in vanadium normalization, but is still a pending issue, regarding the general diffraction measurements.

As a future perspective, the computer code shown in this work will be developed in a user-friendly format, to allow general users to apply it for different systems in diffraction experiments.

9. References

Abdul-Redah, T., Krzystyniak, M. & Chatzidimitriou-Dreismann, C. A. (2005). Neutron compton scattering from water studied with the double-difference technique, *Phys. Rev. B* 72: 052202. URL: *http://link.aps.org/doi/10.1103/PhysRevB.72.052202*

Beckurts, K. & Wirtz, K. (1964). *Neutron Physics, By K.H. Beckurts and K. Wirtz. Translated by L. Dresner*, Springer Verlag OHG Berlin.
URL: *http://books.google.com/books?id=_er8cQAACAAJ*

Bischoff, F. G. (1969). *Generalizad Monte Carlo method for multiple scattering problems in neutron and reactor physics*, PhD thesis, Rensselaer Polytechnic Institute.

Blech, I. A. & Averbach, B. L. (1965). Multiple scattering of neutrons in vanadium and copper, *Phys. Rev.* 137: A1113–A1116.
URL: *http://link.aps.org/doi/10.1103/PhysRev.137.A1113*

Blostein, J. J., Dawidowski, J. & Granada, J. R. (2001). On the analysis of deep inelastic neutron scattering experiments for light nuclei, *Physica B* 304: 357.

Blostein, J. J., Dawidowski, J. & Granada, J. R. (2005). Formalism for obtaining nuclear momentum distributions by the deep inelastic neutron scattering technique, *Phys. Rev. B* 71: 054105.
URL: *http://link.aps.org/doi/10.1103/PhysRevB.71.054105*

Blostein, J. J., Dawidowski, J., Ibáñez, S. A. & Granada, J. R. (2003). Search for anomalous effects in H_2O/D_2O mixtures by neutron total cross section measurements, *Phys. Rev. Lett.* 90(10): 105302.

Blostein, J. J., Rodríguez Palomino, L. A. & Dawidowski, J. (2009). Measurements of the neutron cross sections of hydrogen and deuterium in H_2O/D_2O mixtures using the deep inelastic neutron-scattering technique, *Phys. Rev. Lett.* 102: 097401.
URL: *http://link.aps.org/doi/10.1103/PhysRevLett.102.097401*

Chatzidimitriou-Dreismann, C. A., Abdul Redah, T., Streffer, R. M. F. & Mayers, J. (1997). Anomalous deep inelastic neutron scattering from liquid H_2O/D_2O: Evidence of nuclear quantum entanglement, *Phys. Rev. Lett.* 79: 2839–2842.

Colognesi, D. (2007). Deep inelastic neutron scattering anomalies and quantum decoherence in condensed matter, *Physica B: Condensed Matter* 398 (1): 89–97.

Copley, J. R. D., Verkerk, P., van Well, A. A. & H., F. (1986). Improved monte carlo calculation of multiple scattering effects in thermal neutron scattering experiments, *Computer Physics Communications* 40(2-3): 337 – 357.
URL: *http://www.sciencedirect.com/science/article/pii/0010465586901189*

Cuello, G. & Granada, J. (1997). Thermal neutron scattering by debye solids: A synthetic scattering function, *Annals of Nuclear Energy* 24(10): 763 – 783.
URL: *http://www.sciencedirect.com/science/article/pii/S0306454996000503*

Cuello, G. J. (2008). Structure factor determination of amorphous materials by neutron diffraction, *J. Phys.: Cond. Matter* 20: 244109.

Dawidowski, J. & Cuello, G. J. (2011). Experimental corrections in neutron diffraction of ambient water using h/d isotopic substitution, *Journal of Physics, Conference Series* . In press.

Fischer, H. E., Cuello, G. J., Palleau, P., Feltin, D., Barnes, A. C., Badyal, Y. S. & Simonson, J. M. (2002). D4c: A very high precision diffractometer for disordered materials, *Appl. Phys. A* 74: S160–S162.

Granada, J., Dawidowski, J., Mayer, R. & Gillette, V. (1987). Thermal neutron cross section and transport properties of polyethylene, *Nuclear Instruments and Methods in Physics Research Section A: Accelerators, Spectrometers, Detectors and Associated Equipment* 261(3): 573 – 578.
URL: *http://www.sciencedirect.com/science/article/pii/0168900287903706*

Granada, J. R. (1985). Slow-neutron scattering by molecular gases: A synthetic scattering function, *Phys. Rev. B* 31: 4167–4177.
URL: *http://link.aps.org/doi/10.1103/PhysRevB.31.4167*

Howe, M. A., McGreevy, R. L. & Zetterström, P. (1996). Computer code correct: Correction program for neutron diffraction data. NFL Studsvik.

Kropff, F., Latorre, J. R., Granada, J. R. & Castro Madero, C. (1984). Total neutron cross section of D2O at 20 °C between 0.0005 and 10 ev, *Technical Report EXFOR 30283001*, IAEA.

Lovesey, S. (1986). *The Theory of Neutron Scattering from Condensed Matter*, International series of monographs on physics, Oxford University Press, USA.
URL: *http://books.google.com/books?id=JuupZxrsCTEC*

Mughabghab, S., Divadeenam, M. & Holden, N. (1984). *Neutron Cross Sections: Neutron Resonance Parameters and Thermal Cross Sections*, Neutron Cross Sections Vol. 1, Academic Press. URL: *http://books.google.com/books?id=cgk6AQAAIAAJ*

Paalman, H. H. & Pings, C. J. (1962). Numerical evaluation of X-Ray absorption factors for cylindrical samples and annular sample cells, *J. Appl. Phys.* 33: 2635.

Rodríguez Palomino, L. A., Blostein, J. J. & Dawidowski, J. (2011). Calibration and absolute normalization procedure of a new deep inelastic neutron scattering spectrometer, *Nuclear Instruments and Methods in Physics Research Section A: Accelerators, Spectrometers, Detectors and Associated Equipment* 646(1): 142 – 152.
URL: *http://www.sciencedirect.com/science/article/pii/S0168900211008989*

Rodríguez Palomino, L. A., Dawidowski, J., Blostein, J. J. & Cuello, G. J. (2007). Data processing method for neutron diffraction experiments, *Nuclear Instruments and Methods in Physics Research Section B: Beam Interactions with Materials and Atoms* 258(2): 453 – 470.
URL: *http://www.sciencedirect.com/science/article/pii/S0168583X07003862*

Schmunk, R. E., Randolph, P. D. & Brugger, R. M. (1960). Total cross sections of Ti,V,Y, Ta and W, *Nuclear Science and Engineering* 7: 193–197.

Sears, V. (1975). Slow-neutron multiple scattering, *Advances in Physics* 24(1): 1–45.

Spanier, J. & Gelbard, E. (1969). *Monte Carlo principles and neutron transport problems*, Addison-Wesley series in computer science and information processing, Addison-Wesley Pub. Co. URL: *http://books.google.com/books?id=TP9QAAAAMAAJ*

Squires, G. L. (1978). *Introduction to the theory of thermal neutron scattering*, Cambridge University Press, Cambridge ; New York :.

5

Neutron Diffraction Studies of the Magnetic Oxide Materials

J.B. Yang[1], Q. Cai[2], H.L. Du[1], X.D. Zhou[2], W.B. Yelon[2] and W.J. James[2]
[1]*School of Physics, Peking University*
[2]*Materials Research Center, Missouri University of Science and Technology*
[1]*P.R. China*
[2]*USA*

1. Introduction

Neutron diffraction (ND) is an extremely valuable tool for the investigation of magnetic materials, because of its ability to directly observe periodic magnetic structures, determine magnetic moment directions and magnitudes, to observe light elements (H, C, N, O, F etc.) that are otherwise difficult to locate from x-ray diffraction due to the strong scattering of heavy elements, or to distinguish nearby elements in the periodic chart. Neutron diffraction has proven to be a very useful technique in the study of the magnetic oxides such as perovskites and transition metal oxides.

The perovskite-type ABO_3(A=La, Sr, B=Fe, Co, Ni, Cu, Mn, Ti) perovskites has attracted much attention due to their mixed electronic and oxygen ion conductivity, which shows potential to serve as oxygen separation membranes, oxygen sensors, and solid oxide fuel cell (SOFC), which results from the large number of oxygen vacancies (Kamata et al, 1978, Mizusaki et al., 1991, Kuo et al, 1989.) It is recognized that the physical properties of ABO_3 are largely dependent on the oxygen deficiency (Goodenough, 1955, Srilomsak et al., 1989) . The mixed conductivity can be enhanced through the substitution of La^{3+} by Sr^{2+} at A sites, and the substitution of Fe^{3+} by other transition metal ions at B sites. The charge imbalance and overall charge neutrality can be maintained by the presence of charged oxygen vacancies and mixed valence state ions at the B sites. These point defects are the origin of the mixed electronic and oxygen ion conductivity.

A three-dimensional framework made up of BO_6 octahedra characterizes the ABO_3 perovskite crystal structure. The interstices of this framework are sites for the large A cations, and the small B cations are at the centers of the octahedra. Every two neighboring BO_6 octahedra share a common O atom. The perovskite structure is often stable with both divalent (Sr, Ba, Ca, etc.) and trivalent (e.g. lanthanide) A site occupation, and the charge imbalance and overall charge neutrality will be maintained by the formation of electrons, holes and/or charged oxygen vacancies. These point defects can take part in the electronic and/or oxygen ion conductivity, and this gives the material scientist an opportunity to alter the transport properties of a given oxide.

When perovskite materials that contain transition metal ions at the B sites are heated to sufficiently high temperature, they can equilibrate with the ambient oxygen by oxygen exchange. If the oxygen activities at the two sides of the perovskite are different, the oxygen ions can be transported from the high oxygen activity side to the low one. Therefore perovskites can be used as oxygen separation membranes or electrodes of SOFCs.

Strontium-doped lanthanum manganate is most commonly used as the cathode of SOFCs operated at high temperature. $LaMnO_3$ is a p-type perovskite. At high temperature, it can have oxygen excess, stoichiometry, or oxygen deficiency, depending on the oxygen partial pressure.(Kamata et al, 1978, Mizusaki et al., 1991, Kuo et al, 1989.) For example, at 1200°C, the oxygen content of $LaMnO_3$ ranges from 3.079 to 2.947 under oxygen partial pressures of 1 to 10^{-11} atmosphere pressure.(Kamata et al, 1978) In addition to oxygen nonstoichiometry, $LaMnO_3$ can also have La deficiency or excess. $LaMnO_3$ with La excess may contain La_2O_3 as a second phase, and therefore $LaMnO_3$ with La deficiency is recommended for use in SOFCs.

In recent years, work has focused on SOFCs that can operate at an intermediate temperature (IT) range (600-800°C). In this temperature range, the metal alloy, such as stainless steel, which is cheap and easily fabricated, can be used as the interconnect, and the reliability of SOFCs is thus improved. In order for a SOFC to be operated in the IT range, the electrode kinetics have to be at least as fast as those occurring at high temperature. LSM, the current cathode material, is not suitable for use below 800°C due to its very low oxygen vacancy concentration. Therefore, research has to be done to find new material for use as the cathode of SOFCs operating in the IT range. In general, most cathodes have relied on the B-site cation, and doping on both A-sites and B-sites to improve electrical conductivity and catalytic performance. One approach to finding a new cathode material is to replace the Mn by other transition metals, such as Fe, Co, and Ni etc.

The strontium-doped lanthanum ferrite $La_{1-x}Sr_xFeO_{3-\delta}$ (LSFO) has been intensively studied since it has good mixed conductivity at high temperature, and, thus, can be used as an oxygen membrane and is a candidate for the cathode of the SOFC. Undoped $LaFeO_3$ has a low electrical conductivity and oxygen vacancy concentration, but the electrical conductivity and oxygen vacancy concentration are increased by the substitution of La by Sr. M. V. Patrakeev et al., 2003 studied the electron / hole and ion transport in LSFO in the oxygen partial pressure range of 10^{-19}-0.5 atm and temperatures between 750-950°C, and found that the electronic and ionic conductivity increase with Sr content and attain maximal values at x=0.5. The conductivity can be explained by electron hopping between Fe^{3+} and Fe^{4+} in the high P_{O2} range and between Fe^{3+} and the Fe^{2+} in the low P_{O2} range. The LSFO material is stable when the $P_{O2}>10^{-16}$ atm at 950°C; oxygen vacancies will appear at 600-800°C in air. LSFO has shown excellent cathode performance at 750°C (Sun et al., 2005).

It is important to measure the oxygen vacancy concentration in LSFO. It is well known that neutron powder diffraction coupled with Rietveld refinement(Rodriguez-Carvajal 1998) can be used to determine the oxygen vacancy concentration in many oxides, since the sensitivity of neutron scattering to oxygen is comparable to other atoms. However, the precision of such a determination is unknown, especially at low vacancy content. In addition, neutron diffraction is a very sensitive direct probe of the magnetic moment, and LSFO exhibits magnetic ordering. If the magnetic moment of Fe is sensitive to the oxygen vacancy concentration, it can be used as an indirect probe of the oxygen vacancy concentration. To understand the physical properties of $La_{0.6}Sr_{0.4}FeO_{3-\delta}$ and its behavior in SOFC and to be able to optimize its behavior, it is important to have a good knowledge of the defect chemistry of $La_{0.6}Sr_{0.4}FeO_3$. In the present paper, samples of $La_{0.6}Sr_{0.4}FeO_3$ with different amounts of oxygen vacancies were prepared by heating the samples in N_2, O_2, and CO/CO_2 mixtures. The effect of oxygen vacancies on the structural, magnetic and electronic properties was determined using neutron diffraction, magnetic measurements and Mössbauer spectra. It was found that magnetic ordering, charge disproportionation and charge ordering in these compounds shows strong dependence on the oxygen vacancies. The oxygen vacancies will changes the Fe valence states, the unit cell volume and the Fe-O-Fe bond angle. These dramatically affect the Fe-O-Fe superexchange coupling. Therefore by creating oxygen vacancies or having excess oxygen, the exchange interaction of Fe-O and the valence state of Fe ions are affected, and lead to large changes in the magnetic and transport properties, such as the magnetic ordering temperature, the magnetic moments, the hyperfine interactions and electric conductivity in the pervoskite structure.

2. Experimental methods

The liquid-mixing method (Eror et al., 1986) was used to synthesize fine LSFO powder. An aqueous solution of Fe nitrate was first prepared and thermogravimetrically standardized. The lanthanum carbonate, and strontium carbonate were added to form a clear solution. Citric acid and ethylene glycol were added to the nitrate solution and then heated slowly to form a polymeric precursor, which was heated to 250°C to form an amorphous resin. This resin was calcined at 800°C for 8 hours. The powders were pressed at 207 MPa to form a dense bar. The bar was sintered at 1000–1200°C for 24 h under different environments(air, N_2, O_2 or CO/CO_2), followed by quenching to room temperature. We assume the oxygen nonstoichiometry of the quenched samples is the same as before quenching. The magnetization curves of the samples were measured using a superconducting quantum interference device SQUID magnetometer in a field of up to 6 T from 1.5 K to 800 K. A magnetic field of 50 Oe was used for the field cooling(FC) and zero field cooling(ZFC) process. The crystal phase was identified by x-ray diffraction analysis using Cu *Ka* radiation. The powder neutron diffraction experiments were performed at the University of Missouri-Columbia research reactor using neutrons of wavelength 1.4875 Å. The data for each sample were collected over 24h In order to study the oxygen stoichiometry change in air at high temperature, in-situ neutron diffraction studies were carried out at high temperature (up to 800°C) for the samples unquenched and quenched at 1500°C in air. The Rietveld method was used to refine the data by using the FULLPROF code (Rodriguez-Carvajal 2000), in which the magnetic ordering was modeled by a separate phase.

2.1 Results and discussion

2.1.1 The effect of heat treatment under different gas environments

Figure 1 shows the typical ND patterns of $La_{0.6}Sr_{0.4}FeO_3$ powders at different heat treatment conditions. Similar patterns are observed for all samples, showing them to be single phase. The symmetry of the samples remains rhombohedral (space group $R\bar{3}c$) throughout the series. A cubic structure (space group $Pm\bar{3}m$) was also proposed for these compounds, but it was found that the data could be better fitted in a hexagonal structure (space group $R\bar{3}c$) as confirmed by the following neutron diffraction data refinement. The model is a rhombohedral structure with space group $R\bar{3}c$. The lattice parameters (hexagonal setting) are $a = b \approx \sqrt{2}a_p$, $c \approx 2\sqrt{3}a_p$, $\alpha=\beta=90°$, and $\gamma=120°$, where a_p is the lattice parameter of

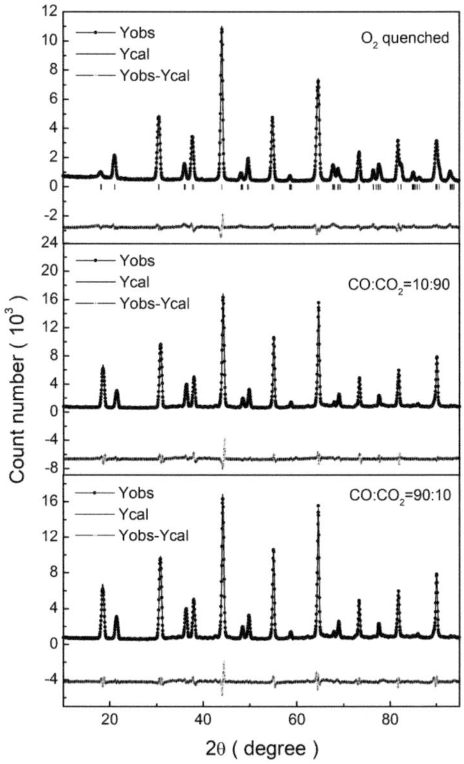

Fig. 1. Neutron diffraction patterns of $La_{0.6}Sr_{0.4}FeO_3$ quenched at O_2 and CO/CO_2. atmospheres

the basic cubic perovskite cell. In this model, the large cations La^{3+}/Sr^{2+} occupy the 6a sites, the small cations Fe^{3+}/Fe^{4+} occupy the 6b sites, and the oxygen ions occupy the 18e sites. The magnetic structure is modeled as having antiferromagnetic ordering with the moments in the hexagonal plane, which reverse between the positions (0, 0, 0) and (0, 0, ½). The magnetic model has the symmetry $R\bar{3}$, and the magnetic moments on the two sites are constrained to be equal. It is difficult to use XRD patterns to determine the structural distortion, and the oxygen vacancy concentration. Accordingly, neutron diffraction was employed to distinguish the differences between the structures of the samples, and to determine the oxygen vacancy concentration.(Yang et al., 2002)

As evidenced from this figure, all samples crystallize in the rhombohedral structure with space group R-3c. The lattice parameters and bond angle are shown in Fig. 2(a) and (b). It can be seen that the unit cell volume increases when the $La_{0.6}Sr_{0.4}FeO_3$ is quenching in the CO/CO_2 atmosphere. The unit cell volume of three CO/CO_2-reduced samples are larger by about 5Å^3 than those of the O_2- and N_2-quenched samples. The ratio of the lattice parameters a/c changes from 0.4113 for the O_2-quenched sample to 0.40840 for the reduced samples, which decreases the distortion from cubicity. This change affects the peak positions and the sharpness of the diffraction peaks. The diffraction peak splitting for the CO/CO_2 treated samples is dramatically reduced compared to the O_2-treated samples. The first diffraction peak is much stronger in the reduced samples than in the O_2-quenched samples.

This peak proves to be purely magnetic and the change reflects a large increase in magnetic moments for the CO/CO_2-reduced samples. An antiferromagnetic structure has been confirmed for all samples. Fe atoms at (0,0,0) couple antiferromagnetically with those at (0,0,1/2) along the c-axis. The Fe atoms show a magnetic moment $3.8\mu_B$ in the CO/CO_2-treated $La_{0.6}Sr_{0.4}FeO_3$ as compared to $1.2\mu_B$ in the O_2-treated samples at 290 K. The extent of oxygen vacancy in these compounds has been obtained by refinements of neutron diffraction data. The O_2-, and N_2-quenched samples show similar patterns, and there are less than 1% vacancies on the oxygen sites. The oxygen vacancy concentration is around 7-10% for the samples quenched in the CO/CO_2 mixtures. It is noticed that these values are about 2% less than those measured from the iodometric method. This difference might cause from the systematic errors of the two techniques.

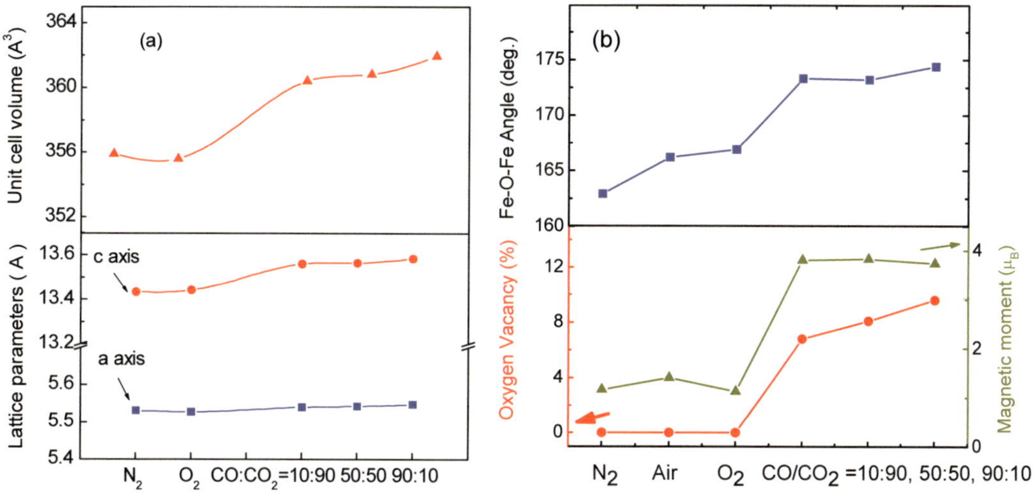

Fig. 2. The lattice parameters and unit cell volumes of $La_{0.6}Sr_{0.4}FeO_3$ treated at different gases(a), the bond angle, oxygen vacancy and magnetic moments of $La_{0.6}Sr_{0.4}FeO_3$ treated at different gases(b)

The typical neutron patterns of the samples (N_2, O_2, and $50\%CO/50\%CO_2$) at 290 K were collected(data not shown here). The N_2-treated sample shows a rhombohedral structure and are refined in the space group $R\bar{3}c$. The diffraction peak half width for the CO/CO2 treated samples is dramatically reduced compared to the N_2-treated samples which is similar to the XRD patterns. The rhombohedral splitting of the peaks is not too obvious for the CO/CO_2-treated samples. Thus, we have used both a cubic cell space group $Pm\bar{3}m$ and a rhombohedral cell, space group $R\bar{3}c$ to refine the structure. The rhombohedral structure gives much better refinement results. For example, χ^2 is 7.56 for the refinement using space group $R\bar{3}c$ and 15.8 for space group $Pm\bar{3}m$ for the N_2-treated sample. Therefore, all samples have been refined with a rhombohedral cell space group $R\bar{3}c$). The first diffraction peak [101] at about 19°) is much stronger in the reduced samples than in the air-, N_2-, and O_2-quenched samples. This peak proves to be purely magnetic and the change reflects a large increase in magnetic moments for the CO/CO_2-reduced samples. The air-, O_2-, and N_2-quenched samples show similar patterns, and there is less than 1%. The ^{57}Fe Mössbauer spectra of $La_{0.6}Sr_{0.4}FeO_{3-\delta}$ treated in N_2 measured at different temperatures are shown in Fig. 3 (a). The hyperfine parameters are listed in Table. I. The Mössbauer spectroscopy of N_2 at room temperature shows a paramagnetic behavior. Two singlets were used in the fitting for the spectrum. The isomer shifts are 0.26mm/s and 0.18mm/s, which are exactly what would be expected for the Fe^{3+} and Fe^{4+} ions. The ratio of Fe^{3+}/Fe^{4+} is 63.9/36.1, corresponding to an oxygen vacancy, δ=0.023. At 130 K and 20 K, the spectra are comprised of three superimposed magnetic hyperfine patterns.

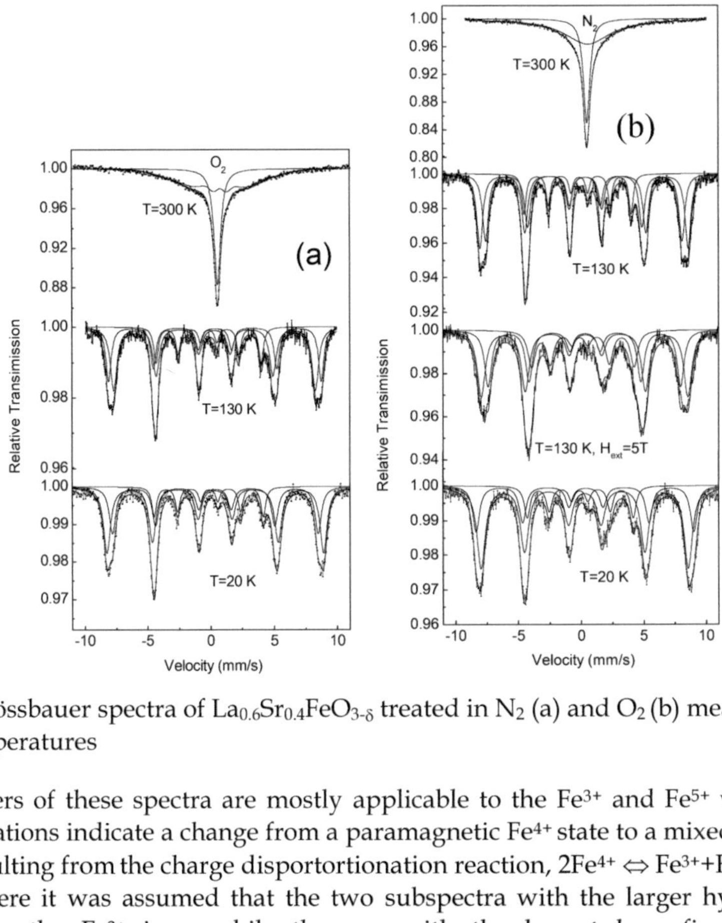

Fig. 3. ^{57}Fe Mössbauer spectra of $La_{0.6}Sr_{0.4}FeO_{3-\delta}$ treated in N_2 (a) and O_2 (b) measured at different temperatures

The parameters of these spectra are mostly applicable to the Fe^{3+} and Fe^{5+} valence states. These observations indicate a change from a paramagnetic Fe^{4+} state to a mixed valence state Fe^{3+}/Fe^{5+} resulting from the charge disportortionation reaction, $2Fe^{4+} \Leftrightarrow Fe^{3+}+Fe^{5+}$ (Takano et al., 1997). Here it was assumed that the two subspectra with the larger hyperfine fields correspond to the Fe^{3+} ions, while the one with the lowest hyperfine field (~26T) corresponds to Fe^{5+} ions, which was used by Dann *et al, 1994.*

Similar spectra were observed for O_2-quenched samples as shown in Fig. 3(b). At room temperature the rhombohedral relaxation of the hyperfine fields appears and therefore a Voigt peak-shaped sextet and a singlet were used in the fitting to account for the Fe^{3+} and Fe^{4+} ions. At low temperature, the charge disportortionation also takes place. The oxygen deficiency obtained from the relative areas of the Mössbauer spectra of the Fe^{3+} and Fe^{4+} ions in O_2-treated sample is nearly zero, which agrees well with the neutron diffraction measurements. The Mössbauer spectra of the CO/CO_2-quenched samples are shown in Fig. 4. They show a typical sextet due to the antiferromagnetic coupling of the Fe atoms, which is

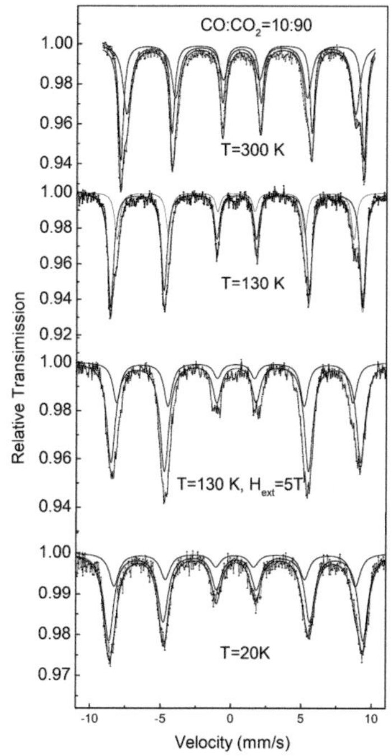

Fig. 4. The Mössbauer spectra of the CO/CO_2-quenched samples at different temperatures

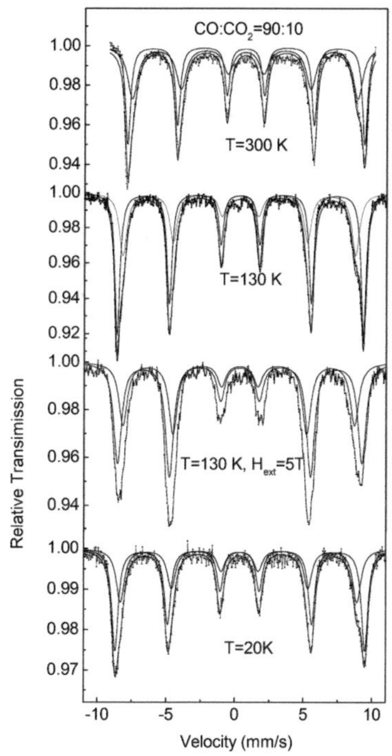

Fig. 5. Mössbauer spectra of the CO/CO_2-quenched $La_{0.6}Sr_{0.4}FeO_3$ samples measured at different temperatures

		IS (mm/s)	QS (mm/s)	B_{hf} (kOe)	Int.(%)
	N$_2$-quenched				
20K	Fe^{3+}	0.442	-.058	540.9	26.3
	Fe^{3+}	0.381	.001	515.0	58.4
	Fe^{5+}	-0.005	0.047	262.3	15.3
130K	Fe^{3+}	0.420	-.106	517.0	43.9
	Fe^{3+}	0.386	-.077	488.9	39.3
	Fe^{5+}	-0.085	-0.118	264.2	16.8
130 K (H= 5T)	Fe^{3+}	0.415	-.051	513.3	44.9
	Fe^{3+}	0.387	-.039	478.0	33.2
	Fe^{5+}	-0.033	0.251	253.4	21.9
RT	Fe^{3+}	0.261	-	-	63.9
	Fe^{4+}	0.180	-	-	36.1
	O$_2$-quenched				
20K	Fe^{3+}	0.422	-0.023	534.3	51.2
	Fe^{3+}	0.399	-0.001	507.1	33.5
	Fe^{5+}	-0.053	-0.034	268.9	15.3
130K	Fe^{3+}	0.427	-0.046	524.6	33.2
	Fe^{3+}	0.399	-0.252	497.1	48.4
	Fe^{5+}	-0.058	-0.015	259.6	18.4
RT	Fe^{3+}	0.331	0.047	191.0	59.9
	Fe^{4+}	0.202			40.1
	CO/CO$_2$=10:90				
20K	$Fe^{(3-x)+}$	0.479	-0.026	561.6	73*
	$Fe^{(3-x)+}$	0.376	0.012	533.1	27*
130K	$Fe^{(3-x)+}$	0.442	0.014	555.8	73*
	$Fe^{(3-x)+}$	0.379	-0.052	522.7	27*
130K (H=5T)	$Fe^{(3-x)+}$	0.435	-0.032	551.4	73*
	$Fe^{(3-x)+}$	0.375	-0.068	521.6	27*
RT	$Fe^{(3-x)+}$	0.324	0.050	535.5	73*
	$Fe^{(3-x)+}$	0.236	0.020	501.1	27*
	CO/CO$_2$=90:10				
20K	$Fe^{(3-x)+}$	0.469	0.024	565.5	60*
	$Fe^{(3-x)+}$	0.417	-0.007	535.4	40*
130K	$Fe^{(3-x)+}$	0.439	0.009	557.4	60*
	$Fe^{(3-x)+}$	0.386	-0.074	526.4	40*
130K (H=5T)	$Fe^{(3-x)+}$	0.434	-0.007	553.1	60*
	$Fe^{(3-x)+}$	0.388	-0.055	524.1	40*
RT	$Fe^{(3-x)+}$	0.324	0.035	535.3	60*
	$Fe^{(3-x)+}$	0.272	-0.063	504.1	40*

Parameter constrained to the given value, and $0<x<1.0$.

Table 1. Hyperfine parameters of the $La_{0.6}Sr_{0.4}FeO_3$ treated at different gases measured at different temperatures.

confirmed by neutron diffraction and magnetic measurements. The best fitting can be reached by using two sextets for the fitting of the entire spectra by assuming two different Fe valence states. The average hyperfine field is 50-53 T at room temperature, which is of the same order as that of the Fe-oxide (such as Fe_2O_3). This large hyperfine field corresponds to a valence state between Fe^{2+} and Fe^{3+}. There is no evidence in the spectra for the presence of any Fe^{5+} at low temperatures. An attempt to find evidence for the presence of distinct Fe^{4+} or Fe^{2+} lines in the spectra also failed. The two sextets have values of an isomer shift somewhere between Fe^{2+} and Fe^{3+} regions. The true oxidation state of Fe seems to be neither Fe^{2+} nor Fe^{3+} but an intermediate state (such as $Fe^{(3-x)+}$,$0<x<1.0$). Because the CO/CO_2-treated samples have much higher oxygen vacancy concentrations, it suggests that some Fe^{4+} or even Fe^{3+} in the non-reduced samples are reduced to $Fe^{(3-x)+}$ ($0<x<1.0$), in order to maintain charge balance in these compounds. It is evident that the change of the Fe valence state from Fe^{4+} to $Fe^{(3-x)+}$ results in a large hyperfine field and a large magnetic moment. The average quadrupole splitting of the CO/CO_2-quenched samples is smaller than that of the non-reduced sample, which indicates a decrease in the distortion from the cubic structure. This is consistent with the neutron diffraction data, which show a decrease of the distortion from cubic. It is found that the CO/CO_2-treated samples have the same a/c ratio (0.40855) as those (a/c=0.40850) of $La_{1-x}Sr_xFeO_{3-\delta}$ (x=0.6 and 0.7), which is, near the boundary between the rhombohedral and cubic structures (Dann et al., 1994). The spectra measured under a magnetic field of 5 T at 130K were also included in Figs. 3-5. As compared with the spectra at 130 K without the magnetic field, it is found that the intensities of the first and sixth line decrease. This indicates that there might be a canted magnetic moment in the magnetic sublattice, which implies that the antiferromagnetic structure is not perfect.

2.1.2 The effect of quenching temperatures

Figure 6 shows the typical ND patterns of $La_{0.6}Sr_{0.4}FeO_3$ powders quenched at different temperatures in the air. The first peak at low angle is a pure magnetic peak; its intensity increases with the quenching temperature, and some of the split peaks merge into one peak when the quench temperature is high.

Figure 7 shows the change of lattice parameters a^* and c^* with the quenching temperature, in which a^* represents $a/\sqrt{2}$ and c^* represents $c/2\sqrt{3}$. The symmetry of L6SF quenched to room temperature from 700 to 1100°C remains rhombohedral, but the rhombohedral distortion becomes small when the quenching temperature is high. When the quenching temperature $T \geq 1200$°C, the unit cell appears to be cubic. The presence of oxygen vacancies can relax the strain in the structure and reduce the distortion.

According to the direct refinement, the oxygen vacancy concentration increases from almost zero for the unquenched sample to about 0.2 for the 1500°C quenched sample. The statistical uncertainty shows that the direct refinement is reliable only when the oxygen vacancy concentration is high.

The unit cell volume increases as the quenching temperature increases by a total of 3.8Å^3 from the unquenched sample to the 1500°C quenched sample. The statistical uncertainty in the unit cell volume is about 0.04Å^3. It shows that, the unit cell volume can be a good metric for the determination of vacancy concentration. However, the unit cell volumes are relative due to the uncertainty in wavelength and sample position. Thus, the volumes of a series of samples can easily be compared, but the unit cell volume of a single specimen cannot be used to estimate its oxygen vacancy concentration. The small downturn in volume at the 1500°C quenched samples appears to be an artifact, since the data on the samples treated in different reducing atmospheres with vacancy concentration up to 9.6% show a continued increase in volume.

The unit cell expansion associated with the formation of oxygen vacancies can be explained by: a): The repulsive force arising between those mutually exposed cations when oxygen ions are absent in the lattice. b): The increase in cation size due to the reduction of Fe ions from a high valence state to a lower valence state, which must occur concurrently with the formation of oxygen vacancies in order to maintain electrical neutrality.

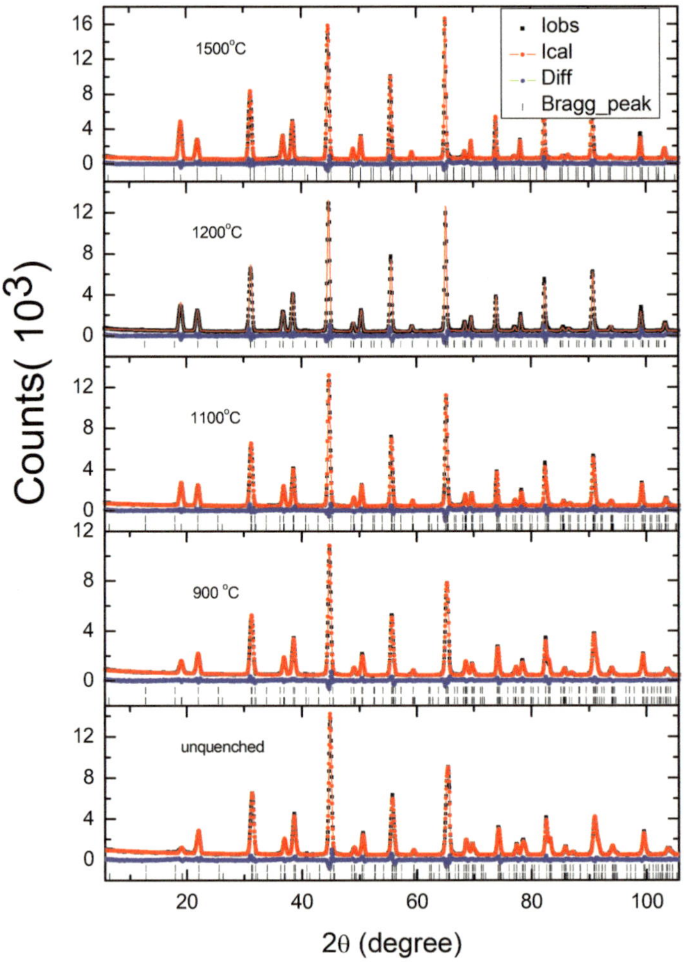

Fig. 6. The neutron diffraction patterns of unquenched, 900°C, 1100°C, 1200°C, and 1500°C air quenched specimens. The black dots are the observed intensity, the solid red line is the calculated intensity, and their difference (blue) is under them. The upper tic marks show Bragg positions for the nuclear phase, and the lower ones are for the magnetic phase.

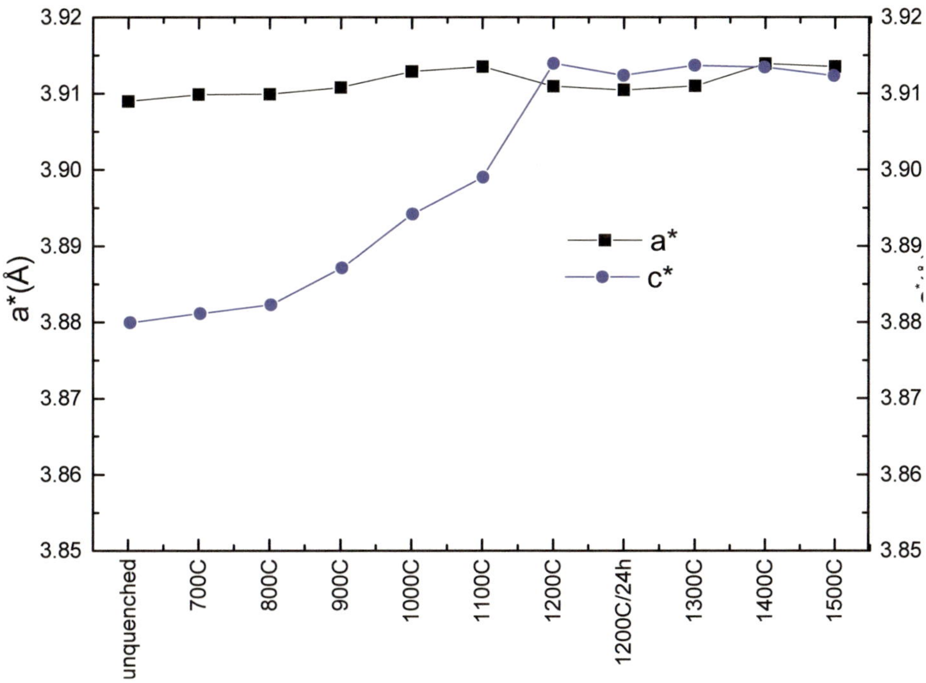

Fig. 7. Lattice parameters of $La_{0.6}Sr_{0.4}FeO_3$ vs. quenching condition.

Quenching Temperature	a (Å)	c (Å)	Vol. (Å^3)	Total O	Mag.(RT) (μ_B)	δ
No Quench	5.52818(17)	13.44049(56)	355.72(2)	2.992(12)	1.31(3)	0.008(12)
700°C	5.52942(15)	13.44457(52)	355.99(2)	2.980(12)	1.42(2)	0.020(12)
800°C	5.52946(15)	13.44860(51)	356.10(2)	2.964(12)	1.67(2)	0.036(12)
900°C	5.53072(16)	13.46527(58)	356.71(2)	2.962(12)	2.07(2)	0.038(12)
1000°C	5.53352(17)	13.48996(65)	357.72(2)	2.936(16)	2.44(2)	0.064(16)
1100°C	5.53456(16)	13.50656(62)	358.30(2)	2.920(14)	2.72(2)	0.080(14)
1200°C	5.53091(41)	13.55839(190)	359.20(6)	2.872(18)	2.98(2)	0.128(18)
1200°C/24h	5.53026(40)	13.55285(191)	358.97(6)	2.856(14)	3.07(2)	0.144(14)
1300°C	5.53098(25)	13.55746(110)	359.18(4)	2.830(16)	3.26(2)	0.170(16)
1400°C	5.53510(29)	13.55662(126)	359.69(4)	2.818(16)	3.39(2)	0.182(16)
1500°C	5.53455(25)	13.55276(108)	359.52(4)	2.804(14)	3.44(2)	0.196(14)

Table 2. The crystal structure parameters of $La_{0.6}Sr_{0.4}FeO_3$ at various quenching temperatures.

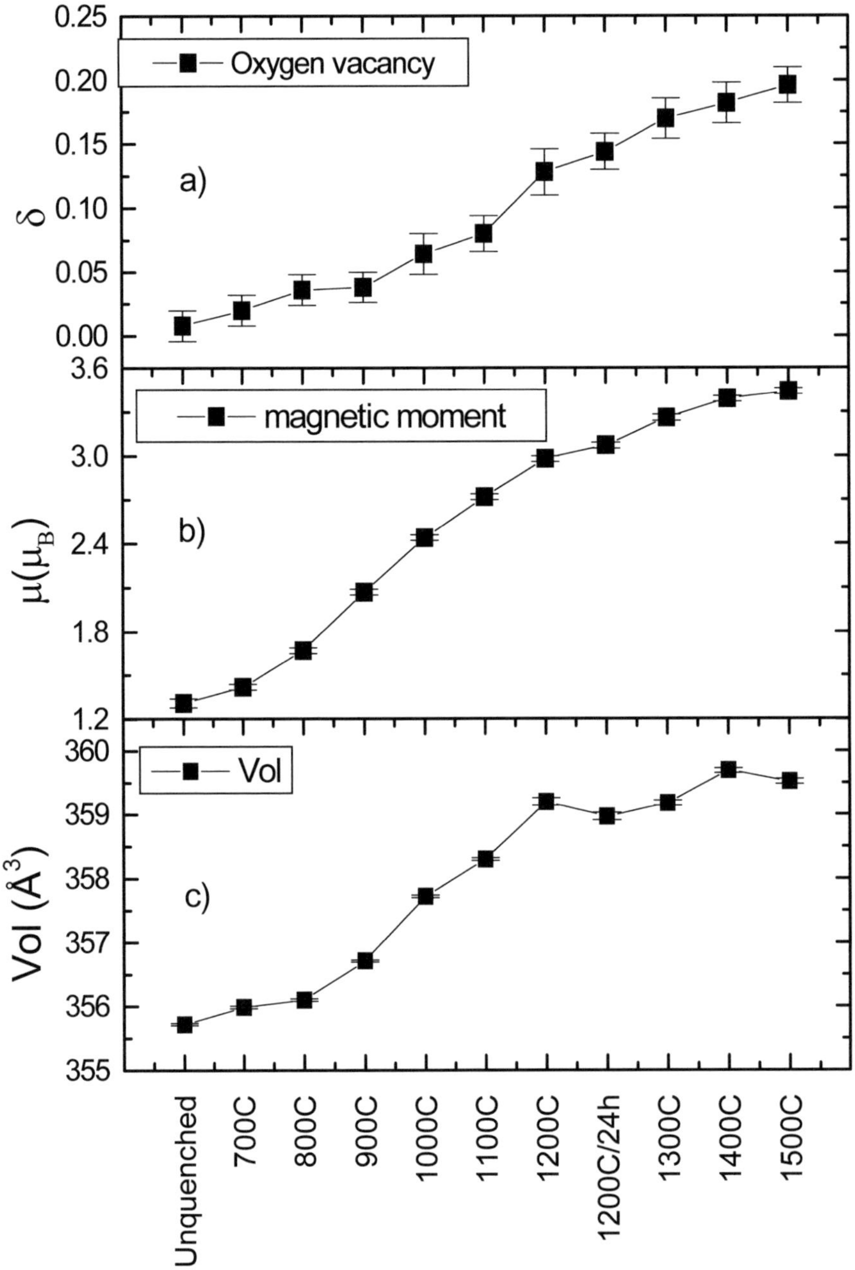

Fig. 8. The changes of oxygen vacancy, magnetic moment, and the unit cell volume with the quench temperature for the $La_{0.6}Sr_{0.4}FeO_3$ samples.

Figure 9 shows the magnetic moment and the unit cell volume of L6SF samples as a function of oxygen vacancy concentration. It appears that, in this range, the magnetic moment shows smoother behavior than does the unit cell volume, and may be the more reliable indirect measurement of the oxygen vacancy concentration.

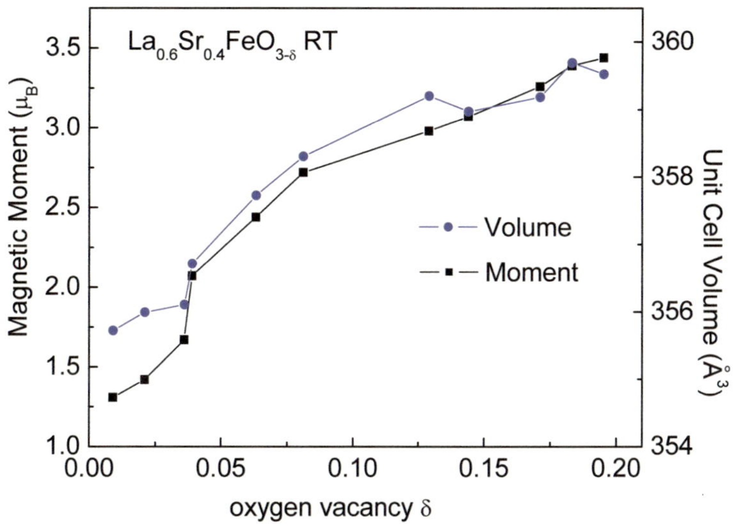

Fig. 9. Magnetic moment and unit cell volume at room temperature vs. oxygen vacancy concentration for $La_{0.6}Sr_{0.4}FeO_3$.

The magnetic moment on the Fe sites at room temperature was found to increase from1.31μ_B for the untreated sample to 3.44μ_B for the sample quenched from 1500°C. The statistical uncertainty in the magnetic moment is less than 2.3% of the total moment, significantly less than the uncertainty in the vacancy concentration directly determined by the crystallographic refinement. The small increase in moment between the untreated and 700°C quenched samples probably reflects the production of a small oxygen vacancy concentration, outside the limit of the direct determination.

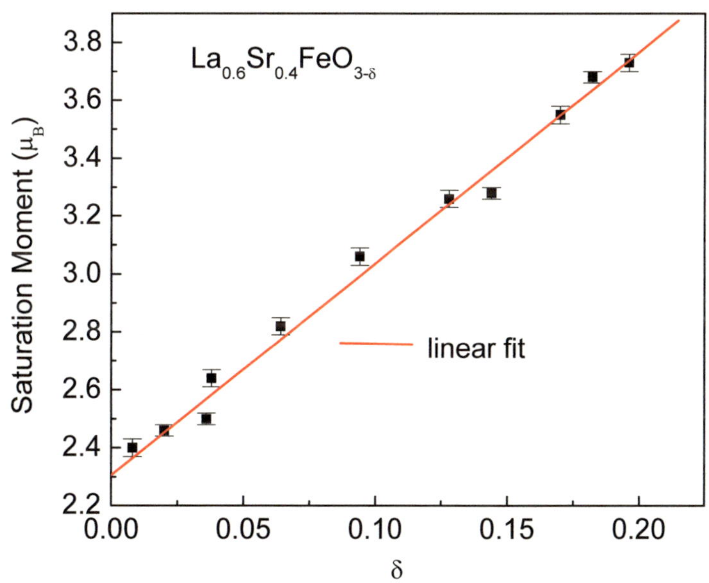

Fig. 10. Oxygen vacancy concentration vs. saturation moment for $La_{0.6}Sr_{0.4}FeO_3$

Fig. 10 shows the saturation moments, determined from the neutron powder diffraction measurement at 10K, as a function of oxygen deficiency. The saturation moment is linear with vacancy concentration, the highest deficiency, $\delta=0.2$, corresponds to a nearly pure Fe^{3+} state, and its magnetic moment is about $3.8\mu_B$, which is equal to the magnetic moment of Fe^{3+} in $LaFeO_3$. The moment when $\delta=0$ is about $2.33\mu_B$, and this value is about 60% of the magnetic moment of Fe^{3+}. This shows the magnetic moments at Fe sites are contributed by Fe^{3+}, and that the moment of Fe^{4+} is almost zero. For the ferrites with high La (50% or more) in an oxidizing environment, it is reasonable to assume that the Fe atoms are in the 3+ and 4+ charge states. The great difference between the saturation moments of the two Fe ions provide a direct determination of the ratio of the two ions and, thus, of the oxygen stoichiometry.

The magnetization as a function of temperature follows the Brillouin curve: saturated at low temperature and decreasing slowly up to about 70% of T_N and more rapidly as T_N is approached. Thus, according the ratio of the magnetic moment at room temperature to the moment at low temperature (Table 3), T_N is a little above room temperature when the quenching temperature is below 900°C.

T(°C)	No Quenching	700	800	900	1000	1100	1200	1300	1400	1500
μ_{RT} (μ_B)	1.31(3)	1.42(2)	1.67(2)	2.07(2)	2.44(2)	2.72(2)	2.98(2)	3.26(2)	3.39(2)	3.44(2)
μ_{10K} (μ_B)	2.40(3)	2.46(2)	2.50(2)	2.64(3)	2.82(3)	3.06(3)	3.26(3)	3.55(3)	3.68(2)	3.73(3)
$\frac{\mu_{RT}}{\mu_{10K}}$	0.546	0.577	0.668	0.784	0.865	0.889	0.914	0.918	0.921	0.922
Fe^{3+}%	61.4	64	67.2	67.4	72.8	76	85.6	94	96.4	98.4

Table 3. The magnetic moment of $La_{0.6}Sr_{0.4}FeO_3$ at different quenching temperature. The mol% of Fe^{3+} is calculated from the oxygen vacancy concentration.

$La_{0.6}Sr_{0.4}FeO_3$ exhibits antiferromagnetic ordering below the Neel temperature. Comparing the magnetic moments at room temperature and 10K, the Neel temperature (T_N) increases with increasing oxygen vacancies alone with the concentration of Fe^{3+}. The magnetic interactions between Fe ions, leading to magnetic ordering in this type of oxide, are predominantly superexchange; exchange that is mediated by polarization of oxygen ions lying between the Fe near neighbors. Since the Fe^{4+} ions have small or zero moments, the exchange interactions in L6SF are expected to be dominated by the Fe^{3+}-O^{2-}- Fe^{3+} interactions. This is the reason why T_N increases with increasing Fe^{3+} concentration, despite the loss of some bonding oxygen atoms.

There are a lot of advantages in using the magnetic moment as a measure of vacancy concentration in $La_{0.6}Sr_{0.4}FeO_3$. First, the saturation moment gives an absolute determination without establishing the room temperature curves, while the room temperature moment may be quickly and reliably determined if the correlations have been established. The uncertainty in magnetic moment is 2% at low vacancy concentration, decreasing to less than 1% when the moment is large, thus, the vacancy concentration should be known, by this indirect determination, to a precision of 1% or 2% over the range of interest. Second, the magnetic moment is not affected by the sample position and the model used for the refinement. Third, it is necessary to collect full data sets in order to accurately refine the vacancy concentration, while it is possible to extract the magnetic moment with the low angle range data only, and such data may be collected in as little as 1 hour. Of course, at very high vacancy concentration (δ>0.2), Fe^{2+} is expected to appear, with a smaller moment than Fe^{3+}, the average moment at Fe sites will become small as δ increases, and the linear relationship between the saturation moment and the oxygen vacancy will break down. At those concentrations, however, the direct determination of oxygen vacancy concentration by crystallographic refinement will become more precise.

On the other hand, collection of a full data set allows us to extract all the three determinations: volume, oxygen occupancy, and magnetic moment. These three determinations are essentially independent. The unit cell volume is based only on peak positions and not peak intensities, and uses the full data set. The oxygen occupancies affect the intensities at all angles and especially uses the high angle data, while the magnetic moment is based primarily on the intensities of magnetic peaks at low angle. Thus use of neutron diffraction provides redundancy and cross checks on the oxygen vacancy determination.

3. Conclusion

$La_{0.6}Sr_{0.4}FeO_3$ shows a very small oxygen vacancy contents prepared by heating in the N_2 and O_2 gases. When treated in a CO/CO_2 mixture, the amount of oxygen vacancies exceeds 7%. Mössbauer spectra of N_2 and O_2-quenched samples show a transition from high temperature paramagnetic Fe^{3+} and Fe^{4+} valence states to a low temperature antiferromagnetic mixed-valence state resulting from nominal charge disproportionation reaction $2Fe^{4+}=Fe^{3+}+Fe^{5+}$. The Fe valence states change from Fe^{4+} to an intermediate valence state, $Fe^{(3-x)+}$, in the CO/CO_2-reduced samples. There is no charge disproportionation throughout the entire temperature range. The change of the valence state in the CO/CO_2-heat treatment increases magnetic moments and the hyperfine fields of the Fe atoms in these compounds. It is found that the antiferromagnetic structure exhibits a small ferromagnetic component, which causes from canted magnetic sublattices $La_{0.6}Sr_{0.4}FeO_3$ spacimens quenched from 700-1500°C exhibit antiferromagnetic ordering at room temperature, the unit cell volume and the oxygen vacancy concentrations increase with increasing quenching temperature. The magnetic ordering is dominated by the Fe^{3+}-O^{2-}-Fe^{3+} interactions, and the Fe^{3+} concentration increases with increasing oxygen vacancy δ. Thus, the magnetic moment and the Neel temperature increase with increasing oxygen vacancy. The saturation moment, determined by neutron diffraction at 10K, is linear with the oxygen vacancy (δ). At high temperature with air flowing, L6SF will absorb oxygen at 303°C-655°C and then lose oxygen when temperature is above 655°C. The presence of oxygen vacancies increases the thermal expansion coefficient.

Neutron diffraction measurement coupled with refinement by the Rietveld method appears to be a reliable, redundant method for determining the oxygen vacancy in L6SF. The unit cell volume, the oxygen occupation, and the magnetic moment can be used to determine the oxygen vacancy. The unit cell volume can be affected by the sample position, but the magnetic moment at room temperature can provide the data more accurately than the other two parameters. The saturation moment (μ_{10K}) is an even more powerful tool which can be used to determine the vacancy concentration.

4. Acknowledgments

This work is supported by the National Natural Science Foundation of China (Grant No. 509701003 and 51171001), the National 973 Project (No. 2010CB833104, MOST of China), and the program for New Century Excellent Talents (NCET-10-1097) and the Scientific Research Foundation for Returned Overseas Chinese Scholars, State Education Ministry.

5. References

Dann E.; Currie D. B.; Weller M. T.; Thomas M. F.& Al-Rawwas A. D. (1994) . The Effect of Oxygen Stoichiometry on Phase Relations and Structure in the System $La_{1-x}Sr_xFeO_{3-\delta}$ ($0 \leq x \leq 1, 0 \leq \delta \leq 0.5$), *J. Solid State Chem.* 109, 134-144.

Eror N. G.; & Anderson H. U.(1986). POLYMERIC PRECURSOR SYNTHESIS OF CERAMIC MATERIALS in "Better ceramics through chemistry II", edited by C J. Briuker, D E. Clark, and D R. Ulrich (Materials Research Society, 1986). Pp571-577. Pittsburgh, PA 1986.

Goodenough J. B. (1955) .Theory of the Role of Covalence in the perovskite-Type Manganites [La,M(II)]MnO_3, *Phys. Rev.*, 100, 564-73 .

Kamata K.; Nakajima T.; Hayashi T.; & Nakamura T. (1978). Nonstoichiometric Behavior and Phase Stability of Rare-Earth Manganites at 1200°C: (1). $LaMnO_3$, *Mater. Res. Bull.*, 13, 49-54.

Kuo J. H.; Anderson H. U., & Sparlin D. M.(1989) . Oxidation-Reduction Behavior of Undoped and Sr-Doped $LaMnO_3$ Nonstoichiometry and defect Structure, *J. Solid State Chem.*, 83, 52-60.

Mizusaki J.; Tagawa H.; Naraya K. & Sasamoto T. (1991) .Nonstoichiometry and Thermochemical Stability of the perovskite-Type $La_{1-x}Sr_xMnO_{3-\delta}$ *Solid State Ionics*, 49, 111-18.

Patrakeev, M. V.; Bahteeva, J. A.; Mitberg, E. B.; Leonidov, I. A.; Kozhevnikov, V. L. & Poeppelmeier, K. R. (2003). Electron/hole and ion transport in $La_{1-x}Sr_xFeO_{3-\delta}$ *Journal of Solid State Chemistry* 172, 219-231.

Rodriguez-Carvajal J.(1998), Program: FULLPROF, Version 3.5d. Program for Rietveld Refinement and Pattern Matching Analysis, France.

Srilomsak S.; Schilling D. P. & Anderson H. U.(1989). Thermal Expansion Studies on Cathode and Interconnect Oxides; pp. 129-40 in *Proceedings of the First International Symposium on Solid Oxide Fuel Cells.* Edited by S. C. Singhai. The Electrochemical Society, Pennington, NJ, 1989.

Sun, L. & Brisard, G.(2005), Fuel Cell and Hydrogen Technologies, *Proceedings of the International Symposium on Fuel Cell and Hydrogen Technologies, 1st*, Calgary, AB, Canada, Aug. 21-24, 85-96. 2005.

Takano, M; Nakanishi N.; Takeda Y.; Naka S., &Takada T. (1997), Charge disproportionation in $CaFeO_3$ studied with the Mössbauer effect, *Mater. Res. Bull.* 12, 923.

Yang, J. B.; Yelon, W. B.; James, W. J.; Chu, Z.; Kornecki, M.; Xie, Y. X.; Zhou, X. D.; Anderson, H. U.; Joshi, Amish G.; & Malik, S. K.;(2002). Crystal structure, magnetic properties, and Mossbauer studies of $La_{0.6}Sr_{0.4}FeO_{3-\delta}$ prepared by quenching in different atmospheres. *Physical Review. B* 66, 184415-1-184415-9.

6

Introduction of Neutron Diffractometers for Mechanical Behavior Studies of Structural Materials

E-Wen Huang[1], Wanchuck Woo[2] and Ji-Jung Kai[3]
[1]*Department of Chemical & Materials Engineering and Center for Neutron Beam Applications, National Central University*
[2]*Neutron Science Division, Korea Atomic Energy Research Institute*
[3]*Department of Engineering and System Science, National Tsing Hua University*
[1,3]*Taiwan*
[2]*South Korea*

1. Introduction

The design of advanced metallic materials for their structural applications requires the understanding of the strengthening mechanisms and property evolution subjected to different types of deformation modes[1]. These metallic systems can interact with their microstructure upon the changes of the environmental conditions, such as strain rate and temperature[2]. While the microstructure has been facilitated for many purposes, this chapter puts forward how to characterize the structural properties with neutron diffractometers. Moreover, nowadays, many neutron diffractometers are equipped with load frames, which advance the diffraction measurements to real-time observations. The chapter considers that deformation mechanisms and their effects on the microstructure are central to the mechanical behavior of structural materials. The main objective of the chapter is to introduce readers how to facilitate the neutron diffractometers to study the mechanical behavior of the structural materials. The reported diffractometers are summarized from the literatures, public information, and on-site visits. Some useful software for diffraction-profile and scattering-intensity analyses is briefly mentioned. The microscopic features are connected with the macroscopic states, such as the applied stresses and temperature evolution to bridge the understanding of the bulk property. What reciprocal information obtained from the diffraction profiles can be inferred to the materials structural parameters will be explained.

2. Neutron diffraction and diffraction-profile-refinement software

Neutrons are elementary particles with a finite mass (m, about 1.67×10^{-27} kg) and spin, without an electron charge. It carries a magnetic moment, and according to *de Broglie* law, the neutrons behave like waves with a wavelength (λ) at the levels of Å and gives rise to diffraction[3]. Neutron diffraction is based on the nuclear interaction between neutrons and the matters and on magnetic interaction with magnetic moments of the atoms due to its magnetic moments. Specifically, in this chapter, we focus only on the application of the elastic neutron scattering. Moreover, for the applications of the neutron diffractormeters to characterize the polycrystalline metallic systems are based on the idea of the powder diffraction. The concept of a powder diffraction experiment is that the sample consists of a large number of small randomly oriented crystallites. If the number is sufficiently large, there are enough crystallites in various diffracting orientations to yield diffraction patterns. Depending on the microstructure arrangements, interferences between the scattered neutrons are constructive when the path difference between diffracted rays differs by an integral number of wavelengths. This is described by the well-known Bragg equation. There are various methods to refine the aforementioned scattering distribution function. The General Structure Analysis System (GSAS)[5] based on full-pattern Rietveld analysis method[6] is one of the most popular peak-profile-refinement software. The users of GSAS[5] can obtain peak-position, intensity, peak profiles and width, respectively with the model fitting. In the following sections, the refinement of the peak position, intensity, width and their related materials microstructure are introduced in Sessions 2.1, 2.2, and 2.3, respectively.

2.1 Peak position, lattice strain, and using lattice strain for mechanical behavior study

With the single-peak refinement of the GSAS, peak positions (d^{hkl}) of specific lattice plane can be refined for the calculations of the lattice strain, ε^{hkl}, as shown in Equation (1):

$$\varepsilon^{hkl} = \frac{d^{hkl} - d^{hkl}{}_0}{d^{hkl}{}_0} \times 10^6 (\mu e) . \tag{1}$$

The refined peak positions of each *hkl* are calculated according to the changes in the *d*-spacing (d^{hkl}). The lattice evolution is relative to the initial *d*-spacing ($d^{hkl}{}_0$) as a function of the deformation levels.

The lattice-strain evolution results can be referred to the generalized Hooke's law to compare the measured results with the classic models. For examples, the relationship between the stress and strain is one of the most fundamental properties of the materials. The Hooke's law presents the strength of interatomic forces between adjacent atoms. The reversible nature of the bulk materials is known as the elastic strain. Hooke's law can be generalized relative to the loading directions. As a stress is applied in one direction (for example, the Y-direction) of the materials, it could yield strains in the X, Y, and Z-directions. Hence, the Poisson's ratio (υ), and the modulus of elasticity (E), and the stiffness can also be determined by knowing the peak position changes upon elastic deformation.

Furthermore, with careful experimental setup, the elastic constants vary as a function of the crystallographic orientation, the strains in each direction subjected to three normal and six shear stress components, can be measured, too. For examples, the cubic crystals, described in Equation 2, have three independent elastic constants and different direction cosines relative to the crystallographic directions.

$$\frac{1}{E} = s_{11} - 2\left[\left(s_{11} - s_{12} \right) - \frac{1}{2} s_{44} \right] \left(l_1^2 l_2^2 + l_2^2 l_3^2 + l_1^2 l_3^2 \right) . \tag{2}$$

l_1,l_2 and l_3 are direction cosines. The direction cosines for principal directions in a cubic lattice can be referred to Hertzbug's book, where the theoretical models are clearly described[7]. This is one of the examples of how to apply neutron diffractometers to calculate the fundamental materials properties. Besides, the fundamental stress-strain relationships, several recent studies show that the lattice strain can be inferred to the thermomechanical behavior[8,9]. Following is a short review of this latest application of neutron peak position data.

2.1.1 Lattice strain for thermomechanical calculations

The state of solids can be described by the strain (ε) or stress (σ) and temperature (T). Under the Hooke's law, the stress- or strain-temperature relation can be derived from the laws of thermodynamics (Stanley and Chan 1985 [10]) as

$$\Delta T = \left(-K \frac{ET}{1-2\upsilon} \Delta\varepsilon \right) \text{ or} \tag{3}$$

$$\Delta T = \left(-KT\Delta\sigma \right). \tag{4}$$

T is the absolute temperature of the current state, and K is a material constant [$K = \overline{\alpha} / \rho C_V$]. Other parameters, $\overline{\alpha}$ is the coefficient of the linear thermal expansion, ρ is the mass density, and C_v is the specific heat at a constant volume. $\Delta\varepsilon$ is the change of the strain, and $\Delta\sigma$ is the change of the stress. υ is the Poisson's ratio. In Equations 3 and 4, the temperature response is negatively proportional to the change in the stress or strain states of a homogenous elastic solid.

It has long been recognized that the mechanical energy can be separated into the stored energy of cold work and thermal energy[11]. Einstein's, Debye's, and Grüneisen's models[12] unify the elastic deformation with thermodynamics. New approaches are proposed to model the stored energy of the deformed microstructure[13-15].

To answer the temperature-lattice strain relationships posed above, Huang et al[9]. applied the lattice strain evolution subjected to cyclic loadings with Wong et al.[13] and Quinn et al.[16]'s thermo-mechanical relationship as described in Equation 5:

$$\Delta T = \frac{-\alpha T}{\rho C} \left(1 - \frac{1}{\alpha E^2} \frac{\partial E}{\partial T} \sigma_m \right) \Delta\sigma \ . \tag{5}$$

ΔT is the temperature response subjected to the mechanical deformation. α is the thermal expansion. C is the heat capacity. E is the Young's modulus. σ_m is the residual stress introduced by the plastic deformation. $\Delta\sigma$ is the stress change due to the elastic deformation, which can be referred in Huang *et al.*'s study[9] as the lattice-strain evolution. Based on their measured lattice strain, the thermo-mechanical responses were estimated with the modified Stanley and Chen's equation[10]:

$$\Delta T = \frac{-\alpha E_{111} T}{\rho C (1-2\upsilon)} \left(\frac{\varepsilon_{111}^{loading} + \varepsilon_{111}^{transverse}}{2} \right) \times \frac{\textit{Volume of the Deformed Specimen}}{\textit{Whole - specimen Volume}} \tag{6}$$

E_{111} is the 111-lattice plane modulus[17]. ν is the Poisson ratio. Because after cyclic loading, there is practical no texture development in the studied system, the lattice strains (ε_{111}) of grains in orthogonal directions (oriented to loading and transverse directions, respectively) can be averaged. Thermal expansion ($\alpha = \frac{\gamma \rho C_V}{3E}$) is a function of Grüneisen parameter ($\gamma \equiv \frac{\partial(ln\,\omega_{Debys})}{\partial(ln V)}$), where ω_{Debys} is Debye frequency and V is the volume. Based on the generalized Hooke's law, the temperature responses can be calculated with the updated volume strain[11] as $\Delta V = (1+\varepsilon_{111})(1-\upsilon\varepsilon_{111})^2 - 1$. Huang *et al.*[9] show that the calculated and the measured temperature-evolution trends qualitative agree with each other. Certainly, there is quantitative discrepancy because during the deformation, the plastic deformation occurred, but not counted in the above elastic-based calculations. Moreover, in Huang *et al.*'s 2010 work[8], their thermal resistivity calculations also validate the above approach of the use of lattice strain to understand the thermomechanical behviors. Besides the above lattice strain applications, to explore the plastic deformation, the following two sections will introduce how to correlate the peak intensity for materials texture and the peak width for microstructure studies, respectively.

2.2 Diffraction-peak intensity and texture

The diffraction is a Fourier transformation from the crystal space into the reciprocal space. Intensities of various *I(hkl)* of the crystalline systems are proportional to the squares of the crystallographic structure factors *F(hkl)*.

h k l are the Miller indices of the unit cell.

The peak intensity can be fitted from the individual *hkl* peaks. The normalized intensity evolution as a function of stress can be calculated using Equation (7) to trace the texture development through the bulk deformation.

$$I_{hkl} = \frac{I^{deformed}_{hkl}}{I^{0}_{hkl}} \quad (7)$$

2.2.1 Combining intensity evolution for texture development and lattice-strain changes for self-consistent modeling

In polycrystalline materials, the orientations of the grains decide the texture of the materials. Under plastic deformation, slips on specific crystallographic planes produce lattice rotations which accumulate texture. Hill's [18] and Hutchinson's[19] self-consistent (SC) models have been applied very successfully to simulate the texture development of polycrystalline materials[20] by the lattice-strain evolution based on neutron-diffraction measurements via the Elasto-plastic self-consistent (EPSC) model for several metallic materials[21, 22]. The intrinsic assumptions of the EPSC model consider the active inelastic-deformation mechanisms, and, hence, the stiffness/compliance constants are important for simulations[23-25]. The present session review Huang *et al.*'s work[17], which extends SC modeling to moderate-to-large deformation strain, considering the grain rotation for describing preferred grain-orientation distributions. Huang *et al.* applied Wang's[26-28] visco-plastic self-consistent (VPSC) model to simulate the texture development, based on their measured lattice strains and macro stress-strain responses from a nickel-based alloy. The VPSC model considers the activity of slip systems and its influence on grain rotation[25]. The VPSC code is based on Wang's fundamental work[26-28]. With the neutron diffractometers, the *in-situ* experimental macro/lattice stress-strain curves can be used as an input to simulate the texture and the probability of the active-slip systems from the multiple *hkl*-reflections. A VPSC model was implemented for estimating the distribution of microstresses and texture evolution in Huang *et al.*'s work[17]. All the parameters were derived by fitting the diffraction profiles of the experimental neutron measurements. The compliances used in the simulations can be calculated from the measurement of lattice-strain (ε_{hkl}) within the elastic deformation. There is good qualitative agreement in the inverse pole figures from the measured intensity evolution to the simulated results. Meanwhile, the quantitative discrepancy suggests that during the plastic deformation of the structural materials, besides the considerations of the lattice evolution and texture development, we need also to take into account the effects of microstructure changes.

2.3 Peak-profile analyses and microstructure changes

Practically, structural polycrystalline materials contain imperfections that influence the intensity of the Bragg reflections distributions. The main deviations observed in the diffraction profiles are from microstructure and from the strain and stress in the sample. The Rietveld method[29] is a breakthrough method to resolve the peak profiles by facilitating the full pattern profiles with the crystal structure refinements. The Rietveld method is a least-square fit of a given profile function to the diffraction pattern by minimizing the difference between the observed and calculated diffraction intensity distributions:

GSAS has three classical profile-shape functions for the users to refine the diffraction profiles. They are the Gaussian, the Lorentzian, and the Pseudo-Voigt-type distribution functions. Typically, the crystalline-size broadening produces Lorentzian distribution-type tails in the peak profile, while microstrains produce Gaussian distribution-type contributions. Full-width-at-half-maximum (FWHM) values are significantly affected with such characteristics.

As a general rule, the determination of a particle size is treated by the Scherrer equation[4]. The Scherrer equation gives a rough estimate of the broadening caused by the crystalline size. The strain broadening can be covered by the Gaussian, which has a width proportional to $\tan(\Theta)$ [30]. Using GSAS, the peak-width can be decomposed into the Gaussian and Lorentzian broadening components simultaneously.

2.3.1 Microstructure for mechanical behavior study

From a mechanistic perspective, the creation of obstacles to the dislocation motion can enhance the strengthening mechanisms. These obstacles provide additional resisting force above the intrinsic lattice friction. The strengthening mechanisms are revealed macroscopically through a larger flow stress[11]. Hence, it is important to examine how the microstructures create the obstacles to the motion of the dislocations.

The cold work on the material can be presented by the microstructural changes at the dislocation level. The increase in the dislocation density can alter the mechanical properties. From the basic observation of generic crystalline solids, the polycrystalline materials are composed of many different grains, each with a particular crystalline orientation and separated by grain boundaries[31]. The grains and the associated boundaries make up the polycrystalline microstructure, such as the grain size and patterned-dislocation structures[32]. In a worked material, the microstructure is characterized by a series of heavily dislocated grains[33].

The microstructures with the defects can significantly change the strengths of the materials[34]. Hence, it is one of the most important structural materials properties for their mechanical applications. Among different perspectives, the material's microstructure must account for the dislocations. In a broad term, the dislocations play the main role of the permanent deformation of crystalline solids largely because the dislocations are the primary means of the plastic deformation[11]. The attempt to monitor such plasticity can be based upon different methods. At the smallest scales, the morphology can be observed by the atomic structure of dislocations via transmission-electron microscopy (TEM) [35].

2.3.2 Microstructure and peak-profile changes in the plastically-deformed materials

To bridge the texture and microstructure developments subjected to plastic deformation, in this session, we continue review Huang *et al.*'s work[17]. Huang *et al.* assume that during the formation of a hierarchical dislocation structure in the plastically-deformed material, some part of a dislocations group within dislocation walls and some part remain randomly distributed. The formation of the hierarchical dislocation structure in many metallic polycrystalline materials under different deformation has been widely reported[35-39]. Readers may refer to Zehetbauer[40], Schafler *et al.*[41], and Hughes and Hansen[42]'s description of the formation of the hierarchical dislocation structure to visualize that with the dislocation walls creating, there are certain misorientations/tilt/twist between the neighboring regions of the same grain and some other random portion of the dislocation population randomly distributed between the walls. There is practically no misorientation/tilt/twist in the interior. For simplicity, the dislocation walls can be assumed to be composed from the equidistant-edge-dislocations. Meanwhile, the screw dislocations remain randomly distributed in the cell interior. Hence, as presented previously[43-46] equidistant-dislocations walls can result in the Lorentzian type of broadening. The randomly-distributed or weakly correlated dislocations in the interior can result in the Gaussian type of broadening. Above all, the readers can base on the microstructure results of their observations and then apply a proper peak-profile function of the GSAS to correlate the refined neutron diffraction results to the microstructure evolution.

2.4 Small-angle neutron scattering and the precipitation strengthening

Finally, in the end of the introduction of the relationship between structural materials microstructure and neutron instruments, the author would like to review how to investigate the precipitates of the metallic alloys. The age-hardening effect of the precipitates has been extensively used to improve the mechanical behavior by impeding the dislocation movements in various metal-based structural materials[47].

If there is a supersaturation of substitutional impurities, subsequent annealing will lead to the formation of precipitates[38]. The precipitates can act as the obstacles to the dislocations. Specifically, the equilibrium of two-phase coexists between the matrix phase and a second phase. The presence of such second-phase particles makes itself known through a substantive change in the mechanical properties of the material. Observations on precipitation hardening make it evident that the key microstructural features in this context are the mean particle radius and the volume fraction of such precipitates[11]. In particular, the variation in the flow stress is a function of particle size. The particle-cutting mechanism is dominant at small radii. The Orowan process is easier at large particle radii[34].

The age-hardening effect was discovered by Wilm[48]. Since then, it has been extensively used for developing various metal-based structural materials for industrial applications. The strengthening mechanism relies on the precipitation in some phases other than that of uniform dispersion. The dislocations are localized and prevented from continued movements by the strain field introduced by the lattice mismatch between the precipitates and the homogeneous matrix. The morphology of the precipitates and their spatial arrangement in the embedded matrix are two known key elements in deciding the mechanical performance.

The small-angle neutron scattering (SANS) approach presents a complementary tool to the microscopy technique. It provides the nano-scale information via the measurement of the Fourier transform of the spatial correlation function. The collected scattering intensity, $I(Q)$, is presented in reciprocal Q space. Therefore, to obtain quantitative real-space information, model fitting is usually required. In this chapter, we review Chen *et al.*'s model[49] for SANS $I(Q)$ study of precipitation strengthening. In general, $I(Q)$ obtained from a system consisted of nonspherical particles can be expressed as

$$I(Q) = \frac{N|\Delta\rho|^2}{V_S} P(Q)\{1+\beta(Q)[S(Q)-1]\} + I_{\mathrm{INC}} \tag{8}$$

where $\Delta\rho$ is the difference in scattering-length densities between the particle and the dispersion medium; V_S, the sample volume illuminated by neutron beam; N, the number of precipitates in V_S; $P(Q)$, the average form factor given by the shape and density profile of particles; $S(Q)$, the effective one-component inter-precipitate structure factor, which is a measure of the inter-particle interference; $\beta(Q)$ the decoupling constant dependent on both the size polydispersity and intra-precipitate density profile[50], and $I_{\mathrm{INC}}$the incoherent background.

In the practical implementation of the introduced model fitting, $\frac{N|\Delta\rho|^2}{V_S}$ is treated as a fitting parameter suggested by Pedersen[51]. The shape of the polydisperse precipitate can be modeled by the form factor, $P(Q)$. The effect of polydispersity can be incorporated through the decoupling approximation and a standard Gaussian law[50].

$$N(x,R,\delta) = \frac{1}{\sqrt{2\pi\delta^2}} \exp\left[\frac{-(x-R)^2}{2\delta^2}\right] \tag{9}$$

where R is chosen to describe the size distribution (δ is the variance).

In Huang *et al.*'s exemplary system[49], the inter-precipitate structure factor, $S(Q)$, is calculated via a stochastic phenomenological model. It is assumed that precipitates are partially ordered and separated from the nearest neighbors with a preferred distance, L, with a deviation measured by the root-mean square denoted by σ. The inter-precipitate structure factor[52], $S(Q)$, is expressed as a function of Q, L, and σ

$$S(Q,L,\sigma)=2\cdot\left\{\frac{1-\exp\left[-\left(Q^2\sigma^2\right)/4\right]\cos(QL)}{1-2\exp\left[-\left(Q^2\sigma^2\right)/4\right]\cos(QL)+\exp\left[-\left(Q^2\sigma^2\right)/2\right]}\right\}-1 \tag{10}$$

Hence, the coherent SANS intensity distributions obtained from the alloys can be fitted by the above model and uniquely refine five parameters: R, ε, δ, L, and σ of the strengthening precipitates.

3. The neutron diffractometers for structural materials research

There are several neutron diffractometers in the world dedicated, but not limited, to the mechanical behavior study. Because this chapter focuses on the mechanical behavior study of structural materials, those diffractometers equipped with the load frames for *in-situ* measurements are summarized. In the following sessions, these neutron diffractometers are categorized into two groups and introduced according to their alphabetic order. The diffractometers located in the spallation neutron source facilities will be introduced in the session 3.1. The others located in the reactor neutron source facilities will be summarized in the session 3.2. The differences between the Spallation Neutron Source and Reactor Source are not the focus of this chapter. Here, only a brief comparison is summarized below.

For the time of flight, Spallation Neutron Source, a pulse of chopped neutrons illuminate the sample at different time. Within this range of the different time, the diffracted neutrons can cover several reflections of the materials within one pulse. With this unique feature, several *hkl* peaks can be revealed at almost "one time" simultaneously. This multiple-*hkl*-peaks-collection capability enables the researchers to easily characterize several phases with a wide range of the q-space at specific environmental at one pulse of the incident neutrons. On the other hand, the Reactor Source Neutron diffractometers can also cover similar range of the q-space, but with the adjusting of the monochromators and analyzers. Hence, it sometimes will take longer time for the Reactor Source to collect the same *hkl* diffractions than that of the Spallation Neutron Source. However, for the structure materials research with known representative *hkl* diffractions, most of the Reactor Source can provide very efficient lattice strain scans with better resolution to map the strain distribution for mechanical behavior study.

3.1 The engineering neutron diffractometers of the spallation neutron sources

3.1.1 ENGIN-X

Engin-X at ISIS, United Kingdom, is one of the most important engineering materials instruments of its kind. The ENGIN-X's two large detectors can collect the diffracted neutron intensity in two orthogonal directions simultaneously. The instrumental parameters can be referred to the ENGIN-X website of (http://www.isis.stfc.ac.uk/instruments/engin-x/engin-x2900.html). With *in-situ* capabilities for sample environments, ENGIN-X enables measurements on small volumes[53] to largesamples[54]. Historically, there are many important milestones research of the structure materials research carried out at the Engin-X[55].

3.1.2 SMARTS

Spectrometer for Materials Research at Temperature and Stress (SMARTS) is located in the beamline of the Los Alamos Neutron Science Center (LANSCE) [56] in the United States. The SMARTS is a third-generation neutron diffractometer constructed at the Lujan Center.

SMARTS provides capability on two important environmental conditions for mechanical behavior study. These two are the load frame and the furnace, which allow measuring the deformation under different types of loading modes and various temperatures. The furnace and load frame suite allows research on materials under extreme loads (250 KN) and at extreme temperatures (1,500°C) (http://lansce.lanl.gov/lujan/instruments/SMARTS/index.html). *In-situ* uniaxial loading on samples up to 1 cm in diameter at stresses of 2 GPa and with lower stresses at temperatures up to 1,500°C are routine. The importance of the two orthogonal detectors is evidenced in a magnesium-alloy research[57]. The appearance and the disappearance intensity evolution of the deformed magnesium-alloy shown in two orthogonal detectors demonstrate the kinetics of the twinning-detwinning. Furthermore, its space capability enables to perform *in situ* observation of temperature and stress evolution during friction-stir welding by Woo et al. [58]. Dr. Bjørn Clausen of the SMARTS create several very useful software for the usres' data deduction. One of them is the SMARTSware. The users can save a lot of time to refine the SMARTS data with SMARTSware.

3.1.3 TAKUMI

TAKUMI is a newly-built Engineering Materials Diffractometer located in Japan-Spallation Neutron Source. The real-space detecting range for the investigated materials covers from 0.5 to 2.7 Å (standard mode). TAKUMI also has a pair of orthogonal scattering-detector banks covering area in both of the horizontal and vertical directions. There large sample table has the load capacity up to 1 ton (http://j-parc.jp/MatLife/en/instrumentation/ns_spec.html). TAKUMI has a loading machine to reach tension up to 50kN and 20kN with compression, respectively. The furnace of TAKUMI is designed to create the high-temperature environment up to 1273K.

3.1.4 VULCAN

VULCAN Engineering Materials Diffractometer at the Spallation Neutron Source of Oak Ridge National Laboratory (http://neutrons.ornl.gov/instruments/SNS/VULCAN/) has created the most various and extreme sample environment for its kind. Although it just opens to the general users since 2009, several new experiments, which were impossible before, have been carried out. New features are summarized below. VULCAN is the latest diffractometer in the world. Dr. Xun-Li Wang and Dr. Ke An build VULCAN's latest data acquisition software. With the most advanced software and flux, VULCAN creates real in-situ experimental environment, which can help the users to investigate the dynamics of the structural evolution subjected to mechanical deformation. VULCAN has the ability to study kinetic behaviors in sub-second times. VULCAN's rapid volumetric (3-dimensional) mapping can narrow a sampling volume down to 1 mm^3 with a measurement time of minutes. VULCAN also has very high spatial resolution (0.1 mm) in one direction with a measurement time of minutes. VULCAN can collect 20 well-defined reflections for in-situ loading studies simultaneously. Moreover, VULCAN is equipped with one of the most sophisticated load frame, which can even perform torsion experiments.

3.2 The engineering neutron strain scanning diffractometers of the reactor neutron sources

The strain scanners are mostly suitable for residual stress measurements but not limited to perform deformation behavior when the detector can be rotated for the appropriate diffraction angle[59-62]. This type of the strain scanning instruments facilitates the concepts introduced in the session 2.1 to perform diffraction experiments. The diffraction peak positions determine the lattice parameters of the phases of the investigated engineering components. The importance of the strain mapping of the neutron diffractometers is that with the high-penetration of the neutrons, the gauged samples can be studied without the destructive specimen preparations. There are a number of research subjects performed with the aid of the strain scanning diffractometers for the understanding of the mechanical properties. Here we introduce a few examples, which have been published recently in literature: (1) Stress variation and crack growth around the fatigue crack tip under loading and overload[63, 64] (2) residual stresses, texture, and tensile behavior in friction stir welding[65, 66] (3) wavelength selection and through-thickness distribution of stresses in a thick weld[67] (4) residual stress determination in a dissimilar weld overlay pipe for the nuclear power plant applications[68] (5) Time-dependent variation of the residual stresses in a severe plastic deformed material[69]. In this session, we also summarize the various strain scanning diffractometers of the reactor neutron source for the mechanical behavior in the manner of alphabetic order.

3.2.1 KOWARI

KOWARI is a Strain Scanner of Australian Nuclear Science and Technology Organisation (ANSTO). KOWARI's advanced sample table can accommodate large objects (up to 1 tonne) and move them around reproducibly to within ~20 mm. With Australia's famous mining industry and the historical heritage of the Bragg Institute, KOWARI is building a program to serve for both of the commercial and academic users.

3.2.2 L3 Spectrometer

L3 Spectrometer of Canadian Neutron Beam Centre is mostly used for strain/stress mapping, crystallographic texture, grain-interaction stresses, precipitation and phase transformations (http://www.nrc-cnrc.gc.ca/eng/facilities/cnbc/spectrometers/l3.html). L3 can also be equipped with a stress rig for examining specimens under uniaxial load. For strain/stress mapping, a large variety of slit dimensions are available. L3 can also be equipped with a variety of sample orientation devices. The sample table with various translation and rotation devices can handle loads of up to 450 kg and provides a large 60 cm × 60 cm (2" × 2") platform. Stress Rig can provide uniaxial tension and compression load. Maximum applied load is 45 kN. L3 Spectrometer is famous for its industrial service and academic user communities for crystallographic texture and grain interaction measurements.

3.2.3 NRSF2

Neutron Residual Stress Mapping Facility (NRSF2) of the High Flux Isotope Reactor (HFIR) Oak Ridge National Laboratory is a strain scanning diffractometer open to the users. The wavelength can be chosen from a variety of monochromator crystal settings with a selection of wavelengths from 1.2 to 2.4 Å (http://www.sns.gov/instruments/HFIR/HB2B/). The high intensity HFIR enable the penetrating power of neutrons of the NRSF2 for scanning residual stresses in engineering materials.

3.2.4 RSI

Residual Stress Instrument (RSI) is a neutron-diffractometer at the High-flux Advanced Neutron Application Reactor (HANARO) of the Korea Atomic Energy Research Institute (KAERI). The diffractometer can measure three-orthogonal-direction strain components with high spatially-resolved sampling. Huang *et al.*'s recent work[70] shows the importance to measure all three directional strains for fatigue study of a stainless steel. RSI has two load frames up to 20 KN for the measurement of strains at room temperature and high temperature up to 800 °C. *In-situ* mapping is also available at the RSI of HANARO.

(1) FRM-II	http://www.frm2.tum.de/en/science/diffraction/stress-spec/index.html
(2) HANARO-RSI	http://nsrc.jaea.go.jp/aonetned/rsi.pdf
(3) HFIR-NRSF2	http://www.sns.gov/instruments/HFIR/HB2B/
(4) CNBC-L3 spectrometer	http://www.nrc-cnrc.gc.ca/eng/facilities/cnbc/spectrometers/l3.html
(5) SMARTS	http://lansce.lanl.gov/lujan/instruments/SMARTS/index.html [72]
(6) ISIS-ENGIN-X	http://www.isis.stfc.ac.uk/instruments/engin-x/engin-x2900.html
(7) ORNL-VULCAN	http://neutrons.ornl.gov/instruments/SNS/VULCAN/
(8) JapanSNS-TAKUMI	http://j-parc.jp/MatLife/en/instrumentation/ns_spec.html
(9) ANSTO-KOWARI	http://www.ansto.gov.au/research/bragg_institute/facilities/instruments/kowari
(10) ANSTO-WOMBAT	http://www.ansto.gov.au/research/bragg_institute/facilities/instruments/wombat

Table 3.1. Some of the website links (http://nsrc.jaea.go.jp/aonetned/index.html) of the neutron diffractometers

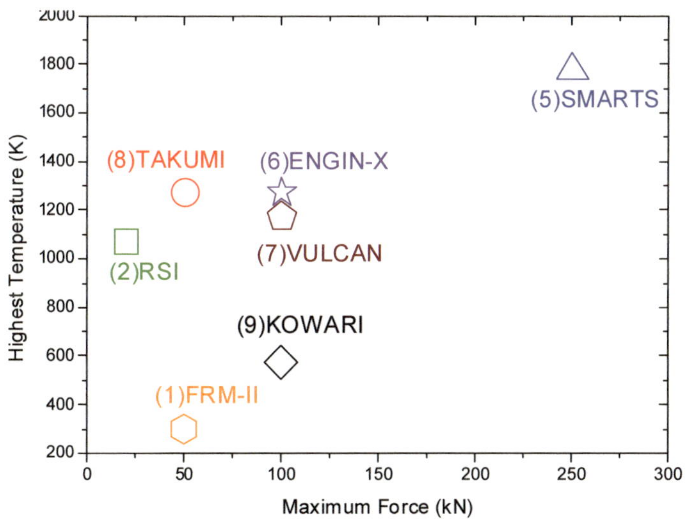

Fig. 3.1. The extreme environmental load and temperature of the neutron diffractometers summarized in Table 3.1..

3.2.5 FRM-II

Stress-Spec of the Forschungs-Neutronenquelle Heinz Maier-Leibnitz (FRM-II) is a diffractometer for residual stress and texture measurements (http://www.frm2.tum.de/en/science/diffraction/stress-spec/index.html). The capability of Stress-Spec is as good as the aforementioned diffractometers of other Reactor Neutron Source. More details can be found in Hofmann *et al.*'s introduction [71].

4. Summary

In this chapter, how to apply the neutron diffractometers to investigate the mechanical behavior of the structure materials is introduced. The applications of the neutron-diffraction experiments can reveal the lattice-strain, texture, and the microstructure evolution upon the deformation. The connections between the macroscopic-mechanical behavior and microscopic characteristics, obtained from the diffraction results, are explained. Moreover, in addition to the concepts, some of the exemplary literatures and the Neutron Engineering Diffractometers are summarized for readers' reference.

5. Acknowledgements

Particular thanks are due to Mr. Kuan-Wei Lee for his patience and careful in editing the manuscript. EWH appreciates the support from National Science Council (NSC)Program 100-2221-E-008- 2 041. JJK, EWH and Kuan-Wei Lee thank the NSC-99-3113-Y-042-001 Program.

6. References

[1] G. B. Olson, Science 277 (5330), 1237-1242 (1997).
[2] A. M.F., Acta Metallurgica 20 (7), 887-897 (1972).
[3] P. Lindner and T. Zemb, *Neutron, X-rays and light: scattering methods applied to soft condensed matter*, 1st ed. (North-Holland, 2002).
[4] G. Will, *Powder Diffraction: The Rietveld Method and the Two Stage Method to Determine and Refine Crystal Structures from Powder Diffraction Data 1st edition.* (Springer, 2006).
[5] A. C. Larson and R. B. Von Dreele, Los Alamos National Laboratory Report LAUR, 86-748 (2004).
[6] H. M. Rietveld, Acta Crystalline 22 (1967).
[7] R. W. Hertzberg, *Deformation and Fracture Mechanics of Engineering Materials*, 2nd ed. (Canada, 1976).
[8] E. W. Huang, R. I. Barabash, B. Clausen, Y.-L. Liu, J.-J. Kai, G. E. Ice, K. P. Woods and P. K. Liaw, International Journal of Plasticity 26 (8), 1124-1137 (2010).
[9] E.-W. Huang, R. I. Barabash, B. Clausen and P. K. Liaw, Metallurgical and Materials Transactions A, in press (2011). (DOI: 10.1007/s11661-011-0972-9)
[10] P. Stanley and W. K. Chan, Journal of Strain Analysis 20, 129-143 (1985).
[11] D. G. E., *Mechanical Metallurgy*, 3rd ed. (McGraw-Hill, New York, 1986).
[12] L. A. Girifalco, *Statistical Mechanics of Solids*. (Oxford University Press, 2000).
[13] A. K. Wong, S. A. Dunn and J. G. Sparrow, Nature 332 (6165), 613-615 (1988).
[14] D. Rittel, Z. G. Wang and M. Merzer, Physical Review Letters 96 (7), 075502 (2006).
[15] M. Huang, P. E. J. Rivera-Díaz-del-Castillo, O. Bouaziz and S. v. d. Zwaag, Acta Materialia 57 (12), 3431-3438 (2009).
[16] S. Quinn, J. M. Dulieu-Barton and J. M. Langlands, Strain 40 (3), 127-133 (2004).

[17] E.-W. Huang, R. Barabash, N. Jia, Y. Wang, G. E. Ice, B. Clausen, J. A. Horton Jr and P. K. Liaw, Journal Name: Metallurgical and Materials Transactions A; Journal Volume: 39A; Journal Issue: 13, 3079-3088 (2008).
[18] H. R, Journal of the Mechanics and Physics of Solids 13 (4), 213-222 (1965).
[19] J. W. Hutchinson, Acta-Scripta Metallurgica Proceedings Series 4, 12 (1989).
[20] T. Holden, C. Tomé and R. Holt, Metallurgical and Materials Transactions A 29 (12), 2967-2973 (1998).
[21] B. Clausen, T. Lorentzen, M. A. M. Bourke and M. R. Daymond, Materials Science and Engineering: A 259 (1), 17-24 (1999).
[22] B. Clausen, T. Lorentzen and T. Leffers, Acta Materialia 46 (9), 3087-3098 (1998).
[23] M. R. Daymond, C. N. Tomé and M. A. M. Bourke, Acta Materialia 48 (2), 553-564 (2000).
[24] M. R. Daymond, M. Preuss and B. Clausen, Acta Materialia 55 (9), 3089-3102 (2007).
[25] U. F. Kocks, C. N. Tome and H.-R. Wenk, *Texture and Anisotropy: Preferred Orientation in Polycrystals and Their Effects on Materials Properties.* (Cambridge University Press, 1998).
[26] Y. D. Wang, R. L. Peng and R. L. McGreevy, Philosophical Magazine Letters 81 (3), 153-163 (2001).
[27] Y. D. Wang, X.-L. Wang, A. D. Stoica, J. D. Almer, U. Lienert and D. R. Haeffner, Journal of Applied Crystallography 35 (6), 684-688 (2002).
[28] Y. D. Wang, R. L. Peng, J. Almer, M. Odén, Y. D. Liu, J. N. Deng, C. S. He, L. Chen, Q. L. Li and L. Zuo, Advanced Materials 17 (10), 1221-1226 (2005).
[29] H. Rietveld, Acta Crystallographica 22 (1), 151-152 (1967).
[30] T. Ungar, I. Groma and M. Wilkens, Journal of Applied Crystallography 22 (1), 26-34 (1989).
[31] F. J. Humphreys, Journal of Materials Science 36 (16), 3833-3854 (2001).
[32] B. R. I. and M. A. Krivoglaz, Physics of Metals 4 (1982).
[33] A. M.F, Acta Metallurgica 37 (5), 1273-1293 (1989).
[34] P. Haasen, *Physical Metallurgy*, 3rd ed. (Cambridge University Press, 1996).
[35] D. Kuhlmann-Wilsdorf, Philosophical Magazine A 79 (4), 955-1008 (1999).
[36] M. H, Acta Metallurgica 31 (9), 1367-1379 (1983).
[37] L. E. Levine, B. C. Larson, W. Yang, M. E. Kassner, J. Z. Tischler, M. A. Delos-Reyes, R. J. Fields and W. Liu, Nat Mater 5 (8), 619-622 (2006).
[38] N. Hansen, Advanced Engineering Materials 7 (9), 815-821 (2005).
[39] W. Nix, J. Gibeling and D. Hughes, Metallurgical and Materials Transactions A 16 (12), 2215-2226 (1985).
[40] Z. M, Acta Metallurgica et Materialia 41 (2), 589-599 (1993).
[41] E. Schafler, K. Simon, S. Bernstorff, P. Hanák, G. Tichy, T. Ungár and M. J. Zehetbauer, Acta Materialia 53 (2), 315-322 (2005).
[42] D. Hughes and N. Hansen, Metallurgical and Materials Transactions A 24 (9), 2022-2037 (1993).
[43] M. A. Krivoglaz, K. P. Ryaboshapka and R. I. Barabash, Physics of Meals and Metallography 30 (1970).
[44] R. I. Barabash and P. Klimanek, Journal of Applied Crystallography 32 (6), 1050-1059 (1999).
[45] R. Barabash, J. Appl. Phys. 93 (3), 1457 (2003).
[46] W. T. Read and W. Shockley, Physical Review 78 (3), 275 (1950).
[47] A. Ardell, Metallurgical and Materials Transactions A 16 (12), 2131-2165 (1985).
[48] A. Wilm, Metallurgie 8, 225-227 (1911).
[49] E. Huang, P. K. Liaw, L. Porcar, Y. Liu, J. Kai and W. Chen, Appl. Phys. Lett. 93 (16), 161904 (2008).
[50] S. H. Chen, Annual Review of Physical Chemistry 37 (1), 351-399 (1986).

[51] J. S. Pedersen, Physical Review B 47 (2), 657 (1993).
[52] R. Giordano, A. Grasso, J. Teixeira, F. Wanderlingh and U. Wanderlingh, Physical Review A 43 (12), 6894 (1991).
[53] S. Y. Lee, H. Choo, P. K. Liaw, E. C. Oliver and A. M. Paradowska, Scripta Materialia 60 (10), 866-869 (2009).
[54] S. Y. Lee, M. A. Gharghouri and J. H. Root, presented at the TMS, San Diego, California, 2011 (unpublished).
[55] E. C. Oliver, M. R. Daymond and P. J. Withers, Acta Materialia 52 (7), 1937-1951 (2004).
[56] M. A. M. Bourke, D. C. Dunand and E. Ustundag, Applied Physics A: Materials Science & Processing 74 (0), s1707-s1709 (2002).
[57] S. R. Agnew, J. A. Horton, T. M. Lillo and D. W. Brown, Scripta Materialia 50 (3), 377-381 (2004).
[58] W. Woo, Z. Feng, X. L. Wang, D. W. Brown, B. Clausen, K. An, H. Choo, C. R. Hubbard and S. A. David, Science and Technology of Welding & Joining 12 (4), 298-303 (2007).
[59] P. J. Withers and H. K. D. H. Bhadeshia, Materials Science and Technology 17 (4), 366-375 (2001).
[60] M. T. Hutchings, P. J. Withers, T. M. Holden and T. Lorentzen, *Introduction to the characterization of residual stress by neutron diffraction*, 1st ed. (Taylor and Francis, London, 2005).
[61] Y. Tomota, H. Tokuda, Y. Adachi, M. Wakita, N. Minakawa, A. Moriai and Y. Morii, Acta Materialia 52 (20), 5737-5745 (2004).
[62] W. Woo, Z. Feng, X. Wang and S. A. David, Science and Technology of Welding & Joining 16 (1), 23-32 (2011).
[63] S. Y. Lee, P. K. Liaw, H. Choo and R. B. Rogge, Acta Materialia 59 (10), 4253 (2011).
[64] S. Y. Lee, H. Choo, P. K. Liaw, K. An and C. R. Hubbard, Acta Materialia 59 (10), 4254 (2011).
[65] W. Woo, H. Choo, M. B. Prime, Z. Feng and B. Clausen, Acta Materialia 56 (8), 1701-1711 (2008).
[66] W. Woo, H. Choo, D. W. Brown, P. K. Liaw and Z. Feng, Scripta Materialia 54 (11), 1859-1864 (2006).
[67] W. Woo, V. Em, B.-S. Seong, E. Shin, P. Mikula, J. Joo and M.-H. Kang, Journal of Applied Crystallography 44 (4), 747-754 (2011).
[68] W. Woo, V. Em, C. R. Hubbard, H.-J. Lee and K. S. Park, Materials Science and Engineering: A 528 (27), 8021-8027 (2011).
[69] W. Woo, Z. Feng, X. L. Wang and C. R. Hubbard, Scripta Materialia 61 (6), 624-627 (2009).
[70] E-W. Huang, S. Y. Lee, W. Woo, and K-W. Lee, Metallurgical and Materials Transactions A (2011). (DOI: 10.1007/s11661-011-0904-8).
[71] M. Hofmann, R. Schneider, G. A. Seidl, J. Rebelo-Kornmeier, R. C. Wimpory, U. Garbe and H. G. Brokmeier, Physica B: Condensed Matter 385-386, Part 2 (0), 1035-1037 (2006).
[72] M. A. M. Bourke, D. C. Dunand and E. Ustundag, Applied Physics A: Materials Science & Processing 74 (0), s1707-s1709 (2002).

7

The Molecular Conformations and Intermolecular Correlations in Positional Isomers 1- and 2- Propanols in Liquid State Through Neutron Diffraction

R.N. Joarder
Jadavpur University
India

1. Introduction

In the case of molecular liquids meaning of structure is two-fold-first is molecular units, their structure or conformation and second is average spatial arrangement of molecules or liquid structure. A liquid can not have structure in the same sense as a crystallographic solid has since the positions of the molecules continuously change and any local configuration will change over a short interval of time. In spite of that there is a possibility that, on average, a particular configuration dominates over the others, especially in H-bonded liquids. Alcohols are good examples where small labile clusters continually break and reform at a mean time interval. During the last few decades, the structural correlations in associative liquids, particularly, those involving H-bonding have been extensively studied, thanks to the successful application of statistical mechanical methods together with improved methods of experimental techniques (Diffraction, Raman, NMR) and Computer simulations. In spite of extensive developments some areas remain unclear even today. For example, the structure and structure related properties of liquid alcohols with large sized molecular species sill remain somewhat controversial. Alcohols are important group of compounds widely used as solvents and many of their properties are due to the existence of intermolecular hydrogen bonds. Though alcohols have amphiphilic molecular structures, the presence of hydrophobic groups do not allow them to form tetrahedral structure like that in water, rather they form H-bonded chains. The diffraction techniques, both x-ray and neutron have yielded useful direct information about the microscopic average structure and near neighbor correlations. The computer simulations, NMR also provide very useful information about their microscopic structure and dynamics. Although all the techniques yield general nature of the liquid structure as chains there is lot of disagreement in the detailed nature of the chain or cluster formed due to H-bonding. In solid and gaseous phases, alcohol structures are now well established. In liquid state however, the situation is still open in view of differences in the results of diffraction and other techniques (Svishchev & Kusalik, 1993; Jorgensen, 1986; Magini et al, 1982; Sarkar & Joarder, 1993, 1994; Benson, 1996).

In the mono-alcohol series two lowest members, methanol and ethanol in liquid phase have been studied quite extensively. The third member of the family, propanol (C_3H_7OH) exists in two stereo-isomeric forms n-propanol (1-propanol or 1P) and isopropanol (2-propanol or 2P) and are relatively less studied even today. The investigation of their molecular conformation and intermolecular correlations, in liquid state is all the more interesting because they have significant differences in thermodynamic and other properties. The significant property differences could be attributed to differences in structures, conformational as well as associational. Therefore, a careful and detailed investigation of these two isomeric liquids is very important. Now, even today, most detailed information about molecular structure and average liquid structure is by diffraction techniques, x-ray or neutron. The neutron diffraction is especially useful because it can yield molecular conformation with correct hydrogen positions and intermolecular correlations. Though x-rays can not see hydrogen positions accurately, the skeleton of the intermolecular structure is however more prominently visible (Sarkar & Joarder, 1993, 1994; Narten & Habenchuss, 1984). So a combination of two techniques is very useful in these studies.

In this short resume we discuss the use of neutron diffraction data sublimented by recent x-ray diffraction data in the detailed investigation of molecular conformation and molecular association of two propanols in liquid state at room temperature (RT). The Monte-Carlo (MC) simulations based on optimized intermolecular potential functions for liquid alcohols including 1- and 2-propanols were carried out long time back (Jorgensen, 1986) and this showed more or less similar chain molecular association for all the members. The molecular dynamics (MD) simulation of 1-propanol in liquid state has also been reported (Akiyama et al, 2004) and the results are comparable with diffraction results. The ab initio calculations of connectivity effects on the clusters of 1-propanol and other alkanol compounds are also available (Sum & Sandler, 2000).

2. 1- and 2-propanols: Thermodynamic and structural differences

In crystalline phase at low temperatures, and also in glassy state, a number of investigations (Talon et al, 2001; Ramos et al, 2003; Talon et al, 2002; Cuello et al, 2002) on the structural differences between the two isomers of propanol and their possible influence on the thermodynamic properties are now available. It is noteworthy that 2-propanol possesses a much larger specific heat than 1-propanol, main reason being the significant larger Debye contribution in 2-propanol.The x-ray and neutron scattering experiments (Talon et al, 2001; Talon et al, 2002; Cuello et al, 2002) have been reported on samples of the glassy and crystalline states. The structural analysis of crystalline state assign a triclinic structure for 1-propanol and a monoclinic structure for 2-propanol. It is possible that the property differences in two propanols in glassy and crystalline states may be attributed to the influence of the position of the hydroxyl (OH^-) group on the elastic constants of the H-bonded network, constituent with the atomic structures of 1- propanol and 2-propanol.

In liquid state too, differences in thermodynamic and other properties are clearly visible (Sahoo, 2011, vide Table-1). Both molecular units however show close values for relevant molecular properties such as van der Waals volumes (124.8 Å^3 for 1-propanol and 127.7 Å^3 for 2-propanol) and molecular electric dipole moments($\sim 1.66\times 10^{-30}$ esu-cm), the overall molecular 'shape 'being the property most affected by such chemical changes. The molecular structural or conformational differences are likely to result in significant differences in basic thermodynamic and other properties. It is therefore indeed interesting to see how the change in the position of a chemical functional group (OH) from one atom to another within the same structural unit changes significantly the thermodynamic and other properties of the liquid. It is true that such comparative studies on the structural correlations of 1- and 2-propanol in liquid state are still rare.

Properties	1-propanol	2-propanol
Density(ρ) in gm/c.c	0.803	0.785
Melting point(T_m°C)	-125	-88
Flash point(F.P°C)	15	11.7
Boiling Point(T_b°C)	97	82
Glass transitiontemperature(T_G°C)	-175	-158
Isothermal Compressibility($\chi_T\times10^{-5}$)/atmos at 25°C	121	112
Heat capacity(C_p) in cal/mol.deg	33.70	36.06
Viscosity(η) in mPa at20°C	1.94	2.37
Entropy of vaporisation at T_b (ΔS_v) in cal/mol	28.43	31.35
Dielectric constant(ε)	20.1(25°C)	20.18(20°C)

Table 1. Properties of liquid 1- and 2-propanols

3. Molecular conformational studies

The propanol is the smallest monoalcohol showing two stereoisomers in crystalline and glassy phases and also likely in liquid phase and so allow us to make comparative study on the effect of rearrangement of the H-bonded structures using diffraction data (both x-ray and neutron). The molecules are however quite large in size and H-bonding being intermolecular, there is considerable amount of overlapping between some intra and intermolecular contributions. The separation of these terms are obviously tricky. Only a few diffraction experiments are so far done on them in liquid state and successfully analyzed. The structures in gaseous phases were obtained by electron diffraction as early as in sixties (Abdel Aziz & Rogowski, 1964).The results showed molecular structures significantly different in two propanols (ccco-chain in trans-configuration for 1-propanol and tetrahedra-angled to middle c-atom in 2-propanol). The first x-ray diffraction on liquid 1-propanol was reported in 1977 (Mikusinska-Planner, 1977) and molecular conformation was shown to be somewhat spherical in shape. The neutron diffraction works on liquid 2-propanol at low temperatures (190K-275K) were done by Howells et al (Zetterström et al, 1991, 1994) with TOF (time of flight) neutron data. The high-q data were directly used to fit the molecular structure of liquid 2-propanol. The molecular parameters like r_{C_2O} , r_{OD} and C_2OD were quite smaller compared to those for gaseous phase (Abdel Aziz & Rogowski, 1964) and also for other alcohols in liquid phase (Sarkar & Joarder, 1993, 1994; Adya et al, 2000). Further the parameters varied considerably over the temperature range though such parameters are expected to be only weakly temperature dependent (Montague & Dore, 1986; Sahoo et al, 2010). In liquid state small variation due to thermal stretching for r_{OD} is not unlikely. The x-ray diffraction works on liquid 1-and 2- propanols (in pure and aqueous solution) were carried out in the recent past (Takamuku et al, 2002, 2004). The molecular parameters were consistent in the two cases though accurate locations of H- positions could not be ascertained. For 1-propanol molecule, the authors considered cis, trans and gauche conformations and gauche conformation was shown to be more favorable though some small population of trans was not ruled out. In the analysis however, several restrictions were imposed e. g. intermolecular O–O distance was kept fixed at 2.752Å (Vide Takamuku et al, 2004). Further for 2-propanol, the authors, being more interested in intermolecular correlations, did not extract molecular parameters, rather estimated them from those of liquid ethanol obtained from neutron diffraction (Tanaka et al, 1984). Recently, neutron diffraction experiments were carried out on deuterated liquid 1- and 2-propanols where the authors have used reactor data (Sahoo et al, 2008, 2010). The authors however claim that they have used modified method on conformational analysis than a direct fitting of high-q data. In literature no other neutron data on liquid 1- propanol is however available. These neutron diffraction studies by Sahoo et al, (2008, 2010) show the conformational difference in 1- and 2-propanols in liquid state at RT quite clearly and it appears that the molecular structures in liquid phase at RT are more or less identical with those in crystalline and gaseous phases. Since the conformational analyses reported by Sahoo et al appear quite convincing, in following paragraphs, some aspects of their method and results discussed.

3.1 Neutron data and their analysis

The neutron scattering experiments on deuteratred liquid propanol samples (99.8% pure) at RT carried out on Hi-Q diffractometer at Dhruva, Bhabha Atomic Research Centre (BARC, India) yielded raw data which have undergone a series of corrections using conventional procedures (Egelstaff, 1987). The corrected neutron cross-section ($d\sigma/\ d\Omega$) data were then extrapolated in the region $0 \leq q \leq 0.3$Å^{-1} and were normalized (on high-q as well) such that graphical extrapolation to $q \rightarrow 0$ limit yields a correct isothermal compressibility of the liquid. The cross-section data are shown in Fig. 1 for both propanols.

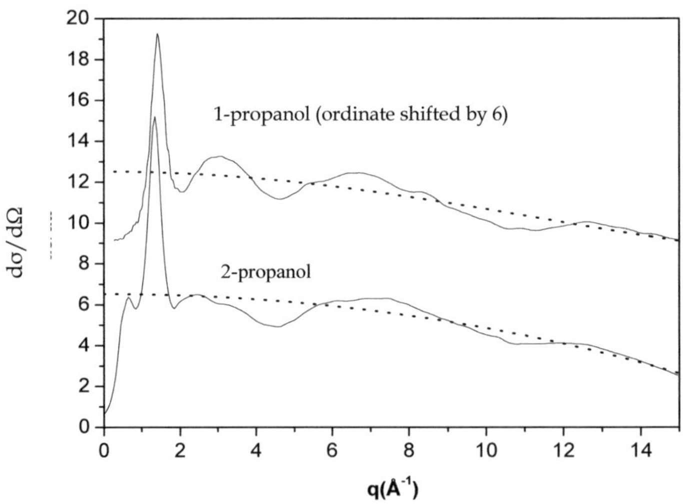

Fig. 1. $d\sigma/d\Omega$ vs. q for liquid D-1-and 2-propanols at RT;—corrected experimental data, -- self term

In 2-propanol, data shows a strong pre-peak at a scattering vector $q \sim 0.7$Å^{-1} similar to one obtained in earlier Howells et al TOF, neutron data (Zetterström et al, 1991, 1994) and also in recent x-ray data (Takamuku et al, 2002). For 1-propanol, however, there is no significant pre-peak at a scattering vector q ~0.7 $Å^{-1}$ similar to earlier x-ray data (Mikusinska-Planner, 1977) though in recent x-ray data there is a weak pre-peak at q ~0.7 $Å^{-1}$ (Takamuku et al, 2004). The cross- section data can be separated into "self" and "interference" terms,

$$d\sigma / d\Omega |_{expt} = d\sigma / d\Omega |_{self} + d\sigma / d\Omega |_{int} \quad (1)$$

$d\sigma/ d\Omega|_{int}$ term oscillates and goes to zero at high q, $d\sigma/ d\Omega|_{expt}$ is represented by appropriate self-scattering term at high q. The cross-section data has a "fall-off" feature at high q and this is due to interaction of incident neutrons with the vibrating scattering sites (deuterium atoms in particular for alcohols). This inelasticity effect modifies the self-scattering term and for alcohols this modification can be represented by a term involving two inelasticity parameters (Montague & Dore, 1986; Champeney et al, 1986; Sahoo et al, 2008, 2010).

Thus,

$$d\sigma / d\Omega |_{self} = [\, 3b_c^2 + b_o^2 + 8b_d^2 + {}^{8\sigma_d^i}\!/_{4\pi}](1 - aq^2 + bq^4) \quad (2)$$

where σ_D^i is the incoherent scattering cross-section for deuterium and a, b are two inelasticity parameters. b_c, b_o, b_d are coherent scattering lengths of carbon, oxygen and deuterium atoms respectively. The inelasticity parameters a, b were estimated by χ^2-fitting between the self scattering term and the experimental cross-section data at high-q values starting from q ≈ 5.5 - 6.0$Å^{-1}$ (vide Fig. 1). The parameters a and b are respectively (3.22 ± 0.04) $\times10^{-3}$ $Å^2$ & (4.05 ± 0.30) $\times10^{-6}$ $Å^4$ for 1-propanol and (1.56 ± 0.04) $\times10^{-3}$ $Å^2$ & (5.04 ± 0.30) $\times10^{-6}$ $Å^4$ for 2-propanol.

Subtracting "self" term from $d\sigma/d\Omega|_{expt}$ in Eqn (1), one gets $d\sigma/ d\Omega|_{int}$ which contains both intra- and inter-molecular contributions. The total structure function, H(q) is defined as

$$H(q) = d\sigma / d\Omega|_{int} / (\Sigma b_\alpha)^2 \quad (3)$$

H(q) is separable into intra- and inter-molecular terms given by

$$H(q) = H_m(q) + H_d(q) \quad (4)$$

where $H_m(q) = (d\sigma / d\Omega)_{int}^{intra} / (\sum b_\alpha)^2$ and $H_d(q) = (d\sigma / d\Omega)_{int}^{inter} / (\sum b_\alpha)^2$:

The explicit expression for $H_m(q)$ is

$$H_m(q) = \sum_{\alpha} \sum_{\alpha \neq \beta} b_\alpha b_\beta j_0(qr_{\alpha\beta}) \exp(-\gamma_{\alpha\beta} q^2) / (\sum_{\alpha} b_\alpha)^2 \quad (5)$$

$r_{\alpha\beta}$ being the mean distance between the sites α and β and $\gamma_{\alpha\beta}$ is the mean square variation in the distance $r_{\alpha\beta}$ with $\gamma_{\alpha\beta} = (1/2)\lambda_0^2 r^2_{\alpha\beta}$, where λ_0 is taken to be a constant for all pairs (similar to Prins relation (Frenkel, 1955)). For Debye-Waller terms same procedure was followed by other workers also (Tanaka et al, 1984). α and β sum independently over 12 atomic sites with $j_0(x) = \sin x/x$ is the zeroth order Spherical Bessel function. $H_m(q)$ is the intra interference term and gives information about the structure of the molecule, while $H_d(q)$ is the distinct structure function gives information about the inter-molecular or liquid structure. In terms of partial structure functions $H_{\alpha\beta}(q)$, $H_d(q)$ is given by

$$H_d(q) = (\sum_{\alpha} b_{\alpha})^{-2} \sum_{\alpha}\sum_{\beta}(2-\delta_{\alpha\beta}) b_{\alpha} b_{\beta} H_{\alpha\beta}(q) \tag{6}$$

The inverse Fourier transform (IFT) of $H_d(q)$ gives the r-weighted neutron intermolecular correlation function *d(r)* and radial distribution function (RDF), $G_d(r)$ given by

$$d(r) = \frac{2}{\pi} \int_0^{\infty} q H_d(q) \sin(qr) dq \tag{6a}$$

and

$$G_d(r) = 1 + d(r) / 4\pi\rho r, \tag{6b}$$

where, ρ is the liquid density. $G_d(r)$ is related to the partial pair distribution functions, $g_{\alpha\beta}(r)$ given by

$$G_d(r) = (\sum_{\alpha} b_{\alpha})^{-2} \sum_{\alpha}\sum_{\beta}(2-\delta_{\alpha\beta}) b_{\alpha} b_{\beta} g_{\alpha\beta}(r) \tag{7a}$$

With

$$g_{\alpha\beta}(r) = 1 + \frac{1}{2\pi^2 \rho r} \int_0^{\propto} q H_{\alpha\beta}(q) \sin(qr) dq \tag{7b}$$

3.2 Modified molecular conformation analysis

In liquid alcohols, the effects of intermolecular hydrogen bonding persist at high q (Narten & Habenchuss, 1984) and as a result $H_d(q)$ continues to exhibit oscillatory behaviour, positive and negative over $H_m(q)$. The function $H_d(q)$ however, tends to vanish more rapidly than the function $H_m(q)$ and so $H(q)$ oscillates around $H_m(q)$ and tends to equalize with $H_m(q)$ at large q. This means that for q greater than some q_{min}, the experimental total structure function $H(q)$ comes primarily from intramolecular part. Assuming a model of the molecule from gas phase electron diffraction analysis (Abdel Aziz & Rogowski, 1964) one can find the atom-atom distances for relevant analysis in liquid phase.

One can find $H_m(q)$ and fit $qH_m(q)$ to experimental $qH(q)$ for $q > q_{min}$ by a χ^2 fitting procedure and refine the distances and angles. One then subtract $H_m(q)$ from experimental $H(q)$ in Eqn. (4) to obtain the first estimate of $H_d(q)$ .The IFT of $H_d(q)$ yields intermolecular radial distribution function $G_d(r)$ by Eqn.(6a & b). Due to limited q-range (q_{max}) available in the experiment, a modification function, $W(q) = \sin(q\pi/q_{max})/(q\pi/q_{max})$ (Champeney et al, 1986) is used in the IFT. Further, one chooses q_{max} such that $G_d(r = 0)$ is almost zero which means that the contribution of the integral in Eqn. (6a) beyond $q = q_{max}$ is almost zero. Again, the function $G_d(r)$ is expected to be zero in the range $0 \le r \le r_0$ where r_0 is about 1.5 Å because the intermolecular distance cannot be less than this value. Setting $G_d(r) = 0$ for this region a FT of the remaining $G_d(r)$ function would yield a new $qH_d(q)$. Subtracting this $qH_d(q)$ from experimental $qH(q)$ one obtains the corrected $qH_m(q)$ function. The difference between the corrected $qH_m(q)$ and original $qH_m(q)$ is generally small. Varying molecular parameters, the subsequent iteration procedure gives the best fitting to this corrected function. A χ^2 fitting is used for the whole range of q data in steps of 0.01Å^{-1}. This modified technique of finding molecular conformation of big molecules like alcohols somewhat resembles the method used by Bertagnolli (Bertagnolli et al, 1976) for neutron data analysis of liquid acetonitrile.

The schematic models chosen from the crystalline and gaseous phases are shown in Fig. 2. The molecules have 12 scattering sites and considering the symmetry of the structure one can minimize the number of parameters to describe the conformation. C_1C_2 is assumed in the X-Z plane and C_1O along the Z-axis. The number of parameters are minimized by assuming all CC and CD distances equal. In 2-propanol the coordinates of C_3 and D are obtained by 120° and 240° rotations of the C_2 coordinates about Z-axis respectively. Denoting by D_1 the hydroxyl deuterium, three rotation angles ϕ_{OD}, ϕ_1,ϕ_2 and in addition to $\angle$ C_1OD_1 angle and three internal distances r_{CC}, r_{CD} and r_{OD} are treated as variable parameters while the methyl backbones are assumed to have tetrahedral geometry with tilt angle of C_1O relative to the methyl groups assumed zero. In 1-propanol the coordinates of C_3 lie in X-Z plane. Here also three rotation angles ϕ_1, ϕ_2,ϕ_3 in addition to $\angle$ C_1OD_1 are treated as variable parameters and so also the r_{CC}, r_{CD} and r_{OD} parameters. The tilt angle of the methyl group is assumed zero, as in gas (Abdel Aziz & Rogowski, 1964) and liquid (x-ray) (Takamuku et al , 2004) works. It is also to be noted that free rotations about C_1C_2 line would generate 'gauche' and 'trans' like conformations and the analysis takes care of these possibilities as well.

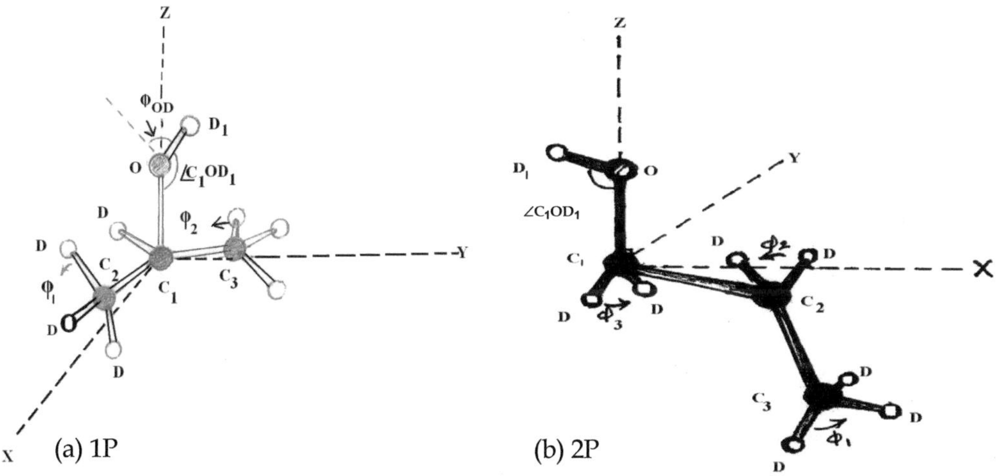

Fig. 2. Assumed models of molecules

3.3 Results

The fitted curves are shown in Figs. (3) and (4) and molecular parameters are listed in Table 2 together with results from other works. The Debye-Waller constant λ_0 were 3.84 $\times 10^{-2}$ Å^{-1}

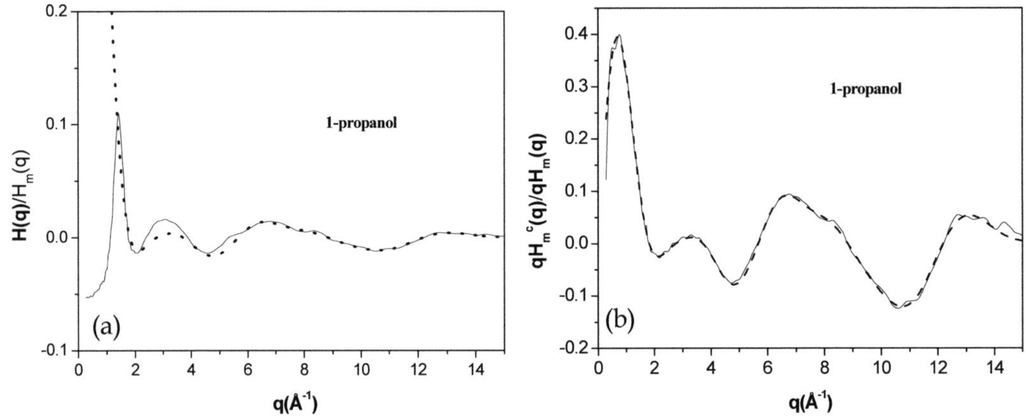

Fig. 3. 1P: (a) Structure function, $H(q)/H_m(q)$ vs q: - - - -$H_m(q)$, — $H(q)$ (Exptl.). (b) $qH_m(q)$ & $qH_m^c(q)$ vs q: - - -$qH_m(q)$, ____$H_m^c(q)$

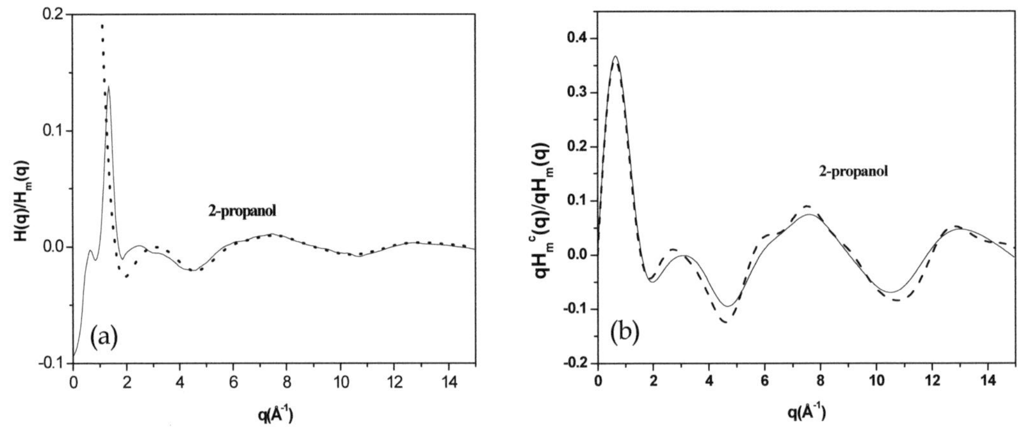

Fig. 4. 2P: (a) Structure function, $H(q)/H_m(q)$ vs q: - - - -$H_m(q)$, — $H(q)$ (Exptl.). (b) $qH_m(q)$ & $qH_m^c(q)$ vs q: - - -$qH_m(q)$, ____$H_m^c(q)$

and 5.53 $\times 10^{-2}$Å^{-1} for 1- and 2-propanols respectively while the agreement factors, χ^2 were respectively 1.35×10^{-5} and 2.56×10^{-5}. In 2-propanol molecule, r_{CC} appears to be little bit more and r_{CO} to be somewhat less but these are not unreasonable if compared with Abdel Aziz & Rogowski's electron diffraction work (1964) for r_{CC} and Howells et al's neutron work(Zetterström et al, 1991) for r_{CO}. The r_{CO} parameter in Howells et al's works was considerably less compared to those for other works (x-ray & electron diffraction). Also in Howells et al's works both r_{OD} and angle C_1OD_1 were abnormally low though their TOF neutron data were extended to as high as 30Å^{-1}. The molecular conformation is however somewhat spherical in both Howells et al and Sahoo et al's neutron works. In 1-propanol molecule, r_{CC} parameter appears to be a little bit less compared to other works. The other parameters are all reasonable. There is however no other neutron diffraction work available in the literature. Anyway, for a large molecule like propanol in liquid state these variations in molecular parameters are not unexpected and differences in r_{CC} and r_{CO} values in two propanols might also result from large conformational differences. What is, however,

Molecular Parameters	Neutrons, Reactor		Neutrons, TOF		X-rays		Electron Diff.(gas)		MD
	2P	1P	2P, SAN-DALS	2P, LAD	2P	1P	2P	1P	1P
r_{CC}(Å)	1.578±0.002	1.477±0.003	1.51±0.02	1.543±0.004	1.526	1.520	1.55-1.56	1.54	1.53
r_{CO}(Å)	1.258±0.007	1.443±0.012	1.37±0.03	1.340±0.03	1.430	1.420	1.40	1.41	1.43
r_{CD}(Å)	1.046±0.001	1.053±0.002	1.09±0.01	1.109±0.007	1.114	1.110	1.09	1.09	_
r_{OD}(Å)	1.029±0.0009	0.986±0.010	0.75±0.05	0.876±0.025	0.965	0.940	0.937	0.937	0.945
$\angle C_1OD_1$ (°)	110.52±0.7	110.41±0.72	76.41±3	82.6±6.8	125	108.5	105.9	105.9	108.5
ϕ_1(°)	-58.00±0.5	14.03±0.03	$\phi_1 = \phi_2$	$\phi_1 = \phi_2$	–	–	–	–	–
ϕ_2(°)	36.14±0.9	0.86±0.03	47.0±4	-14.4±5.5	–	–	–	–	–
ϕ_3(°)/ϕ_{OD}(°)	-44.47±2.0	7.2±0.03	-15±2	0(fixed)	–	–	–	–	–

Table 2. Molecular parameters for propanols at RT

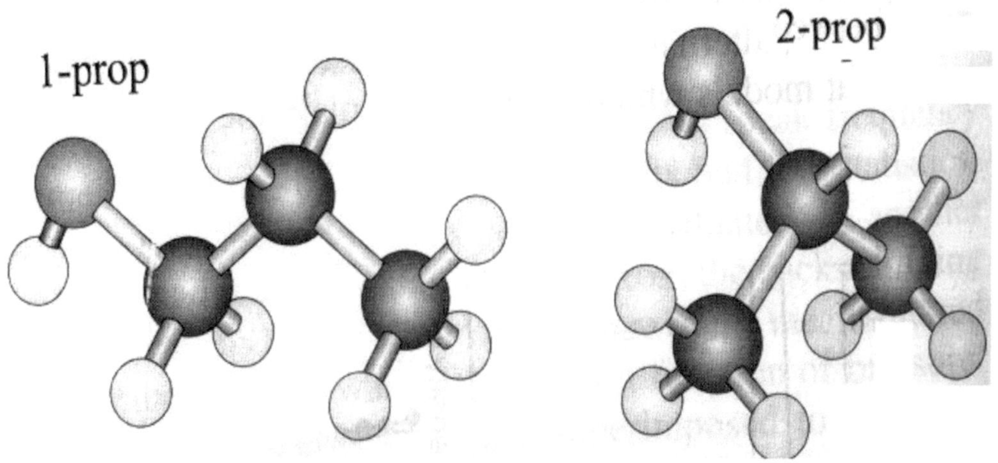

Fig. 5. Conformations of propanols.

important to note here is that the molecular conformation in 1-propanol is considerably elongated more like 'trans' which is also different from recent x-ray result (Takamuku et al, 2004). This difference is probably due to the fact that in x-ray work the positions of hydrogen atoms can not be obtained accurately. It is however to be noted that the earlier x-ray diffraction on liquid 1-propanol (Mikusinska-Planner, 1977) yielded nearly a 'trans' conformation. The molecular conformation of 2-ropanol is almost spherical and in liquid state at RT, two propanols have very different conformations more like that in crystalline solid (Talon et al, 2002).The deduced conformations are shown in Fig. 5. The conformations obtained via neutron data analysis are expected to be more convincing than those obtained through x-ray data analysis. The molecular structural differences result in the differences in intermolecular potential energy functions and radial distribution functions which could lead to significant differences in basic thermodynamic and other properties.

4. Molecular association studies

The hydrogen bonded molecular liquid alcohols (e.g. t-butanol) have very distinct type of intermolecular correlations compared to the corresponding non-hydrogen-bonded molecular liquids (e. g. neopentane) (Sarkar et al, 2000). In normal molecular liquids, the orientation correlations determine the liquid structure and properties (both thermodynamic and transport) of the system. In H-bonded molecular liquids the existence of hydrogen bonds strongly affects the static and dynamic properties of the liquids. The liquid structure is enormously different for the two cases and these structural distinctions are reflected in many physical and some chemical properties of the substances (Pauling, 1967). These changes are likely in the light of the fact that H-bonding may alter the mass, size, shape and arrangement of atoms, as well as the electronic structure of the functional groups resulting in the polymerization or formation of cluster in the liquid state. As a matter of fact, the degree of structural complexity increases as the number of atoms participating in the H-bonding increases. A full understanding of the structural and dynamical properties of these liquids has not been achieved even in ambient conditions. It is however true that the characteristic features of liquid structure and correlations in H-bonded liquids are distinctly exhibited in structure function (presence of pre-peak at $q \sim 0.7$ Å^{-1})and very clear in real space distribution functions obtained by FT of the q-space functions. The lack of long-range order and complexity of intermolecular interactions, however, turn the analysis a complex task except for the simplest molecular systems (e. g. water etc.). The computer simulations (both MC and MD) have been very useful (Jogensen, 1986 ; Svishchev & Kusalik, 1993) but the results often differ with results from diffraction experiments, the reason possibly being the inability of the simulation methods to incorporate the connectivity effects in the intermolecular interactions, so important in H-bonded liquid alcohols (Sum & Sandler, 2000). However, with the advantage of powerful computational tools and modeling algorithms, it is now a realistic possibility to conduct comprehensive 3D models of these complex systems that are consistent with available experimental data (Soper, 1996). Alternatively, a careful analysis via geometrical modeling can also yield satisfactory results in several cases with relatively large molecules (Magini et al, 1982; Sarkar & Joarder, 1993, 1994). Similar efforts have been also made with x-ray data earlier for liquid t-butanol and 1-propanol (Karmakar et al, 1995; Mikusinska-Planner, 1977) and in recent past for liquid 1- and 2-propanols in aqueous solutions (Takamuku et al, 2002, 2004). In view of large conformational differences in liquid 1- and 2-propanols the investigation of their average molecular associations in liquid state is a quite interesting. The investigation using neutron diffraction data reported in recent past (Sahoo et al, 2009, 2010) is briefly described in the following paragraphs.

4.1 Theory of regular molecular association or cluster

(Sarkar & Joarder, 1993, 1994)

The theory is based on a few simplified assumptions:

i. The liquid, on average, is considered to be an aggregate of small clusters of specific size composed of several molecules.
ii. The molecules in different clusters are orientationally uncorrelated.

Now, the general expression for the total structure function H (q) devoid of self term is given by

$$H(q) = MN^{-1}\langle \sum_{i,j} \sum_{\alpha_i,\alpha_j} b_{\alpha_i} b_{\alpha_j} \exp(i\mathbf{q}.\mathbf{r}_{\alpha_i\alpha_j}) \rangle \quad (8)$$

Where $M=(\sum b_{\alpha_i})^{-2}$, the < ... > bracket denotes an ensemble average, i and j label the molecules in the liquid, α_i, the αth atom in the ith molecule, $r_{\alpha i,\alpha j}$ the distance between the atoms α_i and α_j, $b_{\alpha i}$, the scattering length (in neutron case) or q-dependent atomic scattering factor (in x-ray case) of the atom α_i and the summation extends over all the scatterers in N molecules in the liquid.

Considering the assumption (i), on the right hand side of Eqn.(8), the separation of the contribution of atom pair terms within the clusters, called the cluster structure function $H_c^m(q)$, from the inter-cluster contribution is permitted in general. We have then

$$H(q) = H_c^m(q) + MN^{-1}\langle \sum_{\mu\neq\nu} \sum_{1_\mu,1_\nu} \sum_{\alpha_{1_\mu},\alpha_{1_\nu}} b_{\alpha_{1_\mu}} b_{\alpha_{1_\nu}} \exp(iqr_{\alpha_{1_\mu}\alpha_{1_\nu}}) \rangle \quad (9)$$

where

$$H_c^m(q) = MN^{-1}\langle \sum_{\mu} \sum_{1_\mu,1_{\mu'}} \sum_{\alpha_{1_\mu},\alpha_{1_{\mu'}}} b_{\alpha_{1_\mu}} b_{\alpha_{1_{\mu'}}} \exp(iqr_{\alpha_{1_\mu}\alpha_{1_{\mu'}}}) \rangle \quad (10)$$

l and l′ label molecules within a given cluster, and l_μ, the l-th molecule in the cluster μ.

Let us use $r_{c_{1_\mu 1_\nu}}$ to denote the vector distance from the centre of the molecule l_μ to that of the molecule l_ν such that,

$$r_{\alpha_{1_\mu}\alpha_{1_\nu}} = r_{c_{1_\mu 1_\nu}} - r_{c\alpha_{1_\mu}} - r_{c\alpha_{1_\nu}} \quad (11)$$

where $r_{c\alpha_{1_\mu}}$ is the vector distance from the centre of the molecule l_μ to its α-th nucleus within the cluster μ. Eqn.(9) can then be written as

$$H(q) = H_c^m(q) + MN^{-1}\left\langle \sum_{\mu\neq\nu} \sum_{1_\mu,1_\nu} \exp[(iq(r_{c_{1_\mu,1_\nu}})] \times \sum_{\alpha_{1_\mu},\alpha_{1_\nu}} b_{\alpha_{1_\mu}} b_{\alpha_{1_\nu}} \exp[iq(-r_{c\alpha_{1_\mu}} - r_{c\alpha_{1_\nu}})] \right\rangle \quad (12)$$

Eqn.(12) is quite general, the first term can be calculated for a given cluster of molecules but the second term can not be simplified without making some assumption because the orientation of the l_μ molecule (i.e. $r_{c\alpha_{1_\mu}}$) depends on that of the l_ν molecule (i.e. $r_{c\alpha_{1_\nu}}$) and on their separation ($r_{c_{l_\mu l_\nu}}$). If $r_{c_{l_\mu l_\nu}}$ is large, $r_{c\alpha_{1_\mu}}$ and $r_{c\alpha_{1_\nu}}$ are statistically independent. However, an important contribution to H(q) of interest comes from short $r_{c_{l_\mu l_\nu}}$ where orientations are not statistically independent. For this reason, the assumption (ii) is made and then $r_{c\alpha_{1_\mu}}$, $r_{c\alpha_{1_\nu}}$ and $r_{c_{l_\mu l_\nu}}$ all are taken to be statistically independent. One can now simplify Eqn. (12).

Since

$$\langle \exp(i\mathbf{q}.\mathbf{r})\rangle = \langle \exp(-i\mathbf{q}.\mathbf{r})\rangle,$$

Eqn.(12) becomes

$$H(q) = H_c^m(q) + MN^{-1}\langle \sum_{\mu\neq\nu} \sum_{l_\mu, l_\nu} \exp(i\mathbf{q}.\mathbf{r}_{c_{l_\mu l_\nu}})\rangle \times [\sum_\alpha b_\alpha \langle \exp(i\mathbf{q}.\mathbf{r}_{c\alpha})\rangle]^2 \quad (13)$$

In terms of form factor representing completely uncorrelated orientational configuration between molecules one writes

$$F_{2u}(q) = M[\Sigma b_\alpha < \exp(i\mathbf{q}.\,\mathbf{r}_{c\alpha}) >]^2, \quad (14)$$

where $r_{c\alpha}$ is the vector distance from the centre of a molecule to its α-th nucleus.

If one now considers the case of a system consisting of identical clusters, each of which is composed of N_c molecules. Then from Eqn.(13) one can write

$$H(q) = H_c^m(q) + F_{2u}(q)N^{-1}[\langle \sum_{i\neq j} \exp(iq.r_{c_{ij}})\rangle - \frac{N}{N_C}\langle \sum_{l\neq l'} \exp(iq.r_{c_{ll'}})\rangle]$$

or,

$$H(q) = H_c^m(q) + F_{2u}(q)[N^{-1}\langle \sum_{i\neq j} \exp(iq.r_{c_{ij}})\rangle + 1 - N_c^{-1}\langle \sum_{l\neq l'} \exp(iq.r_{c_{ll'}})\rangle - 1] \quad (15)$$

Thus, one writes H(q) in the form,

$$H(q) = H_c^m(q) + F_{2u}(q)\,[S_c(q) - f_3(q) - 1] \quad (16)$$

with

$$S_c(q) = 1 + N^{-1}\langle \sum_{i\neq j} \exp(iq.r_{c_{ij}})\rangle \quad (16a)$$

and

$$f_3(q) = N_c^{-1}\langle \sum_{l\neq l'} \exp(iq.r_{c_{ll'}})\rangle \quad (16b)$$

Here, $S_c(q)$ is the molecular centre structure factor of the liquid and $f_3(q)$, a factor resulting from the molecular centre pairs within the cluster.

Since there is no preferential orientation of the molecule with respect to the direction q, an orientation average may be evaluated and the one writes

$$< \exp(i\mathbf{q}.\mathbf{r}_{ij}) > = j_0(qr_{ij}) = \frac{\sin(qr_{ij})}{qr_{ij}}$$

Including 'Debye-Waller' factors in Eqns. (14), (16a) and (16b) one can write

$$F_{2u}(q) = M[\sum_{\alpha} b_{\alpha} j_0(qr_{c\alpha})\exp(-l_{c\alpha}^2 q^2/2)]^2 \tag{17a}$$

$$S_c(q) = 1 + N^{-1} \sum_{i \neq j} j_0(qr_{c_{ij}})\exp(-l_{c_{ij}}^2 q^2/2) \tag{17b}$$

$$f_3(q) = N_c^{-1} \sum_{l \neq l'} j_0(qr_{c_{ll'}})\exp(-l_{c_{ll'}}^2 q^2/2) \tag{17c}$$

The exponential factor, related to Debye-Waller factors, contains a $l_{\alpha\beta}$ parameter which is the root mean square deviation of the local instantaneous atom-atom separation distance $r_{\alpha\beta}$.

The cluster structure function $H_c^m(q)$ can be separated into the intra-molecular structure function $H_m(q)$ and the inter-molecular structure function $H_c(q)$ within a cluster.

Thus one writes,

$$H_c^m(q) = H_m(q) + H_c(q) \tag{18}$$

where,

$$H_m(q) = M \sum_{\alpha} \sum_{\alpha \neq \beta} b_{\alpha} b_{\beta} j_0(qr_{\alpha\beta})\exp(-l_{\alpha\beta'}^2 q^2/2) \tag{18a}$$

$$H_c(q) = N_c^{-1} M \sum_{l \neq l'} \sum_{\alpha,\beta} b_{\alpha_l} b_{\beta'_{l'}} j_0(qr_{\alpha_l \beta'_{l'}})\exp(-l_{\alpha_l \beta_{l'}}^2 \frac{q^2}{2}) \tag{18b}$$

Eqn. (16) reduces to

$$H(q) = H_m(q) + H_c(q) + F_{2u}(q)\,[S_c(q) - f_3(q) - 1] \tag{19}$$

The second and third term in Eqn.(19) combine to form the conventional inter-molecular "distinct" structure factor $H_d(q)$. So finally one gets

$$H(q) = H_m(q) + H_d(q) \tag{20a}$$

And

$$H_d(q) = H_c(q) + F_{2u}(q)\,[\,S_c(q) - f_3(q) - 1] \tag{20b}$$

The second term in Eqn.(20 b), the inter-molecular contribution, goes to zero for large q and one gets $H_d(q) \rightarrow H_c(q)$, and this is very useful in identifying any average inter-molecular cluster present in the liquid.

4.2 Identifying the average molecular association or cluster from diffraction data

The theory described above is used in this section to identify the presence of average molecular association or cluster in the two liquid propanols at RT using neutron diffraction data. The cluster parameters obtained for the most probable cluster are then used to generate the model x-ray H(q) function and compared with available experimental x-ray data. The good agreement obviously completes the identification of the most probable cluster present in liquid state. The method was earlier successfully applied to liquid methanol, ethanol, t-butanol etc. (Sarkar & Joarder, 1993, 1994; Karmakar et al, 1995). The present application of the method to two propanols in liquid state at RT is to check if there is any significant differences in the nature of most probable cluster or association in them. This investigation is important in view of significant differences in molecular conformation and thermodynamic and other properties of these two isomeric liquids. Between the two, 2-propanol is conformationally simpler and easier to tackle so its associational analysis is reported first.

(a) 2-propanol:

The diffraction data of liquid 2-propanol (both neutron and x-ray (Sahoo et al, 2010; Takamuku et al, 2002) show pre-peak at q~ 0.7-0.8 $Å^{-1}$ which is indicative of molecular association in liquid state. For a large molecule like 2-propanol the center structure factor of the liquid can be approximated by the PY single-site hard-sphere model (Karmakar et al, 1995) with a suitable core diameter. Two widely suggested probable molecular association models for alcohols are considered here namely, a tetramer linear chain (TLC) cluster (Magini et al, 1982) and an hexamer ring chain (HRC) cluster (Sarkar & Joarder, 1993, 1994; Karmakar et al, 1995) of neighboring molecules for liquid 2-propanol at RT. The cluster models are shown in Fig. 6. Considering all O-D…O bonds coordinates of all atomic sites are calculated. For simplification $CD_3(\equiv R)$ is taken as a single scattering unit following group scattering concept with appropriate scattering center and scattering length (Narten & Habenchuss, 1984).

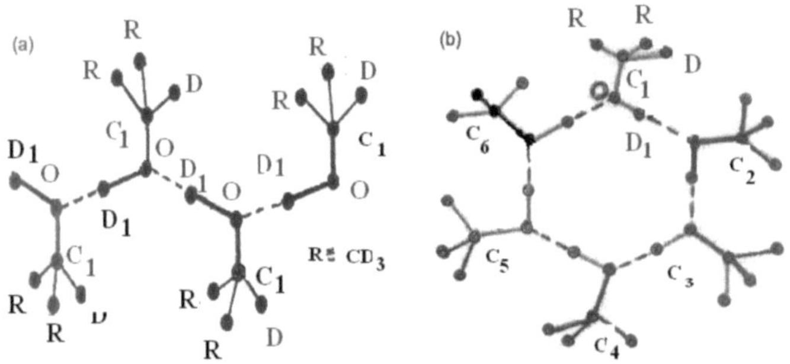

Fig. 6. Model clusters for 2-propanol: (a) Probable TLC structure, (b) Probable HRC structure.

The carbon atom denoted by C (vide Fig. 6) is taken to be centre of the bonded molecule. Varying the parameters like intermolecular O-O distance and the orientation and rotation angles of all C's and R's within the cluster one fits the model $qH_d(q)$ to experimental $qH_d(q)$ function by χ^2 fitting programme. The fitted curves for these two models are shown in Fig. 7(a) and clearly the HRC model is far superior. The model total structure functions H(q) obtained from Eqn. (19) are also shown in Fig.7(b). The pre-peak in the experimental diffraction pattern, characteristic of chain or cluster is satisfactorily generated in the HRC model. The HRC model parameters are shown in Table 3.

It is to be noted that hard core diameter (4.45 Å) and O-O distance parameter (2.68 Å for HRC model) are somewhat smaller than corresponding values for liquid t-butanol (Nath et al, 2002). The O-O parameter is also less compared to 2.736 Å from x-ray data analysis (Takamuku et al, 2002). The smaller O-O distance associated with HRC structure might be attributed to the non-planar nature of hexameric rings which allows the neighbouring molecules to come closer. In x-ray work however, the number of H-bonds per monomer was nearly 2 in agreement with HRC model. The intermolecular $G_d(r)$ obtained from $H_d(q)$ are shown in Fig. 8 for both models and HRC is again seen to be superior. Now using the parameters obtained from neutron data analysis, the model H(q) function for x-ray diffraction can be computed. In the process neutron scattering lengths are replaced by corresponding q-dependent x- ray atomic scattering factors. The HRC model result compare quite favorably with experimental x-ray H(q) data (Fig. 9). As pointed out in by Sahoo et al (2010) much better agreement is possible if the cluster parameters are little bit adjusted. It is however to be noted here that the average molecular association extracted here is very different from the probable chain molecular association suggested in Takamuku et al's x-ray work (2002). In x-ray work authors suggested trimer molecular association in cis and trans form and assumed a continuum beyond nearest neighbours. This association

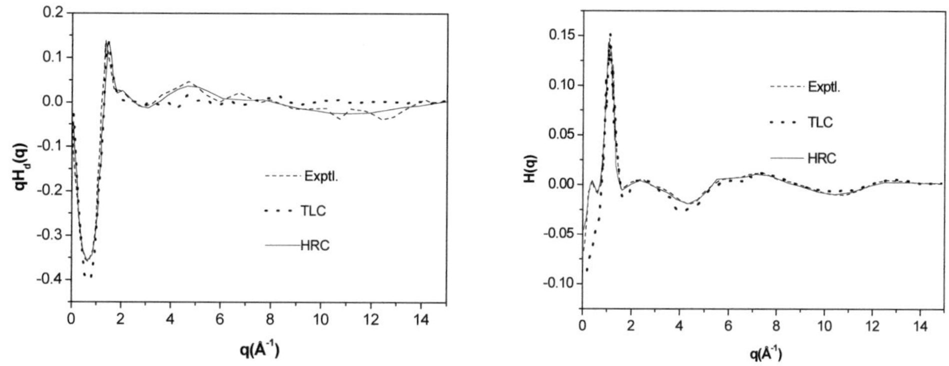

(a) (b)

Fig. 7. (a). $qH_d(q)$ vs q of D-2-propanol, (b). H(q) vs q of D-2-propanol

was, however, not conclusive.

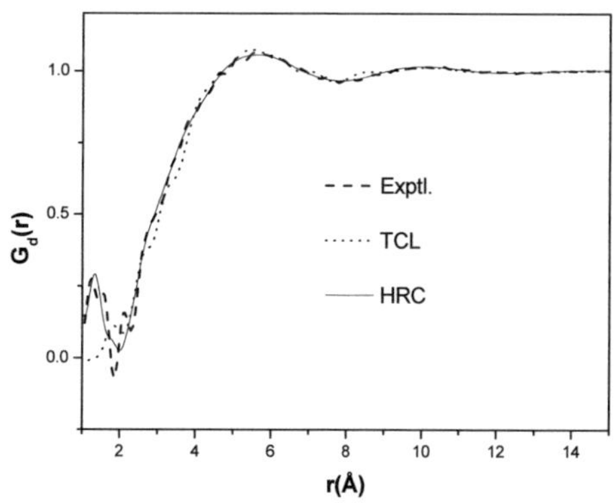

Fig. 8. $G_d(r)$ vs r of D-2-propanol

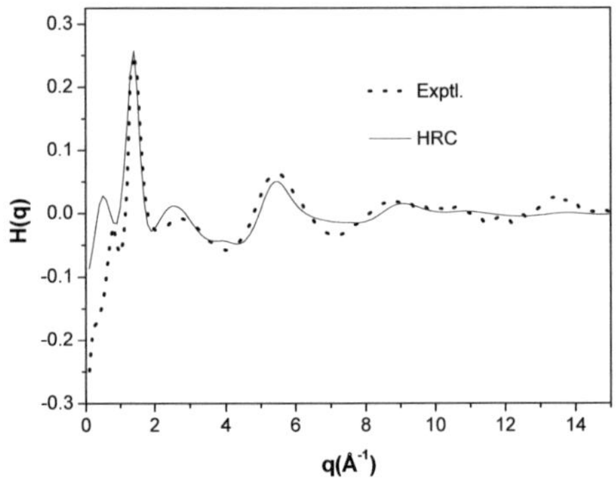

Fig. 9. 2-propanol (x-ray data): H(q) vs q

O-O distance	Rotational angles(deg)of C's about OD line	Rotational angles(deg) of R's about CO line
2.68 Å	(C1) 180.14° (C2) 54.18° (C3) 60.80° (C4) 108.88° (C5) 28.92° (C6) 17.00°	(R1) 18.00° (R2) -114.72° (R3) -11.54° (R4) 86.28° (R5) 1.36° (R6) -11.64°

Hardcore diameter, σ=4.45 Å; Non-planer angle, δ= 3°.

Table 3. HRC model parameters for 2-propanol

(b) 1-propanol:

Since 1-propanol in liquid state at RT has significantly different molecular conformation and properties, its molecular association in liquid state would be interesting. Further, the only neutron data available does not show any significant pre-peak like one for the earlier x-ray diffraction data (Mikusinska-Planner, 1977). The recent x-ray data (Takamuku et al, 2004) however shows a weak pre-peak at q~ 0.7-0.8$Å^{-1}$ but less prominent than one in 2-propanol data. The earlier x-ray diffraction study showed a preference for linear pentamer chain while the recent x-ray diffraction study pointed out a preference for linear trimer chain. Using neutron diffraction data Sahoo et al (2009) tested four model clusters namely, open chain winding 'trimer', 'tetramer', 'pentamer' and also HRC. The model clusters are shown in Fig.10 and in the calculation CD_3(≡R) and CD_2(≡R_1) are considered as single scattering sites with appropriate location of scattering centre and scattering lengths to avoid insertion of many more parameters. This approximation, as pointed out by Sahoo et al (2009), would little affect the intermolecular contribution where, in general, larger distances are involved and also because all D's are symmetrically located with C's in CD_3 and CD_2. This would also mean that when D's are treated separately flexible rotations of D's about corresponding C-C bonds would not contribute very differently to the intermolecular cluster structure function, $H_c(q)$ involving several molecules. Further the size of the molecule being large the centre structure can be approximated by PY single site hard sphere expression with appropriate core diameter and the centre located appropriately.

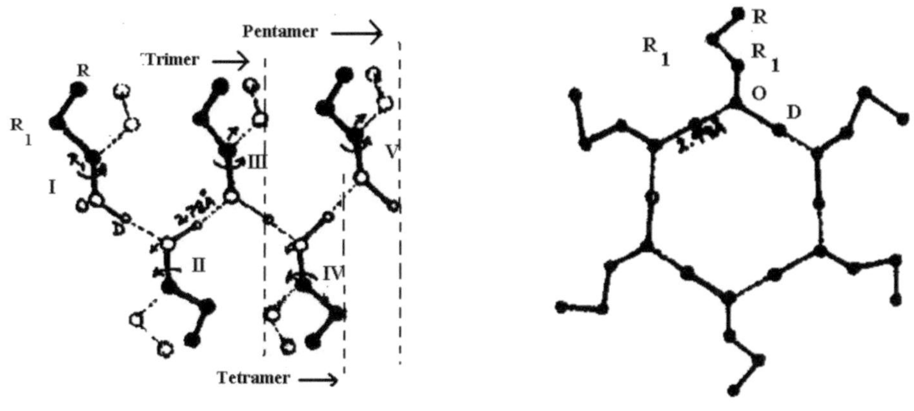

(a) (b)

Fig. 10. Model clusters for 1P, (a) Probable trimer to pentamer structures. (b) Probable HRC structure.

Varying the parameters like the hard-core diameter, inter-molecular O-O distance and in the case of chain clusters the rotational angles of all R_1R_1R and winding of the chain i.e. twist of OR_1's etc. within a cluster and in the case of HRC, the rotational angles of all R_1R_1R about OD axes and all twist angles of R_1R about R_1R_1 within a cluster the model $qH_d(q)$ is fitted to experimental $qH_d(q)$ function by a χ^2-fitting programme. The fitted curves are shown in Fig. 11(a) and (b). Though somewhat surprising (in view of large conformational differences) it is seen that HRC is too the most probable average molecular association in liquid 1-propanol at RT. This is seen to be true for $G_d(r)$ curve also (Fig. 12). The agreement with linear chain models is only so so.

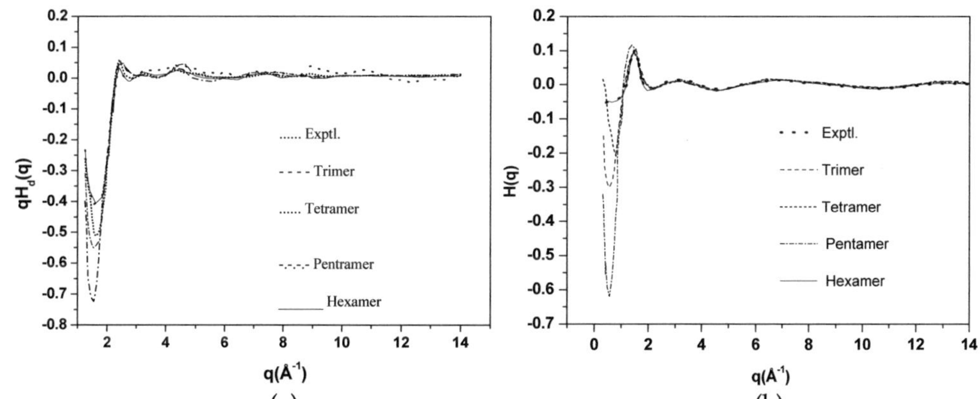

(a) (b)

Fig. 11. (a). $qH_d(q)$ vs q of D-1-propanol, (b). H(q) vs q of D-1- propanol

The HRC model parameters are shown in Table 4. The O-O distance parameter (2.99Å) is somewhat larger compared to that for 2-propanol. Further it is also larger compared to those for x-ray data analysis (Takamuku et al, 2002) and MD work (Akiyama et al, 2004). This larger O-O distance would be probably due to the elongated conformation of 1-propanol molecule and the planar nature of the hexameric rings which might cause stretching on the

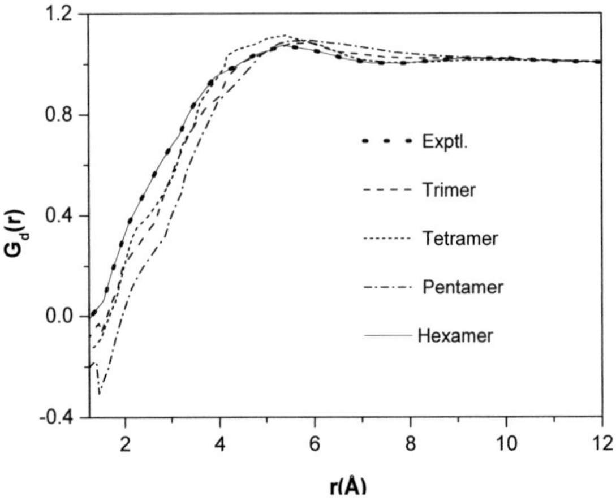

Fig. 12. $G_d(r)$ vs r of D-1-propanol

O-O distance	Rotational angles(deg)of C's about OD line	Rotational angles(deg) of R's about CO line
2.99Å	(C1) 139.90° (C2) 48.62° (C3) 51.38° (C4) 134.24° (C5) 21.38° (C6) 9.962°	(R1) 18.86° (R2) 334.24° (R3) 224.88° (R4) -17.86° (R5) 80.40° (R6) 18.82°

Table 4. HRC model parameters for 1-propanol, Hardcore diameter, σ=4.5Å

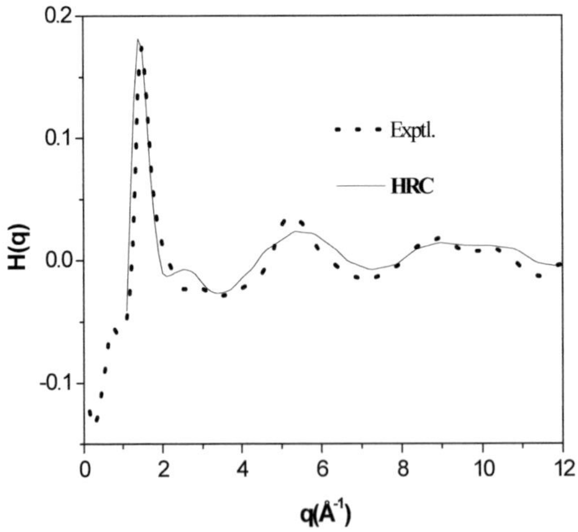

Fig. 13. H(q) vs. q for 1-propanol, (x-ray data)

intermolecular H-bonding. Again, the x-ray H(q) function computed for HRC model using neutron HRC parameters show very reasonable agreement with experimental x-ray H(q) data(Fig. 13). Much better agreement is however possible if the HRC parameters are little bit adjusted (Sahoo, 2011)

4.3 The liquid structure, 1-propanol vs. 2-propanol

In spite of large differences in molecular conformation and other properties the liquid structure is surprisingly similar in two propanols in liquid state at RT and these are most probably hexameric rings. Small but significant existence of other clusters are not ruled out but to show this, very careful examinations are obligatory. The earlier computer simulation studies on several liquid monohydric alcohols including 1-propanol (Jorgensen, 1986) suggested in support of winding linear chain configurations of several molecules and did so on the basic of atom-atom RDF peak locations but these were never conclusive. In 1-propanol, in view of an elongated conformation, a linear chain as average structure could be a possibility but the present result is seen to be otherwise. The very careful recent other studies also show that the presence of hexameric rings as average molecular units is possible in liquid alcohol like methanol (Kastanov et al, 2005, Gonzalez et al, 2007).

Though the average molecular association is similar in two propanols in liquid state there are several differences. The 2-propanol hexamer is not exactly planer. The intermolecular O-O distance is considerably less and most of the rotation angles for R's about CO line are very different. This is not unlikely in view of large conformational differences of two propanols. These differences are also reflected in $G_d(r)$ curves (vide Figs. 8 &12). In spite of apparent similarity there are minor differences in peak locations. It is also to be noted here that using the HRC model it is possible to evaluate the intermolecular pair distribution functions, namely, $g_{OO}(r)$, $g_{OD}(r)$, $g_{DD}(r)$ etc. from the experimental data (Sahoo et al, 2010; Sahoo, 2011). In Table 5 the locations of first and second peaks of these functions for two propanols in liquid state at RT are shown.

Liquids	Temp.	$g_{OO}(r)$		$g_{OD}(r)$		$g_{DD}(r)$	
		1st	2nd	1st	2nd	1st	2nd
D-2-propanol	RT	2.90 (2.75)	4.95 (4.70)	1.95 (1.82)	3.40 (3.30)	2.55 (2.36)	4.55 (4.60)
D-1-propanol	RT	2.95 (2.75) (2.77)	5.20 (4.70) (4.82)	2.04 (1.82) (1.86)	3.60 (3.30) (3.34)	2.66 (2.36) –	4.64 (4.60) –
D_2O	RT	2.87	4.26	2.02	3.45	2.46	3.90

Table 5. Main peak locations of $g_{\alpha\beta}(r)$ in Å. MD simulation data for $g_{DD}(r)$ not available.

In the parenthesis the simulation results (Jorgensen, 1986; Takamuku et al, 2004) are shown. The comparison of distribution curves for 1- and 2-propanols reveals that the peak heights of 2-propanol are in general more while the peak positions are at a little bit smaller r. Another important point to note here is that NDIS (Neutron Diffraction Isotropic Substitution) experimental method enables evaluation of $g_{\alpha\beta}(r)$ for D_2O at RT (Soper et al, 1997). The first peak locations for $g_{OO}(r)$, $g_{OD}(r)$, $g_{DD}(r)$ of D_2O shown in the table agree reasonably well with those for two propanols. This agreement is expected and this also justifies the probable association model suggested here for liquid propanols at RT. Anyway, coming to the case of two propanols the differences in the first peak heights and small differences in peak locations, particularly for the second peaks could be enough to guess that the average intermolecular pair potential, if it could be extracted from experimental data, could be significantly different for the two propanols and this could possibly explain the significant differences in the thermodynamic and other properties of these two isomeric liquids.

5. Conclusion

The detailed molecular conformational and associational structures of liquids with big size molecules like propanols using neutron diffraction have been presented in this chapter. The neutron diffraction analysis is important as H-positions can be accurately seen here unlike x-rays. Still today, the diffraction data analysis happens to be the most powerful tool for molecular as well as liquid structure analyses. The x-ray can not locate H-positions accurately but skeleton of the intermolecular structure is better seen than by neutrons where overlapping of several intra and intermolecular scattering terms occur. So a combined analysis with two diffraction data sets (neutrons and x-rays) is found to be most useful.

The data in the analysis reported here are reactor neutron data extended upto a q not more than 16 $Å^{-1}$, though the data have been claimed to be of good accuracy. It is however to be noted here that all the structural information, molecular and intermolecular are contained in this range. The data at higher q's (say 20 $Å^{-1}$ onwards as in TOF data) do not contain much information other than simple noise. The method of data analysis reported here is consistent and reasonably accurate. The analysis clearly indicates the molecular conformational differences in the two propanols and molecular associational similarities in them at RT. The average molecular association is shown to be hexameric rings. Such molecular association was first proposed for liquid methanol by Pauling (1967) and was supported for methanol and several other alcohols in liquid state by works from this laboratory (Sarkar & Joarder, 1993, 1994; Karmakar et al, 1995; Nath et al, 2002). The justification of such structure for liquid methanol was given by Sarkar & Joarder(1993). The reason for the possibility of such structure in dominant form could be also found in Kashtanov et al's work (2005). It is pointed out there that the molecular orbitals of the six unit ring methanol structure show similarity to those of benzene ring indicating that H-bonding in methanol ring has significant amount of covalent contribution. Further, the electronic structure of methanol dimer is as polarized as the other chain structures and it is completely different from rings and as such the molecular dynamics based on potentials derived from methanol dimer does not predict ring structures. So it is expected that potentials which take into account the unique covalent contribution to H-bonding need to be used in simulation works for generating ring structures like hexameric rings as claimed in the different studies of liquid alcohols. The MC simulation work with refined H-bonding potential does predict the possibility of the presence of hexameric ring structures (Gonzalez et al, 2007).It is also to be noted here that earlier computer simulation (Jorgensen, 1986) yielded 1.91 and 1.92 H-bonds per molecule for 1-and 2-propanols respectively. The x-ray diffraction data (Takamuku et al, 2002, 2004) yield almost similar values (1.92 for 1-propanol and almost 2.0 for 2-propanol). Further, for liquid like methanol, the EPSR (Empirical Potential Structure Refinement) method (Yamaguchi et al, 1999) based on neutron diffraction data predicts 1.95±0.07 & 1.77±0.07 H-bonds per molecule and 6.27±0.7 & 5.5±1 molecules per chain cluster respectively at -80°C & 25°C and these facts approximately agree with HRC model. It is however to be noted here that the other clusters like trimer, tetramer etc. in small proportions are also possible. The analysis for extraction of other structures via molecular cluster theory is in principle possible and could be carried out in near future. Further, provided the experimental data are available, the temperature and pressure effects on the molecular association could also be possibly studied as has been done recently for liquid methanol (Sahoo et al, 2010) and also possibly using MCSQ(Monte Carlo Structure Factor) technique (Sahoo & Joarder, 2010).

6. Acknowledgement

The author thanks his collaborators, Dr. A. Sahoo and Dr. P. S. R. Krishna (BARC, India) for their assistance and contribution.

7. References

Abdel Aziz, N. E. & Rogowski, F. (1964). Structure Estimation of n-Propyl Alcohol, Isopropyl Alcohol, Allylalcohol and Propargylalcohol in Vapor Phase by Electron Diffraction. *Zeitschrift für Naturforschung*, Vol.19b, No. 25 (June, 1964), pp.967-977, ISSN 0932-0776.

Adya, A. K., Bianchi, L. & Wormald, C. J. (2000). The Structure of Liquid Methanol by H/D Substitution Technique of Neutron Diffraction. *Journal of Chemical Physics*, Vol. 112, No. 9(March, 2000), pp. 4231-4241, ISSN 0021-9606.

Akiyama, I., Ogawa, M., Takase, K., Takamuku, T., Yamaguchi, T. & Ohtori, N.(2004). Liquid Structure of 1- Propanol by Molcular Dynamics Simulations and X-ray Scattering. *Journal of Solution Chemistry*, Vol. 33, No. 6/7 (June/ July, 2004), pp. 797-809, ISSN 0095-9782.

Benson, S. W. (1996). Some Observations on the Structures of Liquid Alcohols and Their Heats of Vaporization. *Journal of American Chemical Society*, Vol. 118, No. 43 (October, 1996), pp 10645-10649, ISSN 0002-7863

Bertagnolli, H., Chieux, P. & Zeidler, M. D. (n. d.). A Neutron-diffraction Study of Liquid Acetonitrile.*Molecular Physics,* Vol. 32. No. 3 (September, 1976), pp.759-773, ISSN 0026-8976.

Champeney, D. C., Joarder, R. N. & Dore, J. C. (1986). Structural Study of Liquid D-glycerol by Neutron Diffraction. *Molecular Physics*, Vol. 58, No. 2 (June, 1986), pp.337-347, ISSN 0026-8976.

Cuello, G. J., Talon, C., Bermejo, F. J. & Cabrillo, C. (2002). Chemical Isomeric Effects on Propanol Glassy Sturctures. *Applied Physics A*. Vol. 74(Suppl.), pp. S552-S554, ISSN 0947-8396

Egelstaff, P. A. (1987). Classical Fluids. In: *Methods of Experimental Physics: Neutron Scattering*, D. L. Price & K. Skold(Eds.) , Vol. 23B, Accademic Press, San Diego, USA, ISBN 0-12-475969-6.

Frenkel, J. (1955). *Kinetic Theory of Gases*, Dover, New York, USA.

Gonzalez, M. V., Martin, H. S. & Cobos, J. H.(2007). Liquid Methanol Monte Carlo Simulations with a Refined Potential which includes Polarizability, Nonadditivity, and Intramolecular Relaxation. *Journal of Chemical Physics*, Vol. 127, No. 22(December, 2007), pp.224507(14), ISSN 0021-9606.

Jorgensen, W. L. (1986). Optimized Intermolecular Potential Functions for Liquid Alcohols. *Journal of Physical Chemistry*, Vol.90, No.7 (n. d.), pp. 1276-1284, ISSN 0022-3654

Karmakar, A. K.,, Sarkar, S. & Joarder, R. N. (1995). Molecular Clusters in Liquid tert-Butyl Alcohol at Room Temperature. *Journal of Physical Chemistry*, Vol. 99, No.45 (November, 1995), pp. 16501-16503, ISSN 0022-3659.

Kashtanov, S., Auguston, A., Rubensson, J. -E., Nordgren, J., Agren, H., Guo, J.-H. & Luo, Y. (2005). Chemical and Electronic Structures of Liquid Methanol from X-ray Emission Spectroscopy and Density Functional Theory. *Physical Review B*, Vol. 71, No. 10 (March, 2005), pp. 104205(7), ISSN 1098-0121.

Magini, M., Paschina, G. & Piccaluga, G. (1982). On the Structure of Methyl Alcohol at Room Temperature. *Journal of Chemical Physics*. Vol. 77, No. 4(August, 1982), pp.2051-2056, ISSN 0021-9606.

Mikusinska-Planner, A. (1977). X-ray Diffraction Study of the Structure of 1-Propanol at -25 °C. *Acta Crystallography*, Vol. A33,No. 3(May, 1977), pp. 433-437,ISSN 1600-5724.

Montague, D. G. & Dore, J. C. (1986). Structural Studies of the Liquid Alcohols by Neutron Diffraction, IV CD_3OH and CD_3OD. *Molecular Physics*. Vol. 57, No. 5 (n.d.), pp. 1035-1047, ISSN 0026-8976.

Narten, A. H. & Habenchuss, A. (1984). Hydrogen Bonding in Liquid Methanol and Ethanol determined by X-ray Diffraction. *Journal of Chemical Physics*, Vol. 80, No. 7 (April, 1984), pp.3387-3391, ISSN 0021-9606.

Nath, P. P., Sarker, S., Krishna, P. S. R. & Joarder, R. N. (2002). Intermolecular Structure of Liquid D-tert Butanol by Neutron Diffraction Data. *Applied Physics A*, Vol. 74(Suppl.), pp. S348-S351, ISSN 0947-8396

Pauling, L. (1967). *The nature of Chemical Bond*, 3rd. Edition, Oxford University, Oxford, UK, ISBN 0-8014-0333-2.

Ramos, M. A., Talon, C., Jimenez-Rioboo, R. J. & Vieira, S.(2003). Low Temperature Specific Heat of Structural and Orientational Glasses of Simple Alcohols. *Journal of Physics: Condensed Matter*, Vol. 15 (March, 2003), pp. S1007-S1018, ISSN 0953-8984.

Sahoo, A., Nath. P. P., Bhagat, V., Krishna, P. S. R. & Joarder, R. N. (2010). Effect of Temperature on the Molecular Association in Liquid D-Methanol using Neutron Diffraction Data. *Physics and Chemistry of Liquids*. Vol. 48, No. 4 (August, 2010), pp. 546-559, ISSN 0031-9104.

Sahoo, A., Sarkar, S., Bhagat, V., & Joarder, R. N. (2009).The Probable Molecular Association in Liquid D-1-Propanol through Neutron Diffraction. *Journal of Physical Chemistry A*, Vol. 113, No. 17(April, 2009),pp.5160-5162, ISSN 1089-5639.

Sahoo, A., Sarkar, S., Bhagat, V., & Joarder, R. N. (2008). Molecular Conformation and Structural Correlations of Liquid D-1-Propanol through Neutron Diffraction. *Pramana-Journal of Physics*, Vol. 71, No. 1(July, 2008), pp.133-141, ISSN 0304-4289.

Sahoo, A., Sarkar, S., Krishna, P. S. R, & Joarder, R. N.(2010). Molecular Conformation and Liquid Structure of 2-Propanol through Neutron Diffraction. *Pramana-Journal of Physics*, Vol. 74, No. 5(May, 2010), pp. 765-779, ISSN 0304-4289.

Sahoo, A. & Joarder, R. N.(2010). Monte-Carlo Structure Factor (MCSQ) Calculation for Liquid Structure of H-bonded Molecular Liquids. *Solid State Physics. Proceeding of the 55th DAE -Solid State Physics Symposium, 2010.* AIP Conf. Proc. 1349, Vol. 55 (December, 2010) pp.521-522, ISBN 978-0-7354-0905-7.

Sahoo, A. (2011). Investigation of Molecular Conformation and Association in Some H-bonded Liquids through Neutron Diffraction. *Ph. D. Dissertation,* Jadavpur University(June, 2011), Kolkata, India.

Sarkar, S., & Joarder, R. N. (1993). Molecular Clusters and Correlations in Liquid Methanol at Room Temperature. *Journal of Chemical Physics*, Vol. 99, No. 3(August, 1993), pp.2032-2039; Ibid. Molecular clusters in Liquid Ethanol at Room Temperature.Vol. 100, No.7(April,1994), pp.5118- 5122, ISSN 0021-9606.

Sarkar, S., Nath, P. P., & Joarder, R. N. (2000). Orientation Correlation versus Cluster Correlation in Molecular Liquids-Signature through Diffraction Data. *Physics letters* A, Vol. 275, No. 1-2 (October, 2000)pp.138-141, ISSN 0375-9601.

Soper, A. K. (1996). Empirical Potential Monte Carlo Simulation of Fluid Structure. *Chemical Physics*, Vol. 202. No.2-3 (January, 1996), pp. 295-306, ISSN 0301-0104.

Soper, A. K., Bruni, F. & Ricci, M. A. (1997). "Site-Site Pair Correlation Functions of Water from 25°C to 400°C: Revised Analysis of New and Old Diffraction Data" *Journal of Chemical Physics*, Vol. 106,No.1(January,1997), pp. 247-254, ISSN 0021-9606.

Sum, A. K. & Sandler, S. I. (2000). Ab Initio Calculations of Cooperativity Effects on Clusters of Methanol, Ethanol, 1-Propanol and Methanethiol. *Journal of Physical Chemistry A*, Vol. 104, No. 6 (January, 2000), pp.1121-1129, ISSN 1089-5639.

Svishchev, I. M. & Kusalik, P. G. (1994). Structure in Liquid Methanol from Spatial Distribution Functions. *Journal of Chemical Physics,* Vol. 100, No. 7(April, 1994), pp. 5165-5171, ISSN 0021-9606.

Tanaka, Y., Ohtomo, N., & Arakawa, K.(1984). The Structure of Liquid Alcohols by Neutron Diffraction II. Molecular Structure of Ethylalcohol. *Bulletin of Chemical Society of Japan*, Vol. 57, No. 9(September, 1984), pp.2569-2573, ISSN 0009-2673.

Takamuku, T., Saisho, K., Aoki, S. & Yamaguchi, T. (2002). Large-angle X-ray Scattering Investigation of the Structure of 2-Propanol-Water Mixtures. *Zeitscript für Naturforschung.Vol.* 57a, (n. d.), pp. 982-994, ISSN 0932-0784.

Takamuku, T., Maryyama, H., Watanabe, K., & Yamaguchi, T.,(2004). Structure of 1-Propanol -water Mixtures Investigated by Large-Angle X-ray Scattering Technique. *Journal of Solution Chemistry,* Vol. 33, No. 6/7(June/July2004), pp.641-660, ISSN 0095-9782.

Talon, C., Bermejo, F. J., Cabrillo, C., Cuello, G. S., Gonzalez, M. A., Richardson, J. W., Criado, A., Ramos, M. A., Vieira, S., Combrero, F. L., Gonzalez, L. M.(2002). Chemical Isomerism as a Key to Explore Free-energy Landscapes in Disordered Matter. *Physical Review Letters*, Vol. 88, No.11 (March, 2002), pp. 115506(4), ISSN 0031-9007.

Talon, C., Ramos, M. A., Vieira, S., Symyt'ko, I., Afonikova, N., Criado, A., Madariago, G. & Bermejo, F. J.(2001). Themodynamic and Structural properties of the two Isomers of solid Propanol. *Journal of Non-Crystalline Solids*. Vol. 287,(n. d.) pp. 226-230, ISSN 0022-3093.

Yamaguchi, T., Hidaka, K., & Soper, A. K.(1999). The Structure of Liquid Methanol Revisited: a Neutron Diffraction Experiment at -80°C and +25°C. *Molecular Physics*, Vol. 96, No.8 (April, 1999), pp. 1159-1168; Ibid. ERRATUM The Structure of Liquid Methanol Revisited: a Neutron Diffraction Experiment at -80C and +25 C. vol. 97, No. 4 (August, 1999), pp.603- 605, ISSN 0026-8976.

Zetterström, P. Dalborg, U., Delaplane, R. G. & Howells, W. S. (1991). Neutron Diffraction Studies of Liquid Iso-Propanol. *Physica Sripta*, Vol. 44, (n. d.),pp.56-62, ISSN 0031-8949.

Zetterström, P. Dalborg, U., & Howells, W. S. (1994). A Systematic Study of the Structure of Liquid Iso-Propanol by Time-of-Flight Neutron Diffraction. *Molecular Physics*, Vol. 81, No. 5, (October, 1994), pp.1187-1204, ISSN 0026-8976.

8

Neutron Diffraction Study of Hydrogen Thermoemission Phenomenon from Powder Crystals

I. Khidirov
Institute of Nuclear Physics, Uzbekistan Academy of Science
Uzbekistan

1. Introduction

In hydrogen energy powder metals or compounds that accumulate and desorb hydrogen extensively are used for hydrogen storages. Even in small quantities hydrogen can have stronger effect on structure and properties of some metals and compounds than heavier elements. Study of both hydrogen arrangement in crystal structure and process of its evacuation out of a lattice is of great importance for understanding the reasons and mechanisms of hydrogen influence on material properties. At vacuum hydrogen evacuation, it is possible to find such temperature T_{ev} at which the hydrogen atoms, having low bond energy and high diffusion speed, are eliminated out of the lattice. The configuration of relatively heavy matrix atoms (previously stabilized by hydrogen atoms) does not change because of their insufficient diffusive mobility at this temperature. In this way it is possible to obtain new metastable phases artificially (hydrogen induced phases). In these phases the structure of initial hydrogenous phase can be kept (as if "frozen"), but during heat treatment they can undergo phase transitions uncharacteristic for the initial material. Vacuum removal of hydrogen out of a crystal lattice without change of symmetry is similar to thermoelectronic emission. Therefore we suggest the name of this new phenomenon as "hydrogen thermoemission" from crystals. Thus hydrogen thermoemission is complete removal of hydrogen atoms out of crystal structure without changing the crystal symmetry. The aim of this chapter is generalizing the results of neutron diffraction investigations of low-temperature hydrogen thermoemission phenomenon from powder crystals of Ti-N-H, Zr-N-H, Ti-C-H, Ti-H systems and of rare earth metal trihydroxides $R(OH)_3$, were R is La, Nd or Pr. We will also look into formation of regularities and structural features of hydrogen induced phases.

2. Experiment techniques

2.1 Choice of method of structural research

Neutron diffraction method was chosen as the basic method of structure study which allows to obtain the largest and reliable information in the structural analysis of the alloys consisting of components (metals and hydrogen) with strongly differing order numbers. Neutron diffraction patterns were obtained using the neutron diffractometer mounted at the thermal column of the nuclear reactor of the Institute of Nuclear Physics of Uzbekistan Academy Sciences (λ=0,1085 nm) at room temperature. X-ray diffraction patterns were obtained using the X-ray diffractometer DRON – 3M with CuK_{α} – radiation (λ = 0.15418 nm).

2.2 Calculation methods of crystal structure determination

The observed integral intensity I_k(calc) = I_{hkl} of diffraction reflections from planes with Miller indexes h k l for a cylindrical sample is calculated using the formula (Bacon, 1975):

$$I_{hkl} = I_0(\lambda^3 h / 2\pi r)(V\rho / \rho')(p_{hkl} N^2 | F(hkl)|^2 / \sin\theta \sin 2\theta) A_{hkl} \exp(-2W), \quad (1)$$

where $I_0(\lambda^3 h/2\pi r)(V\rho/\rho')$ = k is the constant dependent on geometry of the device and a sample; p_{hkl} - the multiplicity factor ; F_{hkl}^2 - the structural factor for reflection (h k l); A_{hkl} - the factor of absorption (for the objects of our study the absorption factor is insignificant and, so, in calculations it was neglected), r - distance from a sample up to the counter; h - height of a crack of the counter; N - number of elementary cells in unit volume; V - volume of the studied sample; ρ and ρ' - experimental and calculated density of the studied sample; W - the thermal factor which is determined by equation:

$$W = (8\pi^2 \overline{u^2} / 3)\sin^2\theta / \lambda^2, \quad (2)$$

where $\overline{u^2}$ - atomic mean-square displacement. The structure factor F^2 indicates the functional dependence of reflection intensities upon atom positions in the unit cell. This dependence is the function both of atom coordinates and Miller indexes of the reflex. The general formula of structure factor is:

$$|F|^2 = \left\{ \sum_n b_n \cos 2\pi (hx_n + ky_n + lz_n) \right\}^2 + \left\{ \sum_n b_n \sin 2\pi (hx_n + ky_n + lz_n) \right\}^2, \quad (3)$$

where x_n, y_n, z_n are atom coordinates expressed in fractional units of the corresponding parameter; b_n - amplitude of neutron scattering on the atom nuclei. Summation is made over all atoms of the unit cell. Determination of the structure was carried out by "trial-and-error" method. The method consists in the next: the obtained experimental both angle positions and intensity of diffraction maxima are compared with the values calculated theoretically within the framework of the certain model. The most probable models of structure (taking into account X-ray data) were selected, and for each model the refinement of atom coordinates and position occupancy by least-square method of Rietveld (Rietveld, 1969) was carried out. At that, the refined parameters could be varied both together and separately. The model which yielded the best results (the minimal R-factor) was chosen. The structure model is correct if the experimental positions of maxima coincide with calculated ones and also if discrepancy indexes R are minimal. Last years for the refinement of crystal structures using X-ray and neutron diffraction data, full-profile method offered by Rietveld is widely used. The main point of this method is minimization of R_P by means of full-profile processing of powder neutron diffraction data. At that, over the Bragg's reflections range the discrepancy indexes are calculated using both the full profile (R_P) and weights of each point (R_{WP}) and intensities of Bragg's reflections (R_{Br}). Now there is the software package for calculation and refinement of crystal structures on the basis of this method. In the present work the software packages have been used DBW 3.2 (Young & Wilas, 1982).

2.3 Method of thermal treatment of samples

Powder samples with grain sizes of < 60 μm were studied. Dehydrogenation of hydrogenous compounds was carried out under persistent evacuation (at vacuum < 5.3×10^{-3} Pa). Thermal treatment of the samples under persistent evacuation was carried out in the vacuum furnace of SShVL-type. The principle scheme of this vacuum furnace is shown in Fig. 1. For carrying out annealing the sample (1) was placed in enclosed volume (2) on a spe-cial support (3). For vacuum control the vacuum-measuring block VIT-3 (4) and manometric lamps PMT-2 (5) and PMI-3 (6) were used. By means of mechanical (7) and diffusion (8) pumps in enclosed volume vacuum not worse than 5.3×10^{-3} Pa was made. After that, vacuum heat treatment of a sample began: temperature was slowly raised since the room one up to necessary size with a step of 25 °C; a sample was exposed at each temperature at first during 24 and then during 36 and 48 hours. Temperature of a sample was controlled by potentiometer (9) with help of PPR (platinum - platinum-rhodium) thermocouple (10). Process was continued until the whole hydrogen leaves a sample. It should be noted that we observed vacuum to be not worse. If vacuum became insignificantly worse, temperature was decreased until vacuum was restored up to $5{,}3\times10^{-3}$ Pa. After annealing at each temperature with certain exposure time (24, 36 and 48 h.) neutron diffraction pattern was obtained. Hydrogen quantity in samples was determined as decline of incoherent background caused by incoherent neutron scattering on H nuclei. Since hydrogen nucleus has the largest amplitude of incoherent neutron scatting, its background scatting falls away with increase of Bragg angle (Bacon, 1975). The hydrogen content was by chemical analysis and controlled by the analysis of neutron structure data using the full-profile method of analysis intensities of neutron patterns (Young & Wilas, 1982).

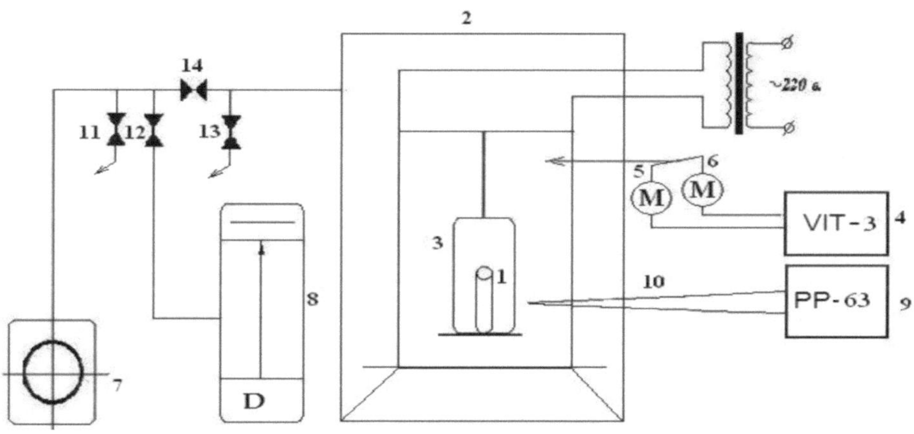

Fig. 1. The principle circuit of the high-temperature vacuum furnace of SShVL-type. 1- powder sample in metal cassette; 2 – enclosed volume; 3 - support for mounting of a sample; 4- vacuum-measuring block VIT-3; 5-manometric lamp PMT-2 for measuring forevacuum; 6-manometric lamp PMI-3 for measuring high vacuum; 7-mechanical pump; 8- diffusion pump; 9-potentiometer PP-63; 10-thermocouple PPR; 11-, 12-, 13-, 14 - vacuum gates.

3. Results

3.1 Hydrogen thermoemission in the Zr-N-H system

The solid solution $ZrN_{0.43}H_{0.38}$ for studying was prepared by self-propagating high-temperature synthesis. This method based on exothermal reaction of initial reagents. As initial materials nitrogen of the "extra pure" brand and Zr powder of the M-41 brand were taken. According to nameplate, the Zr powder contained 0.45 mas % of hydrogen. The pressure of nitrogen in a constant pressure bomb was 7×10^{7} Pa. After synthesis the sample was exposed to homogenizing annealing in evacuated and sealed-off quartz ampoule at temperature 1000 °C during 12 h. Then for fixation of the high-temperature state samples were hardened in air. Powder samples with grain sizes no more than 60 μm were studied. Samples composition was determined by chemical analysis and was controlled by minimizing the R factors of structure determination using neutron diffraction patterns. Dehydrogenation of hydrogenous compounds was carried out under persistent evacuation (at vacuum not more than 5.3×10^{-3} Pa) (Khidirov et al., 2009). After dehydrogenating at every temperature a neutron diffraction pattern was surveyed. Hydrogen quantity in samples was watched as decline of incoherent background caused by incoherent neutron scattering on H nuclei. Since hydrogen nucleus has the largest amplitude of incoherent neutron scattering, its background scattering falls away with the increase of Bragg angle (Bacon, 1975). The hydrogen content was also estimated by the analysis of reflection intensities and sometimes by chemical analysis. As an initial sample the single-phase ordered solid solution $ZrN_{0.43}H_{0,38}$ was taken. According to X-ray structure analysis, the sample was single-phase and homogeneous and had hexagonal close-packed structure with unit cell parameters: a = 0.3274 ± 0.0002; c = 0.5321 ± 0.0003 nm; c/ a =1.625. The neutron diffraction pattern of the initial solid solution $ZrN_{0,43}H_{0,38}$ is shown in Fig. 2,a. Uniform slope of incoherent background points to hydrogen presence. The neutron diffraction data of the solid solution show that nitrogen atoms are ordered over one of two types of octahedral interstices alternating along c axis, and hydrogen atoms − over tetrahedral interstices of hexagonal close-packed (HCP) metal structure; unit cell parameters are: a = 0.3274; c = 0.5321 nm (c/a = 1.625). The metal atoms (z_{Zr}=0.243) are displaced along c axis from their ideal positions (z_{id}=1/4) toward the plane of 1 (a) octahedrons occupied with nitrogen atoms (Table 1). So, the initial sample is the ordered solid solution of nitrogen and hydrogen in HCP structure of α-Zr: it is α' - $Zr_2N_{0.86}H_{0.76}$ phase, space group $P\bar{3}m1$ structure of anti-CdI_2 type on nitrogen. Dehydrogenation of the α'- $Zr_2N_{0.86}H_{0.76}$ phase (of composition $ZrN_{0.43}H_{0.38}$) was carried out by a regime of step annealing within the temperature range 100 °C ≤T<375 °C with steps of 25 °C and exposure time of 24 - 48 h at each temperature. Vacuum in the enclosed volume the vacuum was kept below 5.3×10^{-3} Pa under permanent evacuation. After each vacuum annealing step, neutron diffraction pattern was obtained and quantity of hydrogen in a sample was estimated. Analysis of the neutron patterns shows that such dehydrogenation does not result in marked change of hydrogen content. Hydrogen content is confirmed by conservation of the slope of incoherent background in diffraction patterns. Though according to neutron structure analysis and vacuum extraction data dehydrogenation at T≥375°C during 36 h leads to practically complete hydrogen removal out of the sample (Fig. 2, b) while the quantity of nitrogen is kept. The structural analysis of the dehydrogenated sample showed that the ordered phase structure of anti-CdI_2 type (α'-$Zr_2N_{0,86}$) did not change. The phase was

Atom	Position	Atom coordinates				B, nm^2	ΔB, nm^2	n	Δn
		x	y	z	Δz				
Zr	2 (d)	1/3	2/3	0.243	0.001	0.0039	0.0005	2	
N	1 (a)	0	0	0		0.0072	0.0009	0.86	0.05
H	2 (d)	1/3	2/3	0.619	0.002	0.0107	0.0042	0.76	0.04
R_p = 1.9; R_{wp} = 2.5; R_{Br} = 4.4 %									

Table 1. Structure characteristics and discrepancy indices R of the initial solid solution $ZrN_{0.43}H_{0.38}$ in the space group $P\bar{3}m1$ Notice: B – individual thermal factor; n - occupancy of a position.

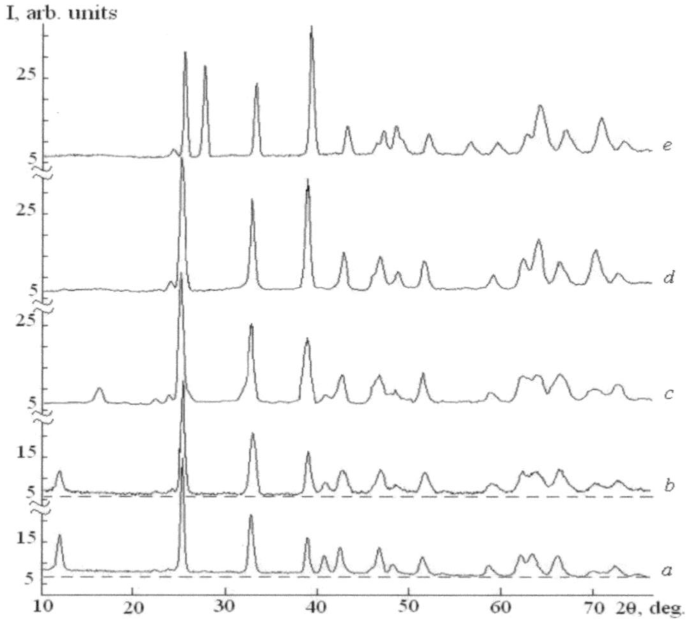

Fig. 2. Neutron diffraction patterns of solid solution: a – $Zr_2N_{0.86}H_{0.76}$ (α'; the space group $P\bar{3}m1$); b - $Zr_2N_{0.86}$ (α'; the space group $P\bar{3}m1$); c - $Zr_2N_{0,86}$ (α''; the space group Pnnm); d - $ZrN_{0,43}$ (L'_3; the space group $P6_3/mmc$); e –mixture of the α and δ phases.

annealed at 1000 °C for 3 h in evacuated and sealed quartz ampoule. As a result of annealing the dehydrogenated sample disintegrated into disordered solid solution of nitrogen in α-Zr – hexagonal α phase (structure of L'_3 type) and a cubic δ-phase (Fig.2*e*) corresponding to the equilibrium phase diagram of Zr-N system (Gusev, 2001). It should be noted that the annealing of hydrogenous $Zr_2N_{0,86}H_{0,76}$ phase under similar conditions does not lead to decay or change of crystal structure. Further annealing of the disintegrated sample (mixture of α+δ phases) at 375 °C for 36 h did not lead to recovery of the phase with anti-CdI_2 structure. As the crystal structure of α′-$Zr_2N_{0,86}$ phase has been stabilized by hydrogen (then all hydrogen atoms are removed) and it is not observed in the phase diagram, this phase is offered to name the hydrogen induced phase, unlike hydrogenous one. For the hydrogen induced phase α′-$Zr_2N_{0,86}$ the least value of divergence factor R_{Br}=6.4% is obtained with nitrogen atoms arranged over 1(a) octahedral positions (space group $P\bar{3}m1$) with the free metal parameter z_{zr}=0.255 diffe-ring from the ideal value z_{id} =1/4 and from z=0.243 for corresponding hydrogenous phase (Table 2). Hence, in the hydrogen induced phase metal atoms displacement directions change: from the plane filled with nitrogen atoms toward the plane of nitrogen vacancies (Fig. 3,*a* and 3,*b*). Hydrogen induced phase α'- $Zr_2N_{0,86}$ has the following lattice parameters: *a* =0.3264; *c*=0.5299 nm (*c/a* = 1.623). These values are less than those of corresponding hydrogenous phase. It should be noted that within the limits of experiment errors the axes relation *c/a* is the same for the hydrogen induced phase and the hydrogenous phase. As a result of hydrogen removal out of hexag-onal lattice its near isotropic compression occurs. Analysis of neutron diffraction pattern of the phase $Zr_2N_{0.86}H_{0.76}$ dehydrogenated at 450-650 °C (Fig. 2,*c*) shows that at these tempera-tures the ordered orthorhombic $Zr_2N_{0.86}$ phase with structure of anti-$CaCl_2$ type (α" phase) is formed. The diffraction pattern of this phase has been indexed in orthorhombic system (space group Pnnm) with the lattice parameters: a = 0.5632 nm ≈ $\sqrt{3}\,a_o$, b = 0.5252 nm ≈ c_o, c = 0.3256 nm ≈ a_o where a_o and c_o are lattice parameters of the initial hexagonal phase. The unit cell of this phase is shown in Fig. 3,*c*. A good agreement between experimental and calculated intensities (R = 6.3 %) is achieved given that four Zr atoms are in positions 4 (g), the nitrogen atoms occupy mainly positions 2 (a) and partially – positions 2 (c) (Table 3). At 650 °C the partial disordering of nitrogen atoms (~20 %), followed by their transition to octahedral interstices of another type – 1 (c), takes place. In neutron diagram of the sample, dehydroge-nated at 800 °C, has been reflections from hexagonal phase with structure of L'_3 - type (α-phase) (Fig. 2, d). The unit cell of the α - phase is of anti - $CaCl_2$ type (α"- phase) is formed.

Atom	Position	Atom coordinates				B, nm²	ΔB, nm²	n	Δn
		x	y	z	Δz				
Zr	2 (d)	1/3	2/3	0.254	0.002	0.0043	0.0008	2	
N	1 (a)	0	0	0		0.0032	0.0012	0.86	0.01
R_p = 2.9; R_{wp} = 3,9; R_{Br} = 6.4 %									

Table 2. Structure characteristics and discrepancy indices R of hydrogen induced phase α'-$Zr_2N_{0.86}$ (obtained at temperature 375 °C) in the space group $P\bar{3}m1$.

The diffraction pattern of this phase has been indexed in orthorhombic system (the space group Pnnm) with the lattice parameters: a = 0.5632 nm ≈ a_o, b = 0.5252 nm ≈ c_o, c = 0.3256 nm ≈ a_o where a_o and c_o are lattice parameters of the initial hexagonal phase. The unit cell of this phase is shown in Fig.3, *c*. A good agreement between experimental and calculated inten-sities (R = 6.3 %) is achieved given that four Zr atoms are in positions 4 (g), nitrogen atoms occupy mainly positions 2 (a) and partially – positions 2 (c) (Table 3). At 650 °C the partial disordering of nitrogen atoms (~20 %), followed by their transition to octahedral interstices of another type – 2 (c), takes place. In neutron diagram of the sample, dehydrogenated at 800 °C, has been reflections from hexagonal phase with structure of L'_3 - type (α phase) (Fig. 2, d). The unit cell of the α-phase is shown in Fig. 3, d. We also investigated mutual transformation of the metastable phases. The annealing of the α' hydrogen induced phase at temperatures 450-650 °C results in formation of α"phase. In other words, in this temperature range phase transition from α′ to α″ is observed. The annealing of the α″ phase at 400- 375 °C during 36 h does not lead to restoration of α′ phase. Hence, $\alpha' \xrightarrow{459^\circ C} \alpha''$ transition is monotropic (nonreversible) and α′ phase can be induced only by hydrogen thermoemission at temperature below that the temperature of α″ phase formation. By increasing the temperature up to 750 -800 °C, the α″ phase passes into α-phase (disordered oversaturated solid solution of nitrogen in a lattice of α-Zr) with structure of L'_3-type (Fig.2, d). This transition is enantiotropic (reversible) because the α″-phase is formed again by annealing the α- $ZrN_{0.43}$ phase at temperatures 650-450 °C. The annealing of the phases at 1000 °C results in their disintegration into to α and δ phases (Fig. 2, e). No any annealing of samples that have been decayed into α and δ phases does not lead to formation of α′, α″ and α phases in a single-phase kind. Therefore, the found phase transitions α′ →α″→ α′ occur in metastable state. The formed metastable phases and phase transitions between them are schematically shown in Fig. 4. Thus, by means of dehydroge-nation of the preliminary hydrogenated sample, three modifications of the oversaturated solid solution $ZrN_{0.43}$ which do not exist in the equilibrium phase diagram have been obtained.

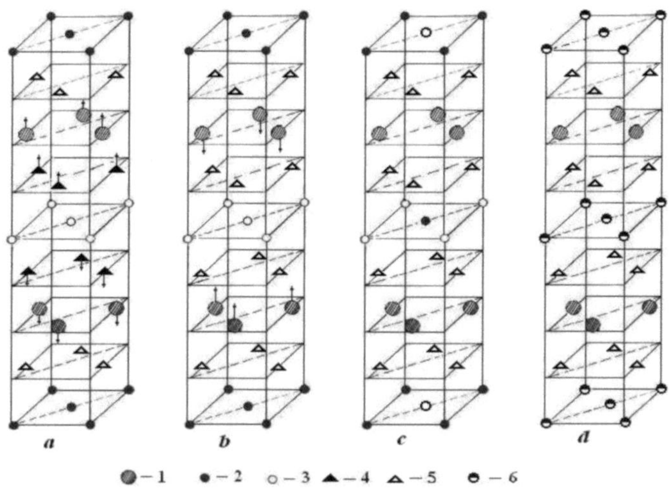

Fig. 3. Unit cells of the hydrogenous solid solution $ZrN_{0.86}H_{0.76}$ and of phases initiated by low-temperature vacuum extraction of hydrogen: 1–metal atoms; 2–nitrogen atoms; 3–nitrogen vacancies; 4–hydrogen atoms; 5–hydrogen vacancies; 6–statistic arrangement of nitrogen atoms.

Atom	Position	Atom coordinates					B, nm²	ΔB, nm²	n	Δn
		x	Δx	y	Δy	z				
Zr	4 (g)	0.324	0.002	0.271	0.001	0	0.39	0.06	4	
N	2 (a)	0		0		0	0.53	0.13	1.38	0.02
N	2 (c)	0		1/2		0	0.53	0.13	0.34	0.02
$R_p = 4.5$; $R_{wp} = 6.0$; $R_{Br} = 6.7$ %										

Table 3. Structure characteristics and discrepancy indices R of the α″-$Zr_2N_{0.86}$ hydrogen induced phase in the space group Pnnm.

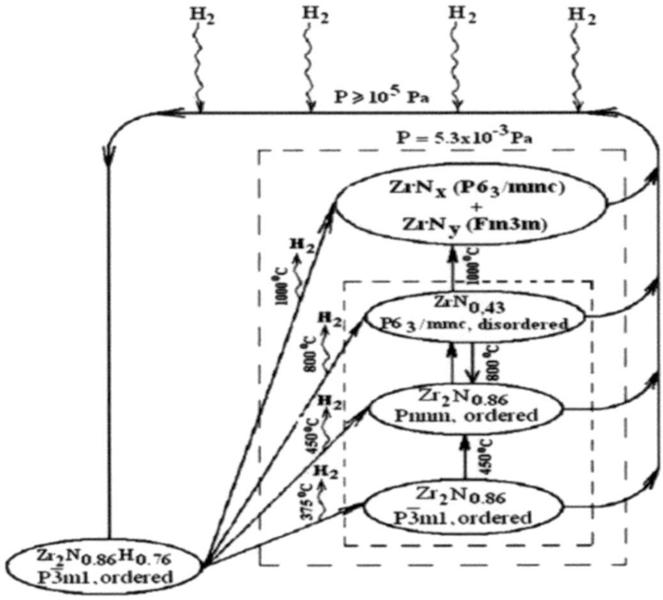

Fig. 4. Scheme of obtaining and phase transformations of metastable oversaturated solid solution in Zr-N system.

In Fig. 4 these modifications are outlined by dashed lines (the small rectangle). Hydrogenation of solid solution of the Zr-N system (Gusev, 2001) reveals that under a hydrogen pressure of P≥10^5 Pa, all samples, including biphase ones, form the sol. s. α'-$Zr_2N_{0.86}H_{0.76}$ (space group $P\bar{3}m1$). This inverse process also is shown in Fig. 4. In these cases, the exposure time required for hydrogen saturation up to the concentration H/Zr = 0.38 will be different at various temperatures. The scheme of reversible and nonreversible phase transformations in metastable state of the $ZrN_{0.43}$ is shown in Fig. 5.

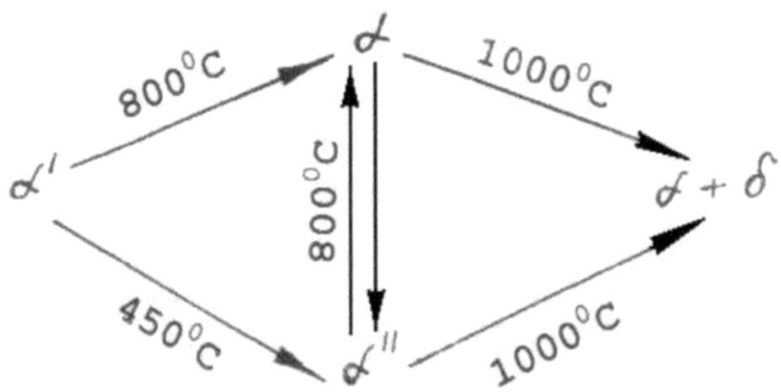

Fig. 5. Scheme of phase transformations on the basis of the hydrogen induced phase α'-$Zr_2N_{0.86}$.

3.2 Hydrogen thermoemission in Ti-N(C)-H systems

Low-temperature vacuum evacuation of hydrogen has been investigated in the ordered solid solution α'-$Ti_2N_{0.52}H_{0.30}$ (space group $P\bar{3}m1$) and the disordered solid solution

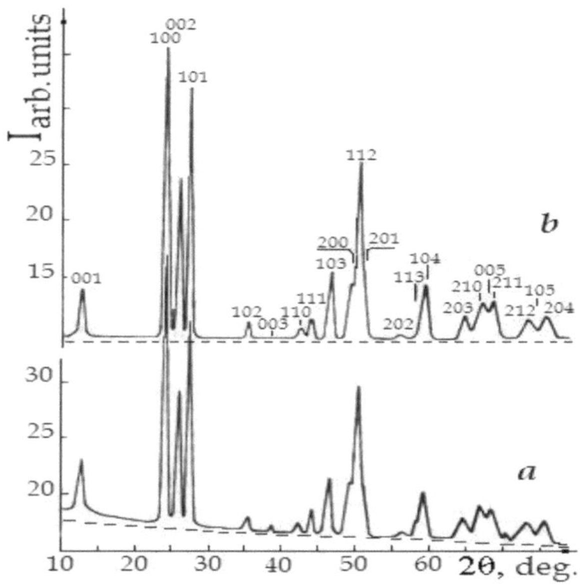

Fig. 6. Neutron diffraction patterns of the α'-$Ti_2N_{0.52}H_{0.30}$ phase (a) and the α'-$Ti_2N_{0.52}$ hydrogen induced phase (b).

α-$TiN_{0.26}H_{0.15}$ (space group $P6_3/mmc$) (Khidirov et al. 1991). In Fig.6, *a*, the neutron diffractogram of the ordered solid solution α'- $Ti_2N_{0.52}H_{0.30}$ is shown. A declining incoherent background is evident. After vacuum annealing at 370 ^{0}C this background disappears (Fig. 6,*b*). This provides evidence for the removal of hydrogen from the lattice, findings are confirmed by chemical analysis. As in the neutron pattern (Fig. 6, *b*), all selective reflections are the same. The ordered structure of the solid solution has not changed. In this way, the "frozen" hydrogen-free metastable phase of Ti-N system – the hydrogen induced α'-$Ti_2N_{0.52}$ phase is obtained. Processing the neutron data by Rietveld method has shown that the value of the free metal parameter z changes in the formed hydrogen induced phase. It changes from z = 0.237 - for the hydrogenous solid solution (that is less than z = ¼ for the ideal position) to z = 0.260 - for the hydrogen induced phase (i.e. greater than z ideal). Annealing the obtained phase in an evacuated and sealed quartz ampoule at T> 370 ^{0}C results in decay corresponding to the equilibrium phase diagram of the Ti-N system (Gusev, 2001). Low-temperature vacuum evacuation of hydrogen was also studied in solid soluti-ons $TiC_{0,45}H_{0,90}$, $TiC_{0,45}H_{0,64}$, and $TiC_{0.35}H_{0.43}$ (Khidirov et al., 2008). In the neutron patterns of the initial solid solutions (Fig. 7), superstructure reflections are visible together with structural reflections and they are indexed within space group $P\bar{3}m1$. According to neu-tron structure analysis of the investigated solid solutions, thermoemission of H is virtu-ally not observed up to 250 °C. In solid solutions with a large carbon concentration, the separation of pure titanium and face-centered cubic phase (most probably carbohydride TiC_xH_y) was found above 275°C. By further increasing the temperature, the quantity of precipitate phases increases; up to 400°C, the incoherent background slope is kept constant. Thus, solid solutions with high concentration of carbon (and hydrogen) – $TiC_{0,45}H_{0,90}$, $TiC_{0,45}H_{0,64}$, are thermostable in vacuum only at temperatures T < 275 °C, and in HCP and face-centered cubic (FCC) phases hydrogen remains up to 400°C. In the neutron diagram of the solid solution $TiC_{0,35}H_{0,43}$ (α'-$Ti_2C_{0.70}H_{0.86}$) after vacuum annealing at 275°C, there are the same reflections which have been observed in the neutron pattern of the initial hydrogenous sample (Fig. 8); incoherent background, however, is completely absent.

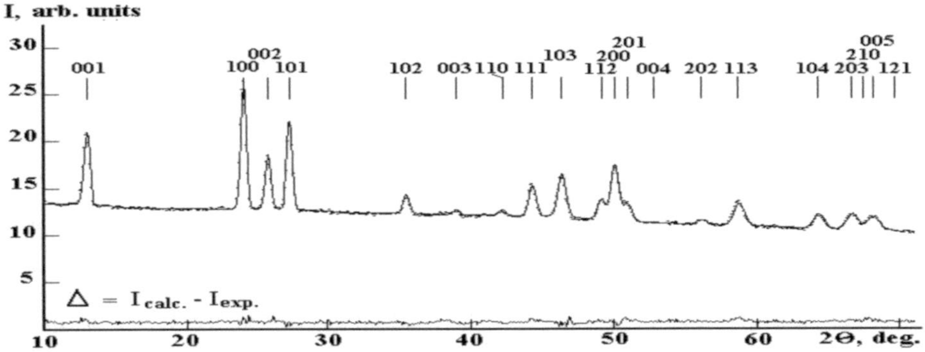

Fig. 7. Neutron diffraction patterns of the α'-$Ti_2C_{0.70}H_{0.86}$ - phase: solid line - calculated, dots - experimental, Δ – differential one. Miller indices are indicated above the reflections.

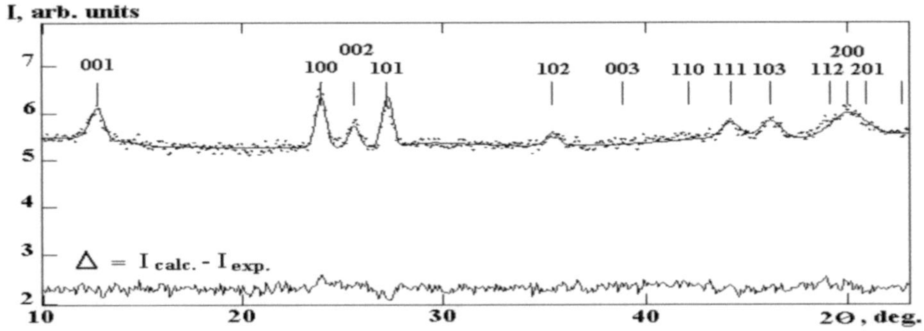

Fig. 8. Neutron diffraction patterns of the ordered α'-$Ti_2C_{0.70}$ – phase. The notations are the same as in Fig. 7. Calculated diagram was obtained in sp. gr. $P\bar{3}m1$ with $z_{Ti} = 0.305$ however, is completely absent.

Analysis of the diffraction compositions corresponds to the formula $TiC_{0,35}$. The initial ordered structure, however, remains unaltered. Therefore, hydrogen thermoemission at 275°C results in the formation of hexagonal ordered α'-$Ti_2C_{0.70}$ - phase in a metastable state. Analysis of the diffraction pattern by the Rietveld method proved a lack of hydrogen and showed that the sample composition corresponds to the formula $TiC_{0,35}$. The initial ordered structure, however, remains unaltered. Therefore, hydrogen thermoemission at 275°C results in formation of the hexagonal ordered α'-$Ti_2C_{0.70}$ - phase in a metastable state. It is necessary to note the large errors in the determination of lattice parameters and the high value of divergence factors (R_{Br} =13,6 %). Evidently, these results are caused by the distortion of peak forms and wide scatter of points in the neutron diffraction pattern; this is because of strong static distortions of the lattice. The large errors do not allow comparison of the lattice parameters of hydrogen induced phase $Ti_2C_{0.70}$ and the initial hydrogenous phase.

In the hydrogen induced α'-$Ti_2C_{0.70}$ phase the sign of metal atoms displacement (z_{Ti} = 0,305) relative to the ideal position (z_{id} = 1/4) changes in comparison with the hydrogenous α'-$Ti_2C_{0.70}H_{0.86}$ phase (z_{Ti} = 0,237). If the value of z_{Ti} in the hydrogen induced phase $Ti_2C_{0.70}$ is assumed to be the same as in the hydrogenous phase α'-$Ti_2C_{0.70}H_{0.86}$, an essential growth of R_{Br} is also observed (up to 30 % at z_{Ti} = 0.250 and 33 % at z_{Ti}=0.237). Thus, in the hydrogen-induced α'-$Ti_2C_{0.70}$-phase, just as in the hydrogen induced α'-$Ti_2N_{0.52}$ phase (Khidirov et al., 1991), the sign of metal atoms displacement relative to their ideal position changes in comparison with corresponding hydrogen-containing phases. As a result of this research, it can be concluded that in solid solutions of the Ti-C(N)-H system, complete thermoemission of hydrogen with a conservation of initial symmetry (caused by hydrogen) is observed in strongly defective structures. Apparently, in strongly defective solid solutions the interaction forces between atoms are weakened, and the temperature of hydrogen evacuation (T_{ev}) is below that of redistribution of interstitial atoms or of recrystallization (T_{recr}): $T_{ev} < T_{recr}$.

4. Neutron diffraction study of tryhydroxides rare-earth metals R(OH)₃ and metasable crystals of trioxides of rare-earth metals R[O₃] (R is La, Pr or Nd)

4.1 Neutron diffraction study of initials samples of tryhydroxides R(OH)₃

The investigated samples were obtained as a result of oxidation in air of the corresponding rare earth metal having 99,95 % purity. The X-Ray phase analyses showed that the samples contained mainly the trihydroxide of the rare earth metal which is consistent with the results of other authors (Portnoy, Timofeeva, 1986). For example, the neutron diffraction pattern of $Pr(OH)_3$ is represented in Fig. 9, *a*) . The steep slope of the background is caused

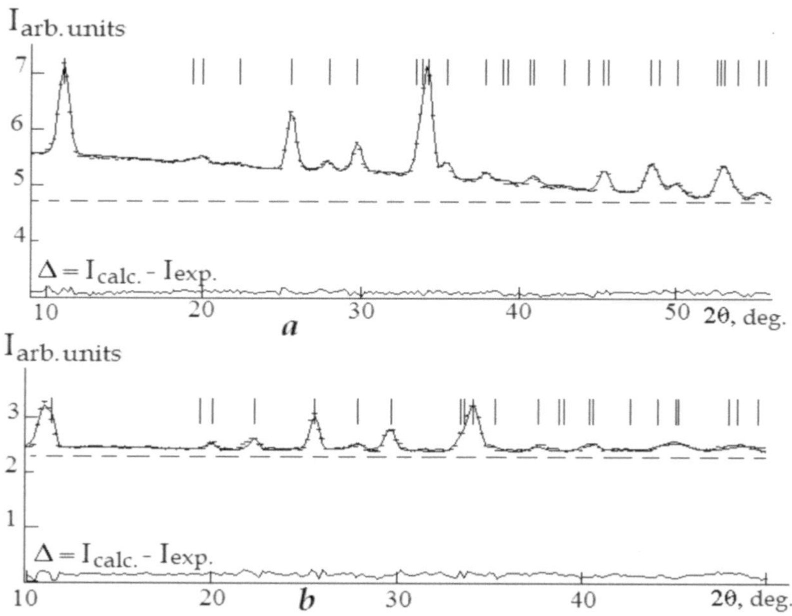

Fig. 9. Neutron diffraction patterns of the $Pr(OH)_3$ and the $Pr[O_3]$ (space group P_3/m). The notations are the same as in Fig. 7.

of neutrons by hydrogen nuclei (Bacon, 1975). Diffraction patterns of other trihydroxides were similar. All rare earth metals from trihydroxides $R(OH)_3$ (R=La to Yb) having UCl_3-type hexagonal structure (Portnoy, Timofeeva, 1986). The structure can be described in $P6_3/m$ space group, where the rare earth element atoms R occupy 2(d) positions and (OH) complexes occupy 6(k) positions having free parameters x=0.29, y=0.38 (Schubert, 1971). The crystal structure of R $(OH)_3$ phases was studied by X-ray diffraction. It was difficult to determine the hydrogen atom coordinates using the X-ray method, because the scattering amplitude of hydrogen atoms is much less than that for rare earth metals. That is why the coordinates of the spherical anions $(OH)^-$ were ascribed to the hydrogen atoms. It seemed reasonable to try to determine the coordinates of hydrogen atoms using neutron diffraction because in case of neutrons the scattering amplitude of hydrogen nuclei ($b_H = -0.374\times10^{-3}$ nm) is comparable with that for rare earth metals ($b_{La} = 0.83\times10^{-3}$ nm, $b_{Pr} = 0.44\times10^{-3}$ nm, $b_{Nd}= 0.72\times10^{-3}$ nm) (Bacon, 1975). The coordinates of hydrogen atoms in rare earth metal trihydroxides R $(OH)_3$ (R is La, Pr, Nd) were determined using the neutron diffraction method (Khidirov & Om, 1991). All diffraction peaks in the neutron diagrams could be indexed in the hexagonal unit cell with the lattice parameters which are given in Table 4.

Sample	*a* (nm)	*c* (nm)	*c/a*
$La(OH)_3$	0.6529 ± 0.0006	0.3852 ± 0.0003	0.5895 ± 0.0003
$Pr(OH)_3$	0.6458 ± 0.0005	0.3771 ± 0.0002	0.5837 ± 0.0004
$Nd(OH)_3$	0.6437 ± 0.0005	0.3747 ± 0.0002	0.5821 ± 0.0004

Table 4. The lattice parameters of $R(OH)_3$.

Atom	Position	x	y	z
2 R	2(d)	1/3	2/3	1/4
6 O	6(k)	0.30 ± 0.002	0.385 ± 0.002	1/4
6 H	6(k)	0.16 ± 0.003	0.288 ± 0.004	1/4

Table 5. Structure characteristics of $R(OH)_3$ in the space group $P6_3/m$.

Sample	R_p , %	R_{wp} , %	R_{Br} , %
$La(OH)_3$	0.74	0.93	6.59
$Pr(OH)_3$	0.54	0.72	7.01
$Nd(OH)_3$	0.88	1.03	7.32

Table 6. R-factors for $R(OH)_3$ in the $P6_3/m$ space group.

The lattice parameters are close to those obtained by other authors (Portnoy & Timofeeva, 1986)] except c of $Nd(OH)_3$. A decrease in the ratio c/a with increasing metal atom number corresponds to the reduction of atomic radius in line of the La –Pr–Nd. The least R–factor was obtained for the model in the $P6_3/m$ space group (Table 5). For the investigated samples this model yielded, the following R – factors are given in Table 6. The R–factor values can be takes into account the high background level of the neutron diffraction patterns caused by incoherent scattering of the hydrogen nuclei and thermal diffusion scattering. The distance between the oxygen and hydrogen atoms in the O–H pairs equals (0.082 ± 0.007) nm. A difference curve (observed minus calculated) is shown in Fig. 9, *a*. It should be noted that if the same coordinates are formally ascribed to the oxygen and hydrogen atoms the R– factor increases considerably. For example, for $La(OH)_3$ R_{Br}=35.7 %. Thus, our results show that, though hydrogen and oxygen atoms occupy the same positions 6(k), their coordinates differ.The unit cell of $R(OH)_3$ is shown in Fig. 10. Oxygen and hydrogen atoms form dumb-bells oriented as is the oxygen atom is attracted and the hydrogen atom is repelled by the nearest metal atoms.

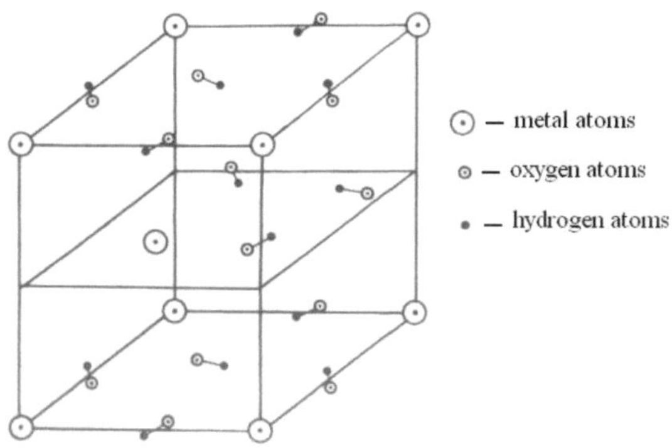

Fig. 10. The unit cell of $R(OH)_3$.

4.2 Neutron diffraction study of hydrogen thermoemission from crystals of tryhydroxides rare-earth metals $R(OH)_3$

For rare-earth metal trihydrooxides $R(OH)_3$ three characteristic temperatures are observed for hydrogen extraction at continuously pumped out high vacuum.

1. The temperature of practically complete hydrogen evacuation from the lattice without change in symmetry of crystal - T_{evac} (140-150 ^{0}C), at which the crystal structure does not change, but is significantly deformed (Fig. 9, *b*). Neutron diffraction calculation of samples after dehydrogenation at 130-150 ^{0}C by means of full-profile Rietveld analysis demonstrates no hydrogen in the sample, though the crystal structure is conserved (Fig. 9, *b*). Significant decrease of diffraction refraction intensities shown by neutron diffraction analysis after vacuum dehydrogenation can be explained, first of all, by lack of hydrogen atoms in lattice; secondly, by lower gross number of dehydrogenated samples. The latter is related to the fact that for acceleration of the process of hydrogen separation from powder $R(OH)_3$ the quantity of samples was taken three times less than that taken for the original powders of $R(OH)_3$. After hydrogen removal one can observe widening of half-width and distortion of diffraction peaks form, which is caused by pronounced deformations in crystal lattice after hydrogen removal. The best agreement between experimental and calculated intensities of neutron-diffraction reflections (Fig. 9, b) and minimal errors in structure determination (R) can be obtained only assumed that these compounds are "trioxides" containing $R[O_3]$. It is worth of mentioning that the temperature of hydrogen evacuation from lattice in all trioxides $R(OH)_3$ within errors in temperature determination ($\Delta T = \pm 12$ C) is almost the same. This can be explained by the fact that rare-earth metals La, Pr and Nd have similar valence electron shells and does not differ in sizes of atoms, where as trihydrooxides $R(OH)_3$ have isomorphic structure. The results of neutron diffraction analysis calculations $Pr[O]_3$ within space group $P6_3/m$ are presented in Table 7 (Khidirov, 2011). One should notice large errors in determination of lattice parameters and large value of R_{Br}.. It is apparently caused by

distortion of peaks forms due to strong statistical lattice deformations appearing after hydrogen removal. Formation of such a compound in R - O systems contradicts to the valence conservation principle. Oxygen is double-valence, and rare-earth metals can be either three or four-valence. But all of them in $Rr(OH)_3$ are three-valence. It is obvious that "trioxides" $R[O_3]$ have broken bonds and unpaired electrons, likewise that of radicals, that is they have excessive negative charge: $R^{3+}[O_3]^{6-}$. This material can be stable only at relatively low temperatures and in medium free of hydrogen. Indeed, annealing of $R[O_3]$ or dehydrogenation of $R(OH)_3$ at temperatures of 150 ^{0}C longer that 24 hours leads to its amorphisation, which is pointed by lack of selective diffraction reflections in neutron diffraction analysis, and formation of small diffuse reflection at angles of $2\theta = 28 - 38$ degrees. Similar phenomenon takes place at even slight increase of temperature (up to 180 ^{0}C). Obtained "trioxide" $R[O_3]$ due to its valence instability "tends" to trap three more protons. This can be accomplished by trapping in lattice of three hydrogen ions at first instance. Therefore, metastable "trioxides" in atmosphere, apparently, interact with water molecules, and retransform step by step back in the trihydrooxide $R(OH)_3$, by compensating the excessive negative charge by three hydrogen atoms. This is demonstrated by the neutron diffraction analysis calculations $R[O_3]$, taken in 30 days after obtaining neutron diffraction patters and by repeated formation of incoherent background on the neutron diffraction patterns. In other words, the neutron diffraction patterns of $R[O_3]$ samples after exposition in atmosphere at temperatures 285 - 290 K (temperature in the reactor hall in winter) within 30 days are both qualitatively and quantitatively become identical to the corresponding neutron diffraction analysis patterns for $R(OH)_3$. Hence, one can conclude that in atmosphere the «trioxide» $R[O_3]$ «self-cures» until complete restoration of the trihydro-oxides of rare earth metals $R(OH)_3$. (Obviously, the reaction rate depends on temperature). Since the hydrogen concentration on Earth's surface is rather low, one can assume that $R[O]_3$ captures hydrogen, mainly from, water vapor in the air, leaving oxygen molecules free:

$$4R^{3+}[O]_3^{6-}(\text{powder metastable crystal}) + 6H_2O = 4R(OH)_3(\text{powder stable crystal}) \rightarrow 3O_2\uparrow$$

Atom	Number	Pozition	X	y	z
Pr	2	2(d)	2/3	1/3	1/4
O	6	6 (k)	0.376±0.002	0.461±0.002	1/4
a= 0.658 ±0.018 HM; *c* = 0.381 ±0.006 HM; R= 0.60 %; R = 0.81 %; R = 13,7%					

Table 7. Structural characteristics and R-factors of the "trioxide" $Pr[O_3]$ the model in the $6_3/m$ space group.

Therefore, «trioxide» $R[O_3]$ in atmosphere can simultaneously be a hydrogen absorber and oxygen generator. Separation (by low-temperature removal) of hydrogen from $4R(OH)_3$ can again lead to restoration of its capabilities to be a simultaneous hydrogen accumulator and oxygen generator in a medium containing water molecules. The cycle described above can be accomplished many times. It is shown in Fig. 11.

2. Amorphisation temperature-$T_{amorph.}$ (180 ^{0}C), at which the selective diffraction reflection disappear completely and diffuse reflection appear in the neutron diffraction pattern (Fig. 12). The experiments revealed that amorphous $R[O_3]$ in atmosphere crystallize spontaneously and transform into trihydrooxides of corresponding rare-earth metals in 1-1.5

months at the temperature of the reactor's hall in winter (15-20 ^{0}C). It was interesting to study the kinetics of this process. We therefore periodically generated the neutron diffraction patters of amorphous La[O_3] (Fig. 13). During this process, the measured sample was located continuously in the 6 mm in diameter vanadium cylinder with an open cap. Fig. 13 shows some peculiar neutron diffraction patterns taken within 50 days.

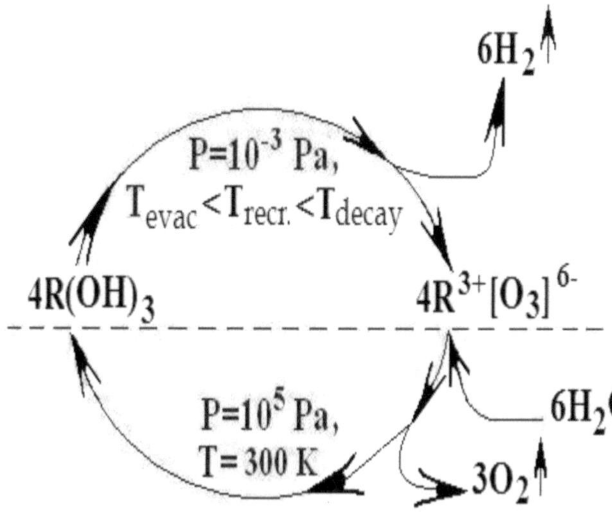

Fig. 11. The scheme of a cycle of reception «trioxides» R [O_3] from R (OH)$_3$ in continuously pumped out high vacuum and repeated formation R (OH) $_3$ from «try-oxides» R [O_3] by self-curing in atmospheric conditions.

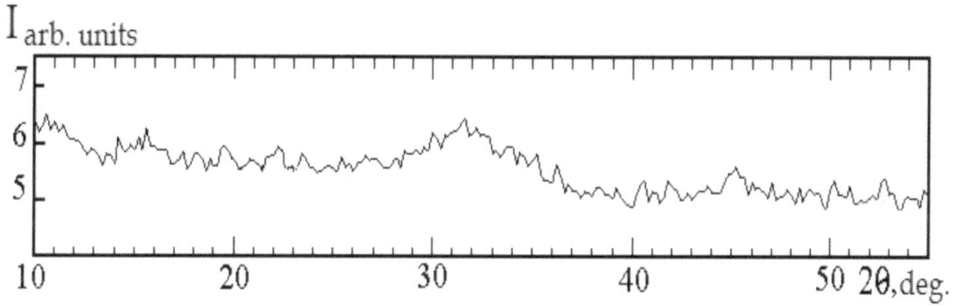

Fig. 12. The neutron diffraction pattern of the amorphous Pr[O_3]

3. Recrystallization temperature - $T_{recr.}$ (≥ 210^{0}C), at which cubic oxide phase of corresponding REM is formed (PrO_{2-x}-phases instead of $Pr(OH)_3$ or La_2O_3 - phases instead of $La(OH)_3$ (Fig. 14). It is natural to assume that at such a temperature the lattice is abandoned not only by hydrogen atoms, but some part of oxygen atoms. In amorphous phase, at the very first stage, the nucleation centers of face-centered cubic oxide phase are formed (Fig. 13). Further, the oxide La_2O_3 - phase intensity increases and diffuse reflection intensities decrease demonstrating that the formed oxide phase grows instead of amorphous phase. It is interesting to observe that on the 5-th day after nucleation of the oxide phase, it does not appear in the neutron diffraction patterns. However, one can observe some small diffraction

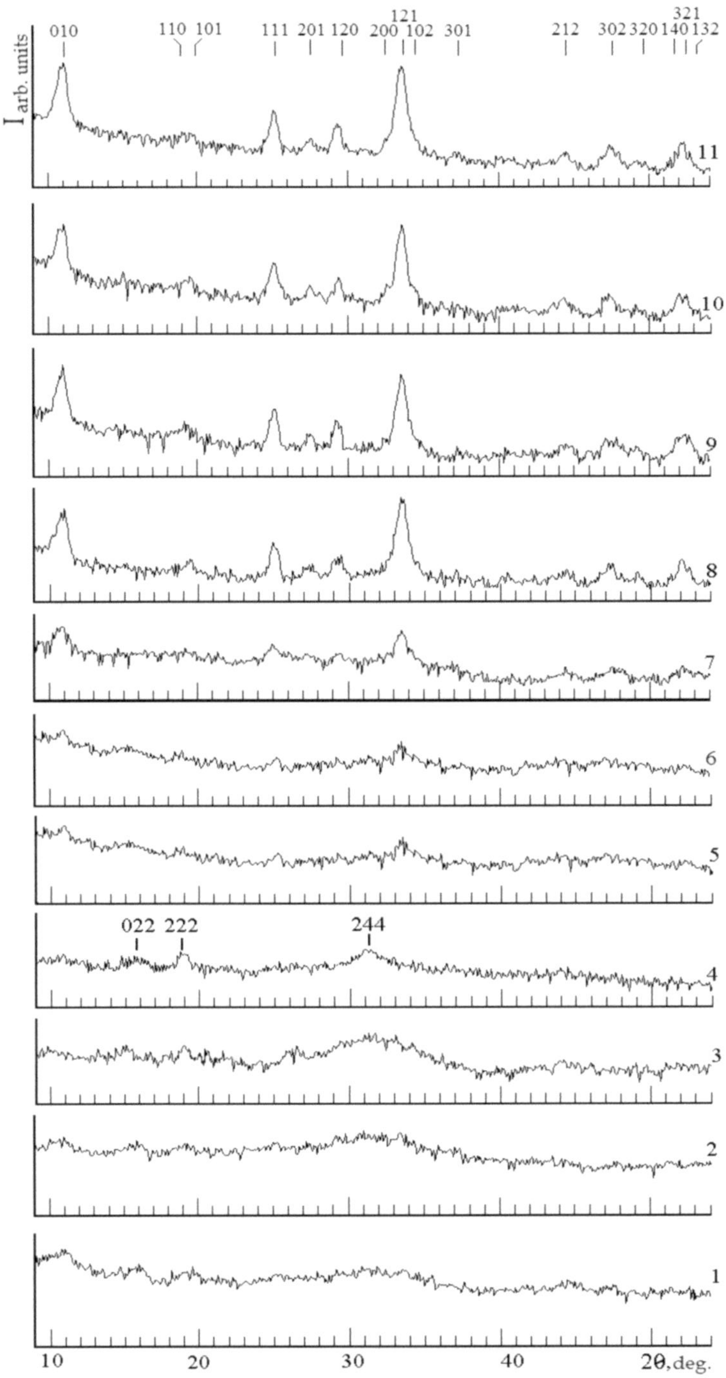

Fig. 13. Neutron diffraction patterns of amorphous La $[O_3]$, removed during spontaneous crystallization and its transformation in La $(OH)_3$: amorphous La $[O_3]$ (1); after its endurance in atmospheric conditions within 5 (2), 10 (3), 15 (4), 20 (5), 25 (6), 30 (7), 35 (8), 40 (9), 45 (10), 50 (11) days, accordingly.

peaks of lanthanum trihydrooxide $La(OH)_3$ (Fig. 13-5). These peaks grow in some time, whereas the diffuse reflections decrease and finally disappear (Fig. 13-8, 13-9, 13-10). In the beginning the diffraction peaks of trihydrooxide $La(OH)_3$ are strongly distorted, but finally they acquire ideal form. Thus, duration of processes of spontaneous crystallizationand complete transformation of amorphous $La[O_3]$ into $La(OH)_3$ in atmosphere at temperatures 15-20 ^{0}C is approximately 45-50 days. Certainly, the observed reaction rate depends on air temperature and humidity. The discovered interesting phenomenon consists of spontaneous crystallization of amorphous matter $R[O_3]$, formed by hydrogen thermoemission, and its reverse transformation into trihydrooxide $R(OH)_3$ in normal atmosphere conditions with selective absorption of hydrogen from water molecules and gaseous environment, having hydrogen partial pressure.

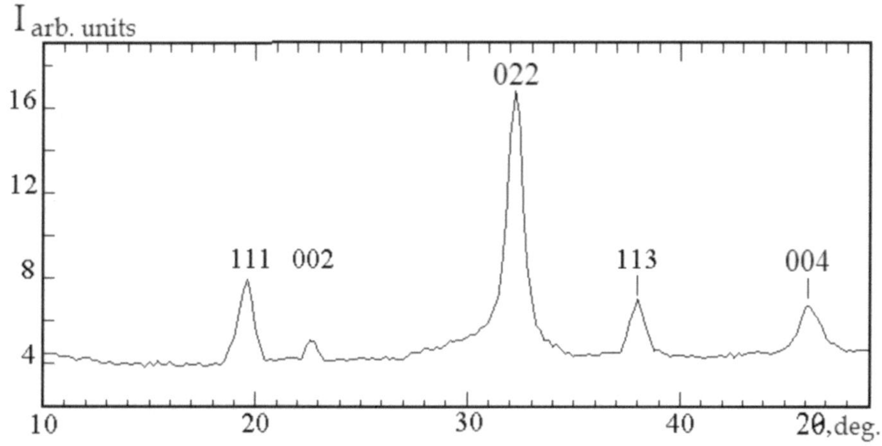

Fig. 14. Neutron diffraction pattern of the cubic PrO_{2-x} phase.

5. Low-temperature vacuum evacuation of hydrogen out of crystal lattice of interstitial alloy TiH_2

We have carried out a neutron diffraction study of the phenomenon of hydrogen thermo-emission out of a lattice of the cubic phase of titanium dihydride $TiH_{1.95}$ at temperatures 50 ÷ 250°C in continuously pumped out vacuum not worse than ~ 5,3 ·10^{-3} Pa. An initial sample $TiH_{1.95}$ has been prepared by direct hydrogenation of a powder of metal by hydrogen at pressure 10^5 Pa. The purity of initial titanium iodide was 99.98 %. For more uniform distribution of hydrogen the samples after saturation were annealed in the pumped out and sealed quartz ampoule during 48 h at temperature 120°C. As X-ray diffraction patterns have shown, the obtained sample was single-phase and homogeneous on composition and had face-centered cubic lattice with parameter a = 0.4454± 0.0001 nm. According to the work (Kornilov, 1975), this lattice parameter corresponds to face-centered cubic phase of compound $TiH_{1,95}$ (δ-phase). Its processing by Rietveld full-profile analysis confirms the data of the work (Kornilov, 1975): the crystal structure of the cubic phase is described within the framework of space group Fm3m. Dehydrogenating of $TiH_{1.95}$ began from the room temperature with a step of 25°C and with exposure time at each temperature not less than 24 h. After dehydrogenating neutron diffraction patterns were surveyed at each temperature,

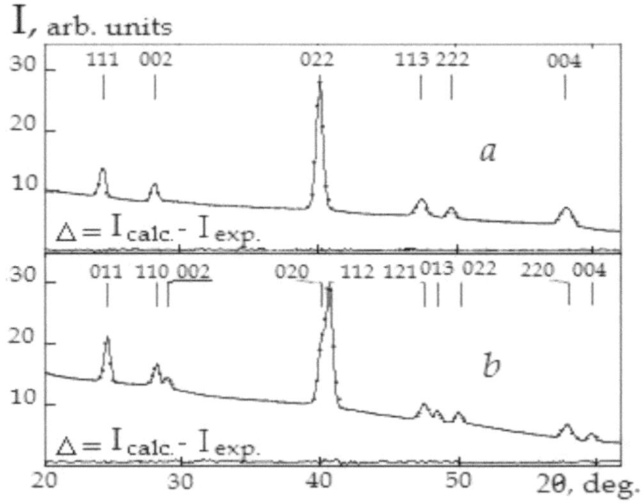

Fig. 15. Neutron diffraction patterns of the sample of titanium hydride $TiH_{1.95}$: (a) – initial and (b) – after vacuum thermal treatment at 200 °C. The notations are the same as in Fig. 7.

and content of hydrogen was controlled by neutron diffraction data using both minimization of the R-factors and the relation of intensities of maxima with even (002) and odd (111) Miller indices of the face-centered cubic phase. Up to temperature 100°C in diffraction pattern neither qualitative, nor quantitative changes were not found. According to neutron structure analysis, at evacuation of hydrogen in the temperature interval of 100 - 175°C the decrease of hydrogen concentration from H/Ti=1,84 up to 1,74 (± 0,04) is observed at practically constant lattice parameter of the initial cubic phase (a = 0.4454±0.0001 nm for $TiH_{1.74}$). When evacuating hydrogen at temperature 200°C the splitting of some diffraction reflections has been found out (Fig. 15). The analysis of character of splitting showed that cubic lattice of dihydride has transformed to tetragonal. According to X-ray structure analysis, the parameters of the new unit cell were a =0.3182 ± 0,0005; c = 0.4413 ± 0,0007 nm, degree of tetragonal alteration is c/a = 1,386. The unit cell parameters of body-centered tetragonal (BCT) phase relate to the initial (cubic) one by the relation: $a_t \approx a_c/\sqrt{2}$; $c_t \approx c_c$. Similar structure was found for compositions $TiH_{1.75\text{-}2.00}$ (Kornilov, 1975) and also was described in (Bashkin et al., 1993) where it was considered in face-centered tetragonal (FCT) aspect. The recalculation of the parameters received by us for face-centered tetragonal unit cell gives: a_{FCT} =0.4418; c_{FCT} = 0.4472 nm; $c/a \approx$ 1.01. The crystal structure of the bode-centered tetragonal phase is described in the framework of space group I4/mmm. The structure characteristic of the bode-centered tetragonal phase, determined from neutron diffraction data and the divergence factors of structure determination are given in Table 8. Preservation of intense incoherent background in the neutron diffraction pattern testifies that in the lattice there was still a lot of hydrogen. If to increase temperature up to 225°C, it is observed the significant reduction of incoherent background in the neutron pattern and also a decay of quasistoichiometric face-centered cubic $TiH_{1.95}$ phase onto nonstoichiometric the FCC cubic $TiH_{2\text{-}x}$ phase and pure Ti. Thus, a change of lattice symmetry occurs before the complete removal of hydrogen; in the $TiH_{1.95}$, recrystallization is lower than temperature of evacuation: $T_{recryst} < T_{evac}$ (at vacuum of P ≈ ~ 5,3·10^{-3} Pa). The obtained results allow one to conclude that vacuum 5,3x10^{-3} Pa (in which TiH_2 dehydrogenating was carried out) is insufficient for removal of hydrogen out of a lattice of metal hydride without a change of crystal symmetry and for obtaining pure titanium with face-centered cubic lattice. For achievement of the given purpose at the temperatures lower than the temperature of recrystallization or decay it is necessary to create high vacuum (about ~ $10^{-9} \div 10^{-12}$ Pa).

AAtoms	Number of atoms	Positions	x/a	y/b	z/c	$B_{eff.}$, nm^2
Ti	2	2 (a)	0	0	0	0.0041± 0,0008
			1/2	1/2	1/2	
H	3.48± 0,04	4 (d)	0	1/2	1/4	
			1/2	0	1/4	
			1/2	0	3/4	
			0	1/2	¾	
R_{wp} = 3,8 %; R_{Br} = 3.7 %; R_F =2.8 %						

Table 8. Structure characteristics within the framework of space group I4/mmm and divergence factors of the bode-centered tetragonal phase of $TiH_{1.74}$ obtained at the temperature 200°C under persistent evacuation.

6. Discussions

The formation of observed hydrogen induced phases and phase transitions in metastable state of Zr-N-H system may be explained as follows. Under 375°C hydrogen atoms are removed out of the lattice but the configuration of relatively heavy atoms of the matrix does not change. As a result it is possible to obtain the "frozen" structure stabilized previously by hydrogen using thermoemission. At temperatures 450°C ≤ T ≤ 650°C the diffusive mobility of nitrogen atoms increases, making possible the ordering type change. Further temperature rise causes large displacements of atoms from lattice sites, leads to disorder in distribution of nitrogen atoms. The disordered state occurs typically for oversaturated solid solutions. This state precedes decay process starting above 800°C. At temperatures below 700°C decay process is strongly hindered, as it requires: nitrogen atoms to move at macroscopic distances and the power barrier to be overcome. It is necessary for formation of nucleus of new phase, namely, cubic zirconium nitride. However, ordering process needs only redistribution of nitrogen atoms and their movement at distances about interatomic space. As a result, during low-temperature annealing (400°C<T<800°C) the oversaturated solid solution does not decay (that would correspond to a minimum of free energy) instead passes into another ordered or disordered state corresponding to a local minimum. Hence, in solid solutions of nitrogen in α-Zr, having high decay temperatures, a number of phase transformations in metastable state can be observed at temperatures below T_{decay}. Fig. 4 shows that depending on the temperature of hydrogen evacuation it is possible to obtain several different phases. The decay temperature for solid solutions ZrN_x is higher than for TiN_x. Therefore in the Zr-N system at $T \leq T_{decay}$ it is possible to change the ordering type (anti-CdI_2→anti-$CaCl_2$). It should be noted that isotropic lattice compression takes place in the induced phases where the structure type of initial hydrogenous phase is preserved (Table 9). Thus, during formation of hydrogen-induced phase two observations were made: isotropic compression of the crystal lattice and change of metal atoms displacement directions. Apparently, such

Hydrogenous phase	System	Space group	Before hydrogen Evacuation			$T_{evac.}$, ⁰C	After hydrogen Evacuation			
			a, nm	*c*, nm	*c/a*		Hydr. induced phase	*a*, nm	*c*, nm	*c/a*
Ordered $TiN_{0,26}H_{0,15}$	HCP	$P\bar{3}m1$	0.2978	0.4795	1.61	370	$Ti_2N_{0,52}$	0.2956	0.4765	1.61
Disordered $TiN_{0,26}H_{0,15}$	HCP	$P6_3/mmc$	0.2978	0.4795	1.61	320	$TiN_{0,26}$	0.2956	0.4765	1.61
$ZrN_{0.43}H_{0,38}$	HCP	$P\bar{3}m1$	0.3276	0.5324	1.63	400	$ZrN_{0,38}$	0.3266	0.5299	1.63
$ZrN_{0,27}H_{0,35}$	HCP	$P\bar{3}m1$	0.3274	0.5240	1.60	350	$ZrN_{0,27}$	0.3266	0.5226	1.60
	HCP	$P\bar{3}m1$	0.3274	0.5332	1.62			0.3270	0.5299	1.62
$TiC_{0,38}H_{0,43}$	HCP	$P\bar{3}m1$	0.3024	0.4894	1.62	275	$Ti_2C_{0,76}$	0.3020	0.4990	1.62
$Pr(OH)_3$	HCP	P_3/m	0.6458	0.3771	0.58	150	$Pr[O_3]$	0.6580	0.381	0.58
$TiH_{1,95}$	FCC	*Fm3m*	0.4454			100	$TiH_{1,85}$	0.4454		
$TiH_{1,95}$	FCC	*Fm3m*	0.4454			170	$TiH_{1,75}$	0.4454		

Table 9. Some structure characteristics of solid solutions before and after hydrogen thermoemission.

effects may be explained by various characters of chemical bond of nitrogen and hydrogen with metal atoms. Hydrogen atoms create isotropic field of elastic stresses resulting in isotropic extension of the structure. One may think that after removing hydrogen atoms isotropic compression of a lattice takes place. According to (Grigorovich & Sheftel', 1980), chemical bonds in Me-N systems are realized so that nitrogen atoms send a part of valence electrons to the conductivity zone of the metal, and thus nitrogen and metal atoms own common valence electrons. It is obvious that positively charged Me and N ions repel each other. According to (Geld, 1985), hydrogen atoms are donors as compared to metal atoms. If this is still rue in presence of nitrogen, metal atoms become negatively charged and will be attracted to nitrogen atoms. Hence the direction of displacement for metal atoms in hydrogenous phase will be the opposite to that of metal atoms in hydrogen induced phase. Hydrogen thermoemission in crystals can be explained as relief of potential wells in crystals (Fig. 16). Heavy atoms and hydrogen atoms have different initial wells. Hence to overcome the potential barrier less energy is required for hydrogen compared to heavier atoms.

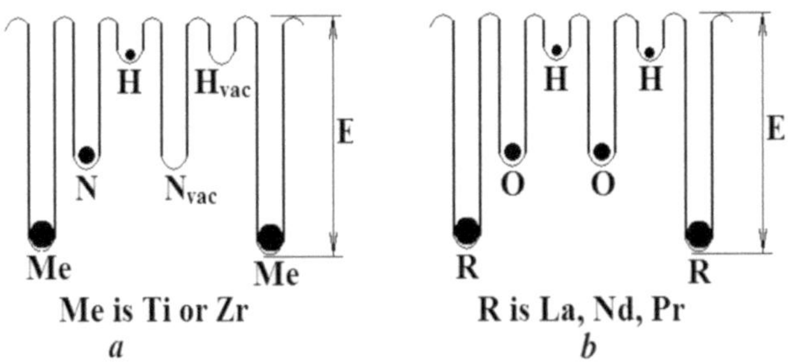

Fig. 16. The relief of potential wells in solid solutions $MeC(N)_xH_y$ (*a*) and in rare earth metal trihydroxides $R(OH)_3$ (*b*).

7. Conclusions

1. It is found a new phenomenon – complete thermoemission of hydrogen out of a H-containing ground crystals which proceeds without altering of crystal symmetry; it is proposed that these artificially-forming metastable phases be named by hydrogen induced phases.

2. At low-temperature (T = 375 °C) high-vacuum evacuations of hydrogen out of lattice of the solid solution $ZrN_{0.43}H_{0.38}$ the new, hydrogen-induced α'- $Zr_2N_{0.86}$ phase was obtained that is absent in the phase diagram of Zr-N system; it is impossible to obtain it by traditional methods.

3. During formation of hydrogen induced phase two observations are made: isotropic compression of the crystal lattice and change for metal atoms displacement directions. Apparently, such effects may be explained by various characters of chemical bond of nitrogen and hydrogen with metal atoms.

4. At rise in temperature of H evacuation two more modifications of the same compound ($Zr_2N0_{.86}$) were obtained but they have somewhat different structure: at T = 450 - 650 °C it is the orthorhombic ordered α'' phase (sp. gr. Pnnm); at T = 750 - 800 °C - disordered hexagonal α phase (sp. gr. $P6_3/mmc$). All the three modifications: α', α'' and α are genetically connected to the initial phase but do not contain hydrogen. Between the found modifications there are reversible and nonreversible phase transitions; conditions and schemes of the transitions are presented.

5. It is established that it is possible to produce the unique metastable hydrogen induced phases, which do not contain hydrogen but keep the structure of initial phase as "frozen", by means of evacuation of hydrogen out of a lattice of some solid solutions of the Zr(Ti) - N - H system using special regimes of vacuum heat treatment. It shows additional opportunities of obtaining new compounds with defined service characteristics by hydrogenating and dehydrogenating.

6. It has been studied the influence of vacuum thermal treatment on stability of the structure of the solid solutions $TiC_{0,45}H_{0,90}$, $TiC_{0,45}H_{0,64}$ and $TiC_{0.35}H_{0.43}$. It is shown that the phenomenon of complete thermoemission of hydrogen is observed in $TiC_{0.35}H_{0.43}$ at temperature 275°C under persistent evacuation not worse than $5.3x10^{-3}$ Pa. In other solid solutions (with the large concentrations of carbon and hydrogen), starting since the same temperature, the decay of the solid solution is observed, α -Ti and FCC titanium carbohydride TiC_xH_y being formed. The conclusion is made that in solid solutions TiC_xH_y thermoemission of H is observed only in strongly defective structures in which interaction forces between atoms are weak and also temperature of hydrogen evacuation T_{evac} is lower than temperature of interstitial atoms redistribution or of recrystallization $T_{recryst}$: $T_{evac} < T_{recryst}$.

7. The crystal structure of metastable hydrogen induced $Ti_2C_{0.70}$ phase is strongly distorted what is appeared in large scatter of points in neutron diagram and also in large errors of determination of lattice parameters and great value of divergence factor $R_{Br.}$. It is found the effect of changing of direction of matrix atoms displacement relatively to their ideal position after complete thermoemission of hydrogen out of lattice of the solid solution $TiC_{0.35}H_{0.43}$; it

may be explained by change of character of interaction between the atoms after "soft" removal of hydrogen.

8. It is shown that at evacuation of hydrogen out of the lattice of titanium hydride $TiH_{1.95}$ in the temperature interval of 100 - 175°C in continuously pumped out vacuum the decrease of hydrogen concentration from H/Ti=1,84 up to 1,74 (± 0,04) is observed at practically constant lattice parameter of the initial cubic phase (sp.gr.Fm3m). So, it is shown the possibility of "soft" regulation of composition of titanium hydride without change of lattice parameter. At temperature 200°C the body-centered tetragonal (BCT) phase is formed with the unit cell parameters a = 0.3182 ± 0,0005; c = 0.4413 ± 0,o007 nm (space group I4/mmm); starting from the temperature of 225°C it is observed the significant reduction in hydrogen content and also a decay of stoichiometric FCC $TiH_{1.95}$ - phase onto nonstoichiometric FCC TiH_{2-x} phase and pure Ti. So, a change of symmetry of the lattice is occurred before complete removal of hydrogen out of lattice of $TiH_{1.95}$, thereby in TiH_2 temperature of recrystallization is lower than temperature of evacuation: $T_{recryst} < T_{evac}$ at pressure of $5{,}3 \cdot 10^{-3}$ Pa and such vacuum is insufficient for obtaining pure titanium with FCC lattice by means of dehydrogenation.

9. The obtained results on study of the phenomenon of hydrogen thermoemission in the solid solutions $TiC_{0,45}H_{0,90}$, $TiC_{0,45}H_{0,64}$ and $TiH_{1.95}$ allow one to conclude that vacuum $5.3 \cdot 10^{-3}$ Pa (in which their dehydrogenating was carried out) is insufficient for removal of hydrogen out of lattice of these phases without a change of crystal symmetry, and for achievement of the given purpose at the temperatures, lower than the temperature of recrystallization or decay, the higher vacuum is necessary.

10. By neutron diffraction is found that by low-temperature ($T_{эвакуации}$ = 130-150 ^{0}C) removal of hydrogen (by thermoemission) from rare earth metal trihydrooxide $R(OH)_3$ (were R is La, Pr or Nd) under continuous high vacuum evacuating, makes possible to obtain metastable "trioxide" $R[O]_3$ of radical type. Existence of such substance contradicts to the valence law (oxygen is bivalent and Pr is trivalent in hydroxides). Such "trioxide" have a superfluous negative charge: $Pr^{3+}[O_3]^{6-}$. So they aspire "to capture" three more protons (hydrogen ions) from a water molecule. Obviously, this substance can be stable at low temperatures and in the mediums, which are not containing hydrogen. In the air at room temperature this substance, most likely, interacting with water molecules, gradually again turns into trihydroxide $Pr(OH)_3$, compensating the superfluous negative charge by three hydrogen atoms: . From this it follows that substance $Pr[O_3]$ can simultaneously be an absorber of hydrogen and generator of oxygen at atmospheric conditions and in any mediums which contains water molecules, without any prior processing like heating or high pressure:

$$4Pr[O]_3(\text{metastable powder crystal}) + 6H_2O = 4Pr(OH)_3(\text{stable powder crystal}) \rightarrow 3O_2 \uparrow$$

Thus, the obtained material, without any prior processing like heating or high pressure, can be simultaneously oxygen generator and hydrogen accumulator in any mediums. Characteristics of $Pr[O_3]$ to transform into stable form $Pr(OH)_3$ by selective bonding of hydrogen from the hydrogen-containing environment allows implication of $Pr[O_3]$ as the hydrogen selective absorber. The cycle described can be accomplished many times.

11. Hydrogen thermoemission is a new method of production for new metastable crystal phases of inorganic compounds which are not observed in phase diagrams and cannot be obtained by conventional methods (quenching, annealing, irradiation, chemical reactions, etc.). Hydrogen is used as a factor of the external system. Comparison of structural peculiarities of hydrogen-induced and corresponding hydrogenous phases opens new sides in understanding of crystal-chemistry of these compounds and crystal-chemical behavior of hydrogen. With such a great variety of hydrogenous inorganic compounds, the clear possibility exists to discover many new metastable crystal phases and materials with new properties that have yet to be explored.

8. References

Bacon, G. E. (1975). *Neutron Diffraction.* Oxford: Clarendon Press, 3rd ed. England.

Bashkin, I. O., Djyueva, I, M., Lityagina, L. M., Malyshev B. Yu. (1993). Concentration dependence of lattice parameter of TiH_2. *Fizika tverdogo tela.* Vol. 35, No 11. (December 1993), PP. 3104-3114, ISSN 0042-1294.

Geld P. V. & Ryabov P. A. (1985). Mokhracheva L. P. Hydrogen and physical properties of metals and alloys. Moscow: Metallurgiya, Russia.

Gusev, A. I. (2001). Order-disorder transformations and phase equilibrium in strongly nonstoichiometric compounds. *Uspekhi fizicheskikh nauk.* Vol. 170, No 1. (January 2001), pp. 3-40, ISSN 0042-1294.

Grigorovich, V. K. & Sheftel', E. N. (1980). Dispersion hardening of refractory metals. Moscow: Nauka, Russia.

Khidirov, I. (2011). Study of the phenomenon of hydrogen thermoemission from crystal lattice of the trihydroxide $Pr(OH)_3$ by neutron diffraction. *International Scientific Journal for Alternative Energy and Ecology,* No 5, (June 2011), pp. 19-22, ISSN 1608-8298.

Khidirov, I., Kurbanov, I. I. & Makhmudov, A. Sh. (1991). Hydrogen induced metastable phase in the Ti-N system. Metallofizika, Vol. 13, No 6, (June 1991), pp.43-47, ISSN 1024-1809.

Khidirov I., Mirzaev, B. B., Mukhtarova N. N., Bagolepov V. A., Savenko A. F., Pishuk V. K. (2008). Influence of vacuum thermal treatment on structure of solid solutioins TiC_xH_y. *In the Book Carbon Nanomaterials in Clean Energy hydrogen Systems.* Pp. 679-686, ISBN 978-1-4020-8896, Dordrecht: Springer, The Netherlands.

Khidirov, I. &, Om, V. T. (1993). Localization of hydrogen Atoms in Rare Earth Metal Trihydroxides $R(OH)_3$. *Physica Status Solidi (a),* Vol. 140, (September 1993), pp. K59-K62, ISSN 1862-6300.

Khidirov, I., Veziroglu, T. N. & Veziroglu A. (2009). Phase transitions in oversaturated solid solution of nitrogen in the α-Zr obtained by using hydrogen termoemission. *International Scientific Journal for Alternative Energy and Ecology,* No 2, (February 2009), pp. 8-14, ISSN 1608-8298.

Kornilov, I. I.(1975). Titan. Moscow: Nauka, Russia.

Portnoy, K. B. & Timofeeva, N. I. (1986). Oxygen connection of rare-earth metals. Moscow: Metallurgiya. Russia.

Rietvild, H. M. (1969). Profile Refinement method for nuclear and magnetic structures.*Journal of Applied Crystallography,*Vol. 2, part 2, (June 1969), pp. 65-71, ISSN 0021-8898.

Schubert, K. (1971). Crystal structures of two-componental phases. Moscow: Metallurgiya, translation from the German language, Russia.

Young, R. A., Wilas, D. B. (1982). Profile Scope Functions in Rietveld Refinements. *Journal of Applied Crystallography,*Vol. 2, (February 1982), pp. 65-71, ISSN 0021-8898.

9

Three-Dimensional Magnetically-Oriented Microcrystal Array: A Large Sample for Neutron Diffraction Analysis

T. Kimura[1], F. Kimura[1], K. Matsumoto[1] and N. Metoki[2]
[1]*Kyoto University*
[2]*Japan Atomic Energy Agency*
Japan

1. Introduction

Structure determination of biomolecules is of great importance because the structure is closely related to biological functions. Some proteins are activated when binding with ligands; a drug molecule functions by binding to a specific protein site (Sousa et al., 2000; Graves et al., 1994). The structure determination is important not only for biomolecules but also for many inorganic, organic, and polymeric crystals that are key materials in materials and pharmaceutical sciences.

Currently, three major methods to solve the structure of molecules are known: solution nuclear magnetic resonance (NMR) (Herrmann et al., 2002), X-ray and neutron single crystal analysis (Blundell et al., 1976; Kendrew et al., 1958; Blake et al., 1965), and X-ray powder diffraction (Margiolaki et al, 2008; Hariss et al., 2004) methods. The solution NMR method has an advantage over diffraction methods because it does not require crystals. However, it can be applied only to proteins with lower molecular weights. X-ray and neutron single-crystal analysis is the most powerful method, but it is sometimes difficult to grow a crystal (Ataka et al., 1986) sufficiently large for conventional or synchrotron single-crystal X-ray measurement. The size requirement is much more severe for neutron diffraction measurements (Niimura et al., 1999). The X-ray powder method can be performed if microcrystalline powders are available, but an appropriate refinement of many parameters is needed to obtain a reliable result. Preferential orientation of a powder sample is utilized to produce single-crystal-like diffraction data (Wessels et al., 1999). However, the data quality strongly depends on the quality of the orientation.

It is well known that feeble magnetic materials such as most biological, organic, polymeric and inorganic materials respond to applied magnetic fields, although the response is weak. A number of studies on the magnetic alignment of such materials have been reported (Maret et al., 1985; Asai et al., 2006). The study of the feeble interaction of diamagnetic materials with an applied magnetic field and its application are now recognized as an immerging area of science and technology, and named "Magneto-Science".

We have recently proposed a method that enables to convert a diamagnetic or paramagnetic microcrystalline powder to a "pseudo single crystal (PSC)" (T. Kimura et al. 2005; T. Kimura,

2009a)(Fig. 1). A PSC is a composite in which microcrystals are oriented three-dimensionally in a resin matrix. A PSC is also referred to as a "three-dimensional magnetically-oriented microcrystal array" (3D-MOMA) (F. Kimura et al., 2011). This method enables us to perform a single-crystal analysis of a material through alignment of its powder sample. The composite is fabricated using magnetic alignment of a microcrystal suspension under an elliptically rotating magnetic field, followed by consolidation of the suspending matrix.

The degree of alignment strongly depends on the diamagnetic anisotropy of the crystal, applied field strength, and size of microcrystals to be aligned. If these conditions are appropriately satisfied, the obtained 3D-MOMA can produce well-separated diffraction spots that allow single-crystal analysis. This method can be applied to biaxial crystals including the triclinic, monoclinic, and orthorhombic systems which have three different magnetic susceptibility values. We only obtain fiber diffraction patterns for uniaxial crystals even if we apply the 3D-MOMA method. This method does not work for the cubic system.

However, the 3D-MOMA method has some drawbacks including the limitation of the suspending medium, the difficulty of recovering a precious sample, and the broadening of diffraction spots by consolidation. A solution to these problems may be measurement of the X-ray diffraction patterns directly from a three-dimensional magnetically-oriented microcrystal suspension (3D-MOMS) without consolidation of the suspending medium. X-ray diffraction from magnetically oriented solutions of macromolecular assemblies was reported (Glucksman et al., 1986; Samulski et al., 1986). Also, small-angle X-ray scattering of colloidal platelets (van der Beek et al., 2006) and molecular aggregates (Gielen et al., 2009) under magnetic fields were reported. *In-situ* X-ray (Kohama et al., 2007) and neutron (Terada et al., 2008) diffraction measurements under applied magnetic fields were reported. We showed a preliminary result (Matsumoto et al., 2011) of the X-ray diffraction from a MOMS achieved with and without sample rotation in a static magnetic field.

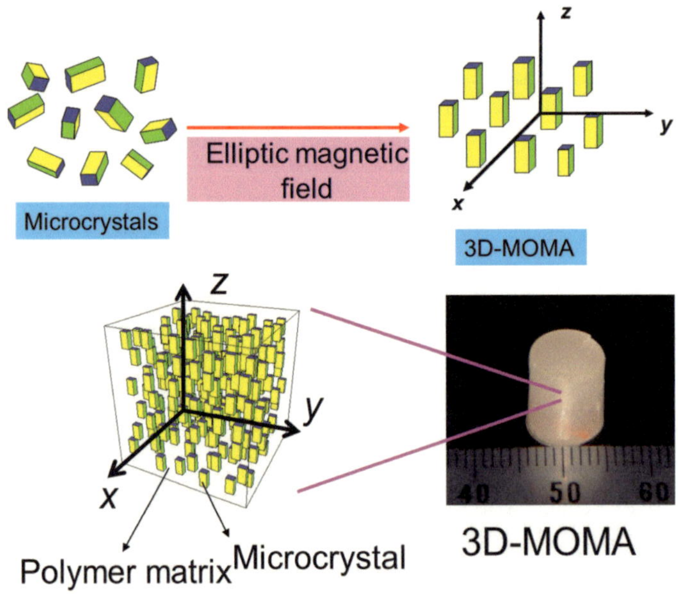

Fig. 1. Schematic of conversion of microcrystals to "three-dimensional magnetically-oriented microcrystal array" (3D-MOMA) or "pseudo single crystal" (PSC).

We believe that the magnetic method we introduce here is of considerable use for the crystal structure determination of materials that do not grow large sizes necessary for the diffraction measurement. Especially, our method has advantages when applied to single crystal neutron diffraction analyses of proteins where a single crystal sample much larger than that needed for the X-ray diffraction measurement is required.

2. Theoretical background of preparation of a 3D-MOMA

A crystal has a diamagnetic susceptibility tensor χ, which is expressed in terms of the principal diamagnetic susceptibility axes ($\chi_3<\chi_2<\chi_1<0$). When the crystal is exposed to a magnetic field **B**, it has a magnetic energy, expressed with respect to the laboratory coordinate system as follows

$$E = -(V/2\mu_0)^{t}\mathbf{B}(^{t}\mathbf{A}\,\chi\,\mathbf{A})\mathbf{B} \tag{1}$$

where μ_0 is the magnetic permeability of the vacuum, V is the volume of the crystallite, **A** is the transformation matrix defined by the Eulerian angles (θ, ϕ, φ), and the superscript t indicates the transpose. We consider an applied elliptical magnetic field that periodically changes in direction and intensity on the xy plane as follows:

$$\mathbf{B} = B\,^{t}(b_x \cos\omega t,\ b_y \sin\omega t,\ 0) \tag{2}$$

Here, ω is the angular velocity of the field rotation, B is the intensity of the field, and b_x and b_y are the longer and shorter axes of the ellipse ($b_y < b_x$), respectively. If ω is far larger than a certain rate τ^{-1} (the intrinsic rate of magnetic response of a particle exposed to the static field of the intensity B) then the easy magnetization axis, χ_1, cannot follow the rotation of the field. This condition is called "Rapid Rotation Regime (RRR)", defined as $|\omega\tau|>>1/2$, where τ is expressed as $\tau = 6\,\mu_0\eta/\chi_a B^2$ in the case that the particle shape is spherical (T. Kimura et al., 2000) where χ_a and η are magnetic anisotropy and the viscosity of the surrounding medium.

In the RRR, a crystal could be assumed to be placed in an time-averaged magnetic potential expressed by the following equation:

$$E_{av} = -(\omega/2\pi)\int_0^{2\pi/\omega}(V/2\mu_0)^{t}\mathbf{B}(^{t}\mathbf{A}\chi\mathbf{A})\mathbf{B}\,\mathrm{d}t \tag{3}$$

Inserting eq (2) into eq (3) and performing the integration, we obtain E_{av} as an analytical function of the Eulerian angles. Expanding eq (3) around $\theta=\phi=\varphi=0$ and truncating the higher terms, we obtain (T. Kimura et al., 2005)

$$E_{av} \approx (V/4\mu_0)B^2 b_y^{\,2}(\chi_2-\chi_3)\theta^2 + (V/4\mu_0)B^2\left(b_x^{\,2}-b_y^{\,2}\right)(\chi_1-\chi_2)(\phi+\varphi)^2 \tag{4}$$

where the constant terms are not shown.

Fluctuations of θ^2 and $(\phi+\varphi)^2$ around the minimum are calculated using the Boltzmann distribution to obtain the expression

$$<(\Delta\theta)^2> \approx 2\mu_0 k_B T/[B^2 b_y^{\,2} V(\chi_2-\chi_3)] \tag{5}$$

$$<(\Delta(\phi+\varphi))^2> \approx 2\mu_0 k_B T / [B^2(b_x^2 - b_y^2)V(\chi_1 - \chi_2)] \quad (6)$$

where k_B is the Boltzmann constant and T is the temperature. To obtain a good 3D-MOMA, we should minimize and equalize the values of $\langle (\Delta\theta)^2 \rangle$ and $\langle (\Delta(\phi+\varphi))^2 \rangle$ by appropriately choosing the values of b_x and b_y. The crystallite volume V is also important. The smaller the V, the larger the fluctuations. Typically, the size necessary is about the order of micrometers, depending on the field strength used and the magnetic anisotropy of the crystal. To obtain a high quality diffraction pattern, the rotation speed ω is also important. It should satisfy $|\omega\tau| >> 1/2$ (RRR) (T. Kimura et al., 2005).

In general, the susceptibility axes of a biaxial crystal do not necessarily coincide with the crystallographic *a*, *b*, *c* axes, except the orthorhombic system (Fig. 2a) (F. Kimura et al., 2010a). The orthorhombic system includes three point groups, that is, 222, *mm*2, and *mmm*. Both for 222 and *mmm*, three crystallographic axes are 2-fold axes that coincide with the susceptibility axes. As a result, the rotation about the susceptibility axes through an angle of π does not create new crystal orientations. On the other hand, *mm*2 has just one 2-fold axis. Therefore, the rotation about the susceptibility axes can produce a new orientation.

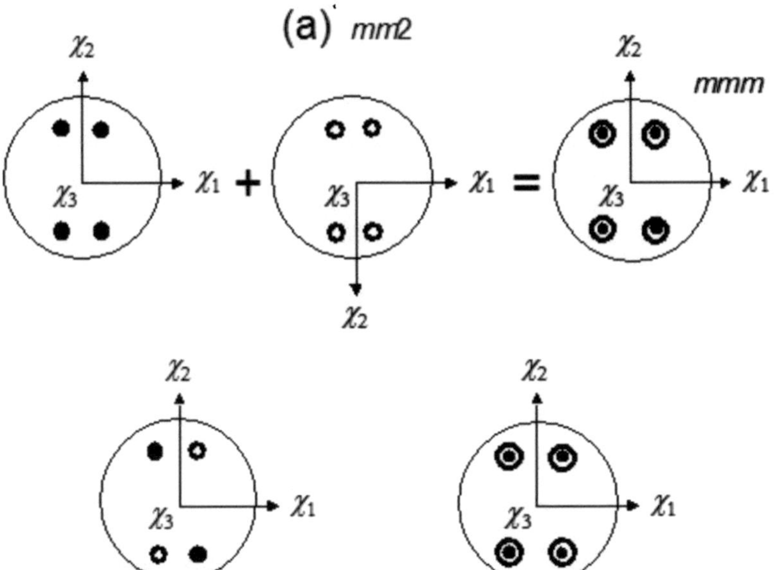

Fig. 2a. Crystal orientations obtained by a rotation about the susceptibility axis by angle of π are magnetically equivalent. The orthorhombic system includes three point groups, *mm*2, 222, and *mmm*, as displayed with stereo diagrams. In the case of (a) *mm*2, an orientation different from the original one is produced by rotation about χ_1 or χ_2. In a 3D-MOMA, these orientations coexist. For (b) 222 and (c) *mmm*, the rotations do not alter the original crystal orientation, resulting in a single orientation in a 3D-MOMA. (Reprinted from (F. Kimura et al., 2010a): Cryst. Growth Des. Vol.10(2010) pp. 48-51, No. 1, DOI: 10.1021/cg90132h. Copyright 2010 American Chemical Society)

For the monoclinic system (point group 2, space group $P2_1$), its twofold axis (b axis) coincides with one of the magnetic susceptibility axes (Nye, 1985). The crystallographic and magnetic axes are shown schematically in Fig. 2b, where the b axis and χ axis are placed perpendicular to the plane of the diagram (F. Kimura et al., 2010b). Since the three magnetic susceptibility axes are mutually perpendicular, the other two magnetic susceptibility axes are placed on the plane of the diagram. The a and c axes are on the same plane. However, there is no rule to relate the χ axes with the a and c axes. If the crystal is rotated by an angle of π around each of the magnetic susceptibility axes, four different crystal orientations are obtained, as shown in Figs. 2b (a)–(d). These four orientations have the same magnetic energy when they are placed in a given magnetic field. Because of the twofold symmetry (point group 2) along the b axis, the crystal orientations of Figs. 2b(a) and 2b(c) are identical; those of Figs. 2b(b) and 2b(d) are also identical. Thus, there are two different orientations with the same magnetic energy. In Table 1, possible crystal orientations with an equal magnetic energy are summarized for crystals belonging to the orthorhombic, monoclinic, and triclinic systems.

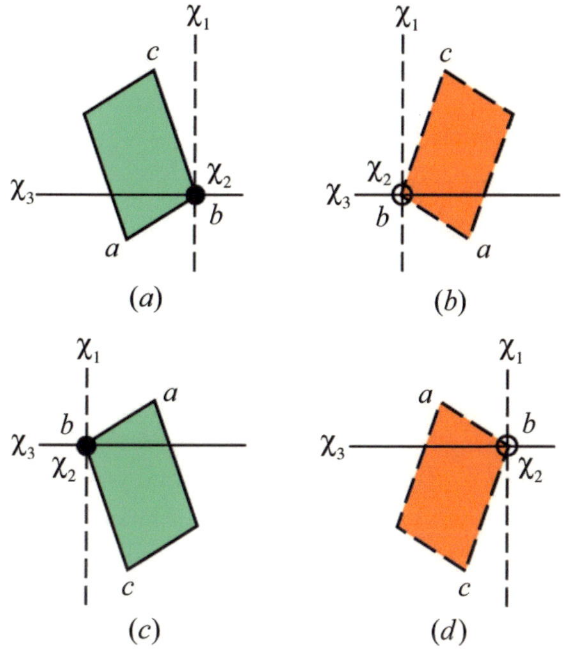

Fig. 2b. Schematic diagram of a crystal belonging to the monoclinic system (point group 2) showing crystallographic axes and magnetic susceptibility axes. (a) The b axis (twofold axis) and χ_2 axis (its direction coincides with that of the b axis) are placed perpendicular to the plane of the diagram. The other crystallographic axes (a and c axes) and magnetic susceptibility axes (χ_1 and χ_3 axes) are placed in the plane. The crystal (a) is rotated by π around each of susceptibility axes to produce three additional possible orientations, (b), (c), and (d). All four orientations have the same magnetic energy under a given magnetic field. Because of the symmetry (point group 2), the crystal orientations of (a) and (c) are identical, and those of (b) and (d) are also identical. (Reprinted from (F. Kimura et al., 2010b): *J. Appl. Cryst.* Vol.143(2010) pp. 151-153, Part 1, DOI: 10.1107/S0021889809048006. Copyright 2010 International Union of Crystallography)

Crystal system	Point group of single crystal	Number of orientations	Point group of PSC
Triclinic	1	4	222
	$\bar{1}$	4	*mmm*
Monoclinic	2	2	222
	m	4	*mmm*
	2/*m*	2	*mmm*
Orthorhombic	222	1	222
	*mm*2	2	*mmm*
	mmm	1	*mmm*

Table 1. Number of possible orientations and the resultant point symmetry of Pseudo Single Crystal (PSC) for biaxial crystal systems with different point groups

3. Magnetic field used to prepare 3D-MOMA

3.1 Generation of modulated magnetic field

A 3D-MOMA of a biaxial crystal was prepared under modulated magnetic fields. In the previous section, the calculation was performed using the elliptical field. In the actual experiment, there are various versions for the elliptical field. We used two different dynamic magnetic fields (T. Kimura et al., 2009b). The first one is an amplitude modulated dynamic field (T. Kimura et al., 2005). This is generated by a four-pole electromagnet, in which one pair of poles generates a B_x field strength sinusoidally oscillating at ω, and the other generates B_y field strength oscillating at the same frequency as the previous one but with a phase shift of $\pi/2$. This combination created a magnetic field expressed as $(B_x \cos\omega t, B_y \sin\omega t, 0)$, shown in Fig. 3. The second one is a frequency modulated dynamic magnetic field (T. Kimura et al., 2006). In this setup, a sample is non-uniformly rotated in a uniform static magnetic field. A sample-rotating unit was placed in a uniform horizontal magnetic field, shown in Fig. 4. The rotation axis was vertical. The rotation was not uniform. In the figure, the *x*-, *y*-, and *z*-axes are laboratory coordinates and the *x'*-, *y'*-, and *z'*-axes are imbedded in the rotating plate. The *x*-axis is parallel to the magnetic field. The *z'*- and *z*-axes are parallel to vertical direction. The angle between the *x'*-axis before rotating and *x*-axis is $\alpha/2$. The *x'*-axis was rotating at the angular velocity ω_s between $360-\alpha/2$ and $\alpha/2$ degree, and between $180-\alpha/2$ and $180+\alpha/2$ degree. On the other hand, it was rotating at the angular verocity $\omega_q(>\omega_s)$ between $\alpha/2$ and $180-\alpha/2$ degree, and between $180+\alpha/2$ and $360-\alpha/2$ degree.

The three parameters, ω_s, ω_q, and α, must be selected appropriately in order to obtain sharp diffraction spots. The half-width of a spot is a result of the fluctuations of the χ_1 and χ_3 axes about the *x'* and *z'* axes, respectively. In terms of diffraction analysis, it is advantageous that the magnitudes of the two fluctuations are equal.

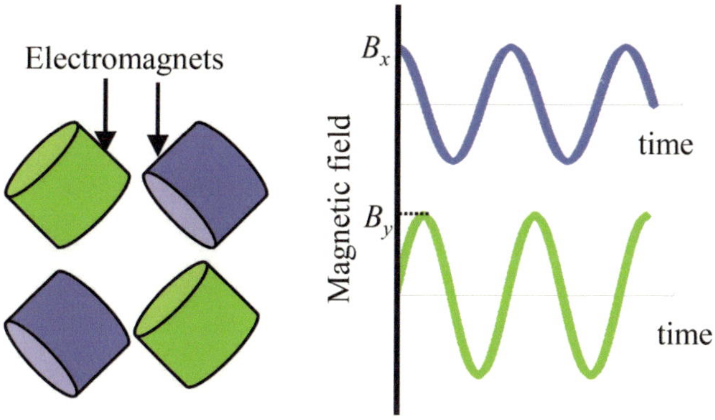

Fig. 3. Generation of amplitude modulated magnetic field. Two pairs of magnets generate $B_x \cos\omega t$ and $B_y \sin\omega t$ resulting in an elliptical magnetic field.

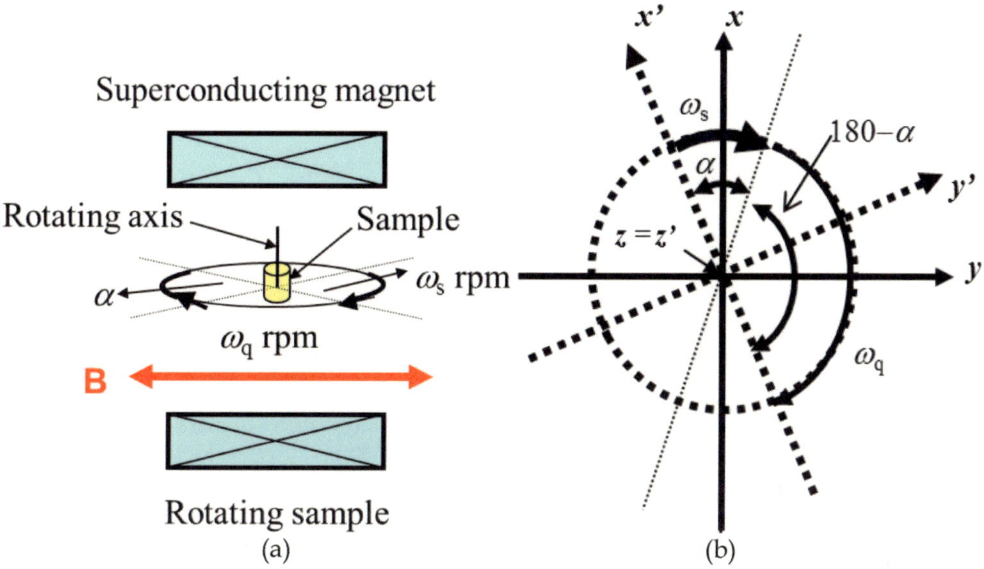

Fig. 4. (a) Generation of modulated magnetic field by rotating a sample in a static field. Rotation speed is switched. (b) Diagram illustrating the frequency-modulated sample rotation. The x-, y-, and z-axes are laboratory coordinates; the x- and y-axes are on the horizontal plane and the z axis is on the vertical plane. The x'-, y'-, and z'-axes are imbedded in the sample that is rotated around the z'-axis (the z-axis coincides with the z'-axis). The rotation is performed non-uniformly so that the condition $\omega_s < \omega_q$ is satisfied. The x-axis coincides with the direction of the magnetic field.

4. Sample preparation

4.1 Preparation of 3D - MOMA

Micro-crystallites were mixed with a UV (Ultra Violet) curable monomer. The concentration of the crystallites in the monomer is 10-30 v/v%. The suspension is poured into any size of plastic container, exposed to a dynamic magnetic field, followed by UV light irradiation to polymerize the resin precursor to fix the alignment. See experimental details in Table 2.

4.2 Preparation of MOMS

Typical experimental setting (KU model χ10-1) is shown in Fig. 5. Sample suspension is poured into a capillary, and then it is placed on a rotating unit equipped with a pair of magnets.

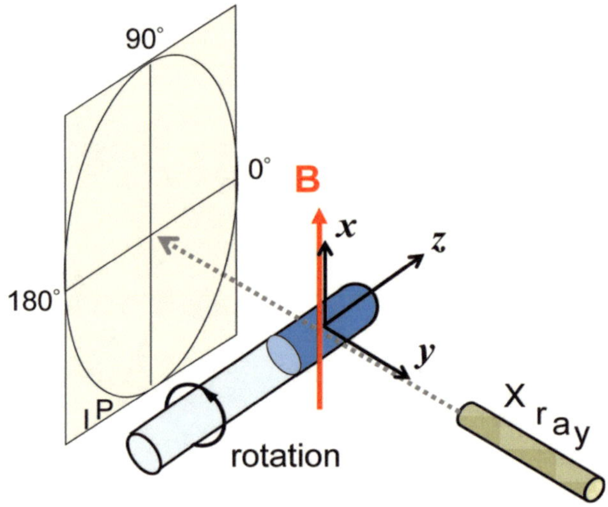

Fig. 5. Experimental setting of *in-situ* X-ray diffraction measurement of a microcrystal suspension. A glass capillary containing the suspension is rotated at the rotation speed ω about the z-axis. The magnetic field **B** is applied parallel to the x-axis. The X-ray beam is impinged from the y direction, and the diffractions are detected by an imaging plate (IP). The azimuthal β angle is indicated. (Reprinted from (Matsumoto et al., 2011): Cryst. Growth Des. Vol.11(2011) pp. 945-948, No. 4, DOI: 10.1021/cg200090u. Copyright 2011 American Chemical Society)

5. X-ray results

We fabricated 3D-MOMAs using frequency modulated magnetic field except $LiCoPO_4$. We fabricated 3D-MOMAs of $LiCoPO_4$ using both amplitude and frequency modulated magnetic fields. Alignment of these 3D-MOMAs are almost the same. Here, we only show the results obtained from the MOMA using frequency modulated magnetic field.

5.1 Alanine (T. Kimura et al., 2006)

The top figure in Fig. 6 shows the XRD powder pattern obtained from the experiment in which magnetic field was not applied. A sparse ring pattern is observed in this case. Figures 6(a), (b) and (c) show the XRD patterns of the obtained pseudo single crystal. Patterns in (a), (b), and (c) verify the alignment of the a, b, and c axes, respectively. The y', z', and x' axes imbedded on the sample correspond to the a, b, and c axes, respectively. This indicates that the b and c axes correspond to the hard and easy magnetization axes, respectively.

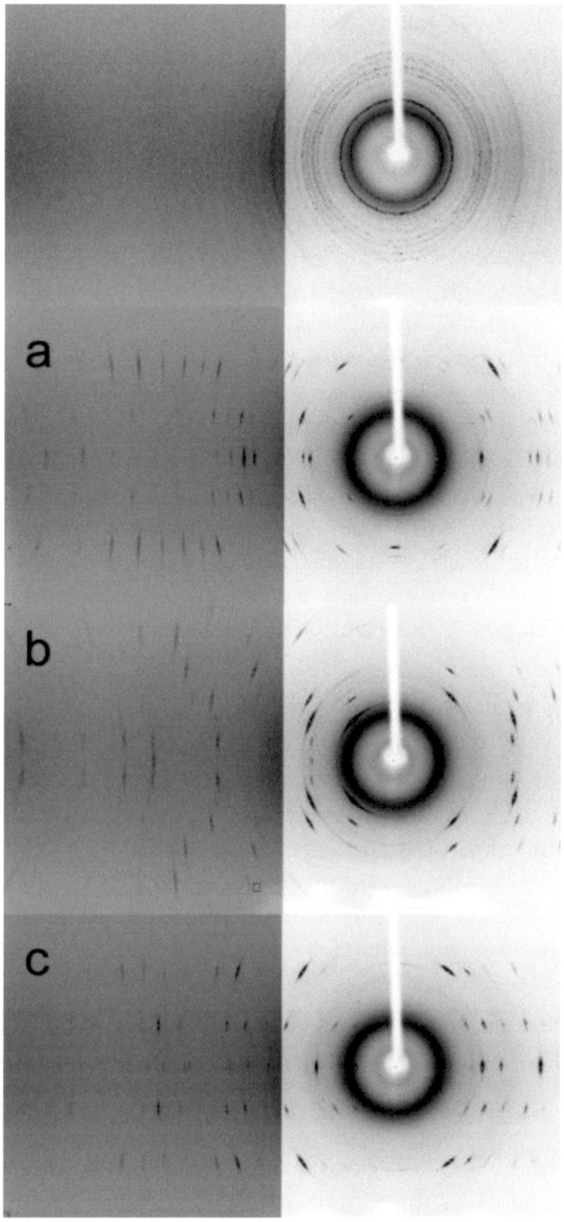

Fig. 6. X-ray diffraction pattern obtained for the sample prepared without the application of a magnetic field(top). X-ray diffraction patterns obtained for the L-alanine pseudo-single crystal sample. Patterns in (a), (b), and (c) were obtained with the alignment of the *a*, *b*, and *c* axes, respectively, perpendicular to the X-ray beam; this is achieved by using an automatic crystal axis alignment system. An oscillation of 10° was applied around the aligned axis. Some of the assignments were performed using a software program. Contrasts are different between left and right halves. (Reprinted from (T. Kimura et al., 2006): *Langmuir* Vol.22(2006) pp. 3464-3466. No. 8, DOI: 10.1021/la053479m. Copyright 2006 American Chemical Society)

The diffraction spots are broad compared to those from a real single crystal, but they are sufficiently distinguishable to be assigned by the software program. Since the lattice parameter is the largest for the *b* axis, the spots in Fig. 6(b) are relatively less resolved. The half widths of the spots obtained by using an azimuthal plot are ca. 3° for most of the spots in all of the patterns in Figs. 6 (a), (b), and (c). Incidentally, the half widths for an actual original single crystal are ca. 0.5°. The fluctuations in θ and $\varphi+\psi$ are complex functions in the case of the frequency-modulated magnetic field used in the present study. However, they can be roughly estimated by using eqs (5) and (6). Using the values of B=5 T, b_x=1, b_y=0.5, T=300 K, and a rough estimation of $\chi_1-\chi_2$ and $\chi_2-\chi_3$ =10^{-7}, we can obtain the estimation for the fluctuation of θ and $\phi+\varphi$ ca. 1° for V =1000 μm^3. This value should be compared with the half widths of the diffraction spots. The experimental values are larger than those obtained by the theoretical estimation. Using the values of the viscosity (η =1.2 Pa s) and the shape factor (F = 0.064, corresponding to the assumed aspect ratio of 11 for crystallites), we can obtain the estimation, τ=1 min. The time required for the completion of the alignment might be estimated as five times of τ, i.e., ca. 5 min. This value is far shorter than 2 h actually applied in the experiment. These discrepancies have several possible explanations, including the existence of smaller crystallites, imperfect dispersion, shrinkage of the resin during the UV cure, the flow caused by sample rotation, insufficient rotation speed, imperfect switching from one rotation to the other, etc.

5.2 $LiCoPO_4$ (T. Kimura et al., 2009b)

A typical diffraction of the 3D-MOMA image is shown in Fig. 7. The estimated mosaicity is 3.9°, which is larger than that observed for normal single crystals on the same instrument (around 0.8°). The cell constants and an orientation matrix for data collection corresponded to a primitive orthorhombic cell with dimensions a = 10.202 (6), b = 5.918 (3), c = 4.709 (2) Å , V = 284.3(3) Å^3 and Z = 4. The space group was determined to be *Pnma* (No. 62). These results are the same as those reported by Kubel (1994), who used a 0.090 × 0.108 × 0.158 mm single crystal. The structure was solved by direct methods and expanded using Fourier techniques. The present result was compared with that from the literature, showing that the atomic coordinates determined in this study are in excellent agreement with those determined using a traditional single crystal. The $R1$ and $wR2$ values were 6.59 and 16.8%, respectively.

5.3 Sucrose (F. Kimura et al., 2010b)

The sucrose crystal belongs to the monoclinic system (point group 2, space group $P2_1$). Assuming an initial random orientation of crystallites in a suspension, the probability of finding two different orientations with the same magnetic energy is equal. These two orientations produce a diffraction pattern similar to that produced by a twin crystal. The diffraction image was analyzed using software designed for twin structures. The cell constants correspond to a primitive monoclinic cell with dimensions a = 7.7735(12), b = 8.7169(13), c = 10.8765(17) Å , β = 102.936(4)°, V = 718.29(19) Å^3 and Z = 2. The space group was determined to be $P2_1$ (No. 4). The structure was solved by direct methods and expanded using Fourier techniques. The $R1$ and $wR2$ values were 7.88 and 17.25%, respectively.

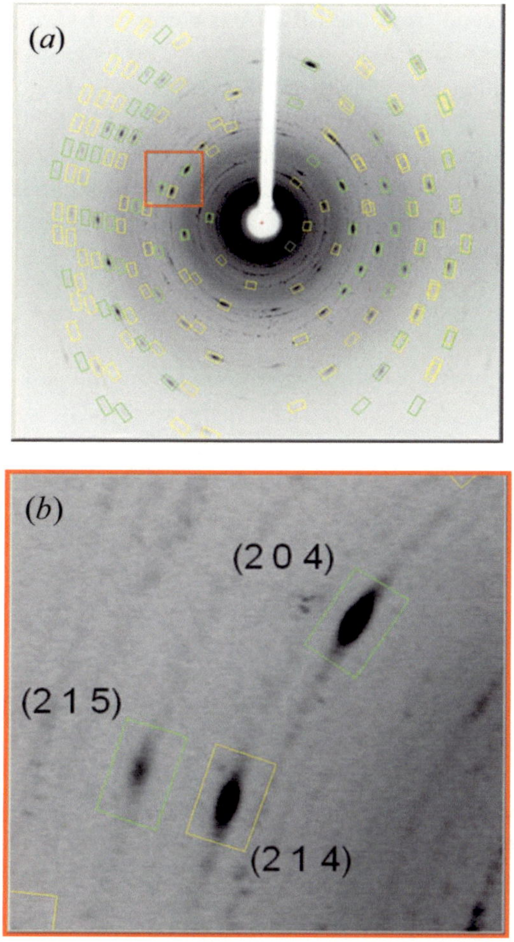

Fig. 7. (a) A typical X-ray diffraction image (oscillation angle 5°) of a pseudo single crystal of $LiCoPO_4$ is shown with diffraction spots enclosed by prediction rectangles. (b) An enlarged view of some diffraction spots is shown with Miller indices. (Reprinted from (T. Kimura et al., 2009b): *J.* Appl. Cryst. Vol.142(2009) pp. 535-537, Part 3, DOI: 10.1107/S0021889809013430. Copyright 2009 International Union of Crystallography)

	applied magnetic field / T	α /degree	ω_s /rpm	ω_q /rpm
L-alanine for X-ray	5	90	5	25
L-alanine for neutron	8	90	40	132
$LiCoPO_4$	0.3	90	10	60
Sucrose	12	90	10	60
Lysozyme		10	10	60
	8	10	30	100
		20	10	60

Table 2. Preparation condition of MOMAs

In Fig. 8, the structure solved from the MOMA is compared with that reported previously (Hanson et al., 1973). They show a good agreement. The twin structure is shown in Fig. 9. The magnetic susceptibility axes, as discussed previously (Fig. 9), are also shown. In principle, these magnetic susceptibility axes are attributed to each of the easy (largest) χ_1, hard (smallest) χ_3, and intermediate (medium) χ_2 magnetic susceptibility axes. The angle between the magnetic axis, χ_1, and crystallographic c axis is determined to be ca 16°. This value is different from a previously reported value of -1° 5′ (Finke, 1909). The reason for the difference is unclear at present.

5.4 Lyzozyme (F. Kimura et al., 2011)

The fluctuations of the χ_1 and χ_3 axes were estimated from the half-widths determined from the rocking curve and the azimuthal β scan, respectively. The rotation axis of the ω scan coincided with the χ_3 axis. The data was collected using ω scans at 1.0° steps. The three parameters, ω_s, ω_q, and α, must be selected appropriately in order to obtain sharp diffraction spots. The half-width of a spot is a result of the fluctuations of the χ_1 and χ_3 axes about the x' and z'- axes, respectively. In terms of diffraction analysis, it is advantageous that the magnitudes of the two fluctuations are equal. Sample rotation speed $\omega_s > 1/(2\tau)$should be met (Rapid Rotation Regime (RRR)). Next, an appropriate choice of α andω_q is necessary in order to minimize and equalize the fluctuations of the χ_1 and χ_3 axes.

In alanine, $LiCoPO_4$, and sucrose studies describing the single-crystal analysis of MOMAs, α=90° was appropriate. However, a theoretical study (T. Kimura, 2009a) shows that the equalization cannot be achieved for some sets of (χ_1, χ_2, χ_3) if α is fixed to 90°. Since the three values of the magnetic susceptibility of lysozyme are unknown, we need to find out an appropriate value of α by trial and error. Table 3 summarizes the half-widths of a diffraction spot ((400) diffraction) obtained from MOMAs fabricated with some sets of parameters (ω_s, ω_q, α).

α /degree	ω_s /rpm	ω_q /rpm	Fluctuation of χ_1 /degree	Fluctuation of χ_3 /degree
10	10	60	4.5	4.3
10	30	100	5.7	NA
20	10	60	6	5.1

Table 3. Fluctuation of χ_1 and χ_3 for three sets of parameters(ω_s, ω_q, α), determined using the (400) diffraction spot.

We tested several combinations of the parameters, and α =10°, ω_s =10 rpm, and ω_q =60 rpm was found to be the most suitable regarding minimization and equalization of the half-widths. This condition was therefore chosen for the preparation of a MOMA for X-ray analysis. In Fig. 10, photograph and microphotograph of fabricated lysozyme MOMA are shown. In Fig. 11, diffraction patterns taken from three directions are shown. Well-separated diffraction spots were obtained. The directions of the magnetic axes as deduced from the preparation procedure are indicated in the figure. From the results of the indexing described in the next paragraph, the crystal belongs to the orthorhombic system. For this crystal

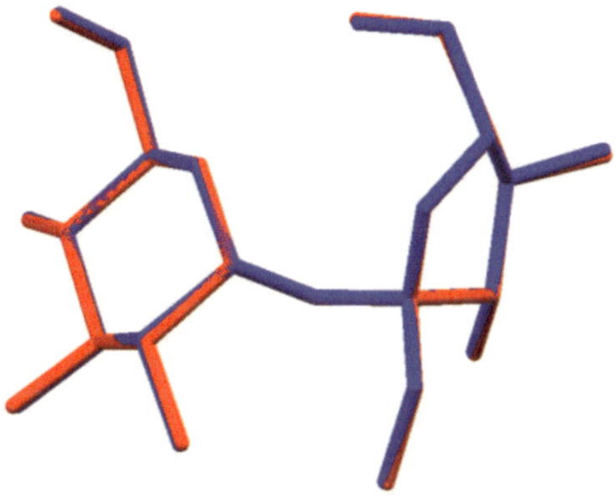

Fig. 8. Comparison of the structure determined in this study (blue) and the structure reported previously (red). (Reprinted from (F. Kimura et al., 2010b): *J.* Appl. Cryst. Vol.143(2010) pp. 151-153, Part 1, DOI: 10.1107/S0021889809048006. Copyright 2010 International Union of Crystallography)

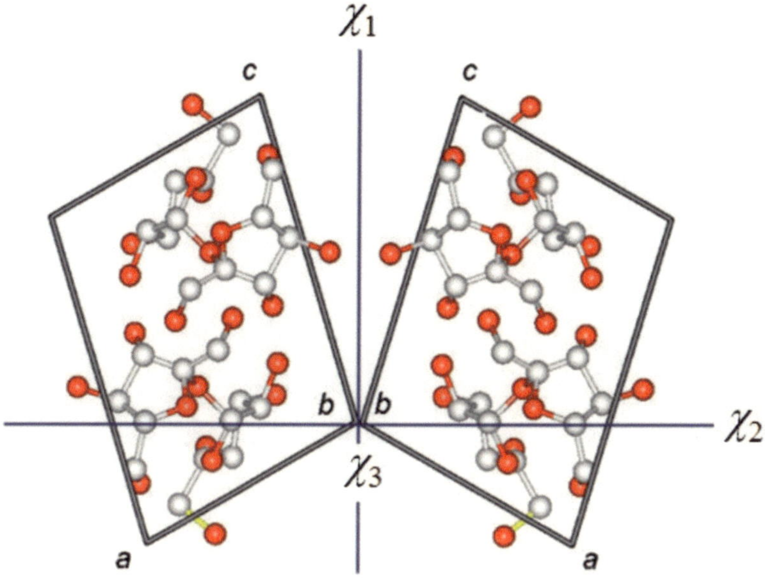

Fig. 9. Twin structure of a 3D-MOMA of sucrose. The *b* axis and one of the susceptibility axes (here denoted as χ_3) are placed perpendicular to the plane of the diagram. The other axes (χ_1 and χ_2) are placed in the plane. The angle between the c axis and the χ_1 axis is ca 16°. (Reprinted from (F. Kimura et al., 2010b): *J.* Appl. Cryst. Vol.143(2010) pp. 151-153, Part 1, DOI: 10.1107/S0021889809048006. Copyright 2010 International Union of Crystallography)

system, the magnetic axes correspond to the crystallographic axes. From the figure, we find that the cell dimensions increase in the order χ_1, χ_2, and χ_3. Combining the indexing results, we conclude that $\chi_1 = c$, $\chi_2 = a$, and $\chi_3 = b$. The alignment of the *c* axis parallel to the magnetic field has been reported for orthorhombic lysozyme crystals.(Sato et al., 2000). Figure 12 shows a detailed diffraction pattern with resolution rings.

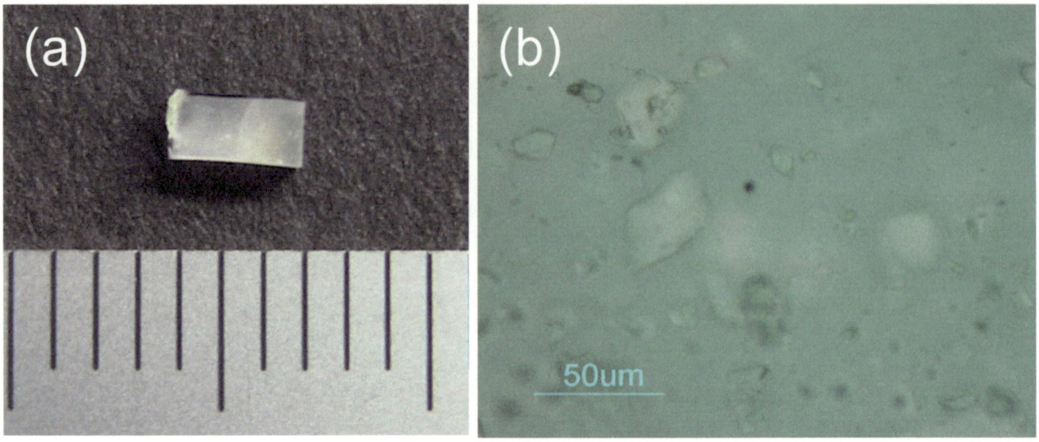

Fig. 10. (a) Photograph (a division: 1 mm) of a magnetically oriented microcrystal array (MOMA) of lysozyme. (b) Microphotograph of MOMA that is composed of various sizes of microcrystals. (Reprinted from (F. Kimura et al., 2011): Cryst. Growth Des. Vol.11(2011) pp. 12-15, No. 1, DOI: 10.1021/cg100790r. Copyright 2010 American Chemical Society)

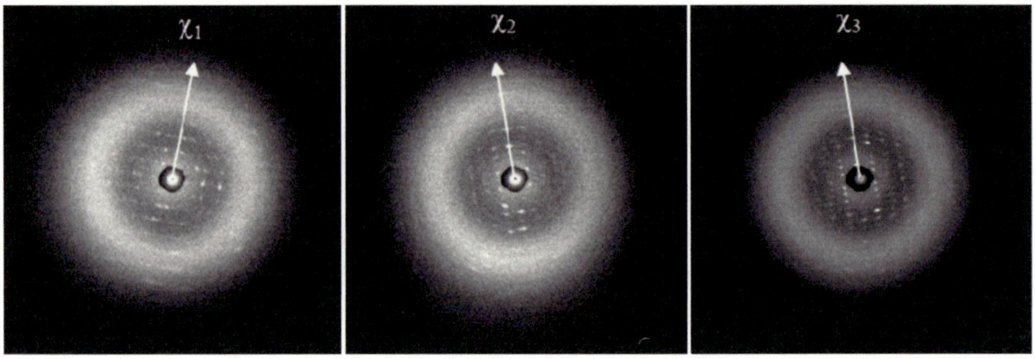

Fig. 11. X-ray diffraction of a lysozyme MOMA taken from three different orthogonal directions. (Reprinted from (F. Kimura et al., 2011): Cryst. Growth Des. Vol.11(2011) pp. 12-15, No. 1, DOI: 10.1021/cg100790r. Copyright 2010 American Chemical Society)

The X-ray results are summarized in Table 4. The indexing was determined as follows: space group, $P2_12_12_1$; lattice constants, a=51.26 Å, b=59.79 Å , c=29.95 Å , and V=91803 Å^3. This data was compared with that reported on the lysozyme single crystal (PDB code 2ZQ4) in Table 5. The cell dimensions obtained in this study are shorter than those reported in the literature. The shrinkage was attributed to the dehydration of the crystal. X-ray diffraction of dehydrated lysozyme crystals in triclinic (Kachalova et al., 1991) and monoclinic (Nagendra et al., 1998) forms has been reported. The graphical display is shown in Fig. 13 for easy comparison of the MOMA with the reported structure (PDB code 2ZQ4). A comparison of the CR positions between the present result and 2ZQ4 gave rmsd=0.755 Å , indicating that the shrinkage of the lattice was mainly due to the loss of water molecules and that the protein chain conformation remained essentially unchanged.

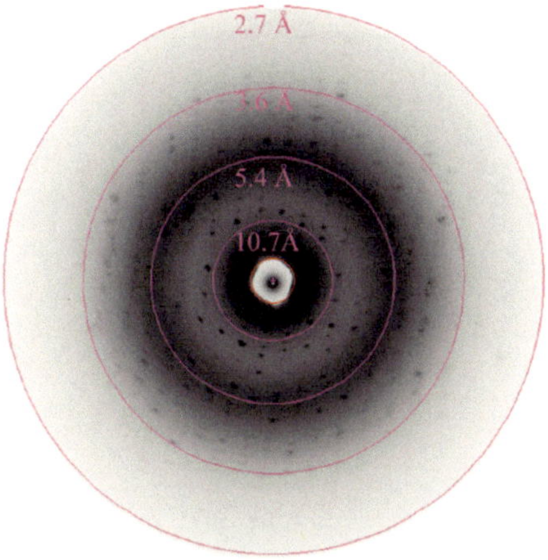

Fig. 12. X-ray diffraction image of the lysozyme MOMAwith resolution rings indicated. (Reprinted from (F. Kimura et al., 2011): Cryst. Growth Des. Vol.11(2011) pp. 12-15, No. 1, DOI: 10.1021/cg100790r. Copyright 2010 American Chemical Society)

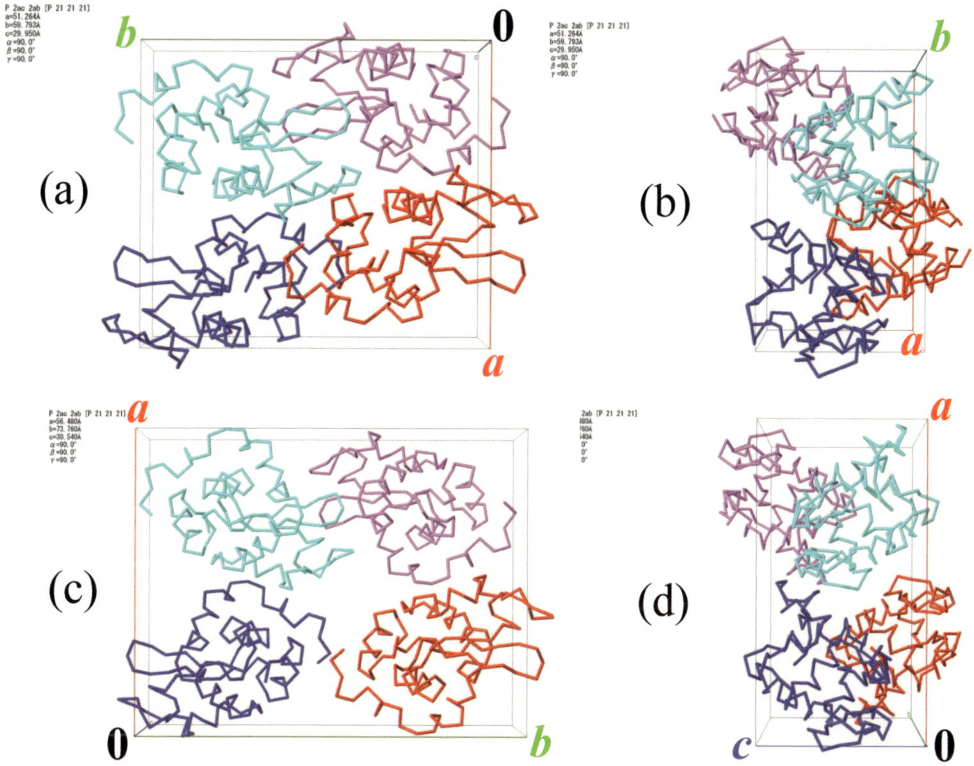

Fig. 13. Comparison of the crystal structures of lysozyme: (a and b) the structure determined through a MOMA prepared in the present study; (c and d) the structure reproduced from the database (PDB code 2ZQ4). (Reprinted from (F. Kimura et al., 2011): Cryst. Growth Des. Vol.11(2011) pp. 12-15, No. 1, DOI: 10.1021/cg100790r. Copyright 2010 American Chemical Society)

A. Crystal data	
Wavelength (Å)	1.5418
Space group	$P2_12_12_1$
Cell dimensions	
a (Å)	51.26
b (Å)	59.79
c (Å)	29.95
V (Å^3)	91, 803
Observed reflections	10, 604
Resolution (Å)	26.77 - 3.00 (3.16 - 3.00)
Independent reflections	1, 828 (250)
completeness (%)	89.6 (88.6)
R_{sym} (%)	19.4 (48.7)
Redundancy	5.8 (5.5)
Mean I/σ	9.3 (2.7)
B. Refinement statistics	
Resolution limits (Å)	25.86 - 3.00 (3.78-3.00)
No. of reflections used	2, 989 (1, 423)
completeness (%)	84.4 (81)
No. of protein atoms	1, 001
No. of solvent molecules	0
Final R-factor	0.215 (0.253)
Free R value	0.270 (0.292)
Average B-factor (Å^2)	51.0
r.m.s. deviation from ideal geometry	
Bond distances (Å)	0.004
Bond angles (deg.)	0.743
Dihedrals (deg.)	22.596

Table 4. Summary of data collection and refinement for lysozyme MOMA. The values in highest resolution bin were indicated in the parentheses.

6. Neutron diffraction data

6.1 Alanine (F. Kimura et al., 2010a)

In measuring the pole figure, a 3D-MOMA sample was set on a goniometer approximately in the direction shown in Fig. 14(a). The angles χ and φ correspond to the operation angles of the goniometer. The angle φ runs from 0 to 180° at a given value of the angle χ that runs from 0 to -180°. At this setting, the diffractions corresponding to the (040), (002), (120), etc. are expected to appear at locations shown in the figure. In Fig. 14(b), the measured pole figure is displayed. Here, all the diffraction spots are displayed in the same figure. A half sphere was scanned so that spots for {120} were all observed. These spots appear around χ =0, -180° and φ =45, 135° because $a/b = 1/2$. On the other hand, a spot for (040) was only scanned on one side. Instead, a close examination of (040) was made by scanning in the vicinity of χ =0° and φ =90°. A contour plot of the intensity of (040) is shown in Fig. 15 as a function of χ and φ. The average fwhm over two angles is ca. 4°. A 2θ profile for the (040) diffraction is shown in Fig. 16 as a typical example. The peak is clearly distinguished from the background. This high signal-to-noise ratio was unexpected because no care was taken to reduce the background incoherent scattering when choosing the resin precursor. This might be attributed to the fact that the coherent diffraction is extremely increased because the microcrystals are aligned three-dimensionally.

(a)

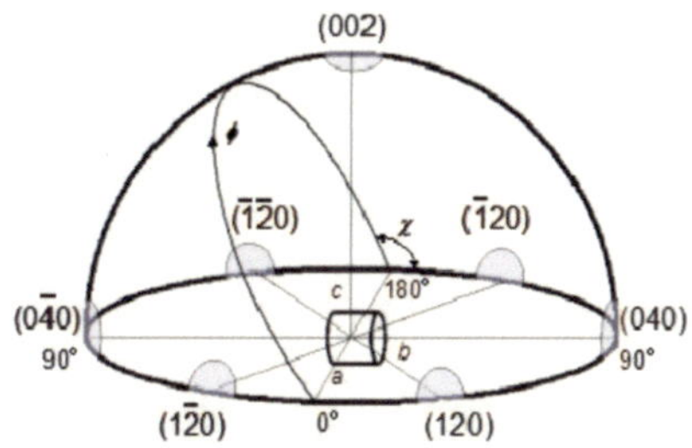

(b)

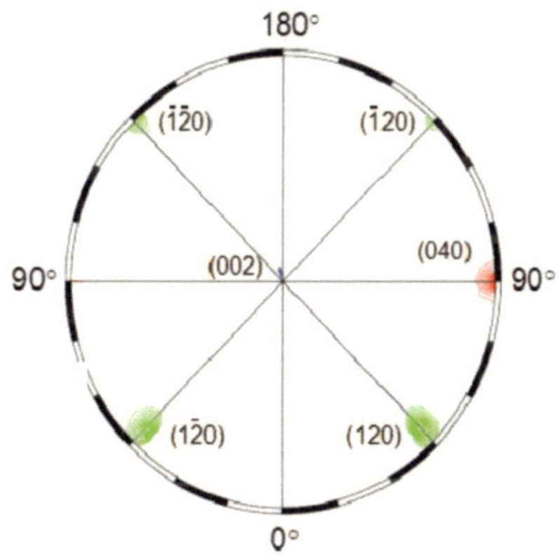

Fig. 14. (a) Sample setting of the L-alanine PSC and the expected diffraction spots. Angle χ runs from 0 to -180° and φ runs from 0 to 180°. (b) Pole figure obtained for the PSC to be compared with (a). (Reprinted from (F. Kimura et al., 2010a): *Cryst. Growth Des.*Vol.10(2010) pp. 48-51. No. 1, DOI: 10.1021/cg90132h. Copyright 2010 American Chemical Society)

Sample	MOMA	Single crystal
PDB Code	present work	2ZQ4
Space group	$P2_12_12_1$	$P2_12_12_1$
Cell dimensions		
a (Å)	51.26	56.48
b (Å)	59.79	73.76
c (Å)	29.95	30.54
V (Å^3)	91, 803	127, 229

Table 5. Comparison of crystal data of lysozyme.

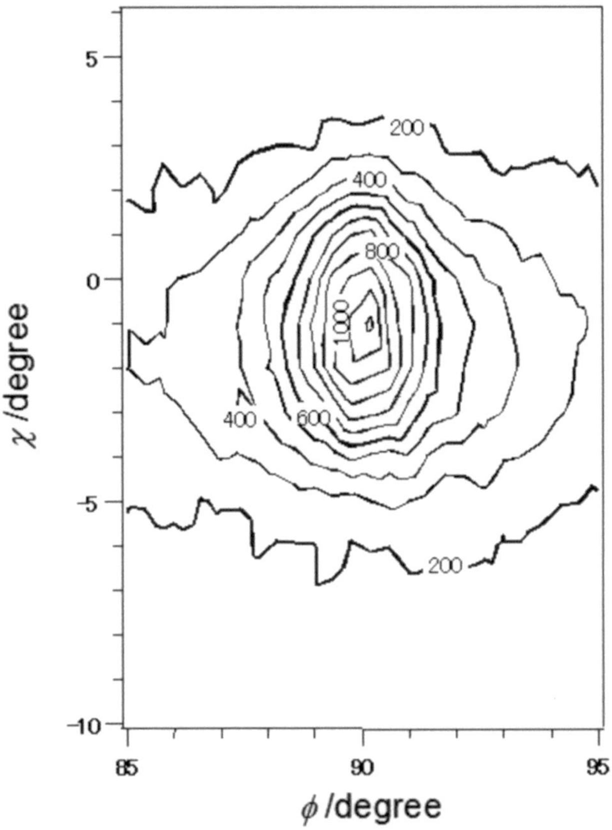

Fig. 15. Contour plot of the diffraction spot for (040) shown in Fig. 4(b) as a function of χ and ϕ. Half maximum of the intensity is ca. 600 from which the average fwhm is estimated to be ca. 4°. (Reprinted from (F. Kimura et al., 2010a): Cryst. Growth Des. Vol.10(2010) pp. 48-51, No. 1, DOI: 10.1021/cg90132h. Copyright 2010 American Chemical Society)

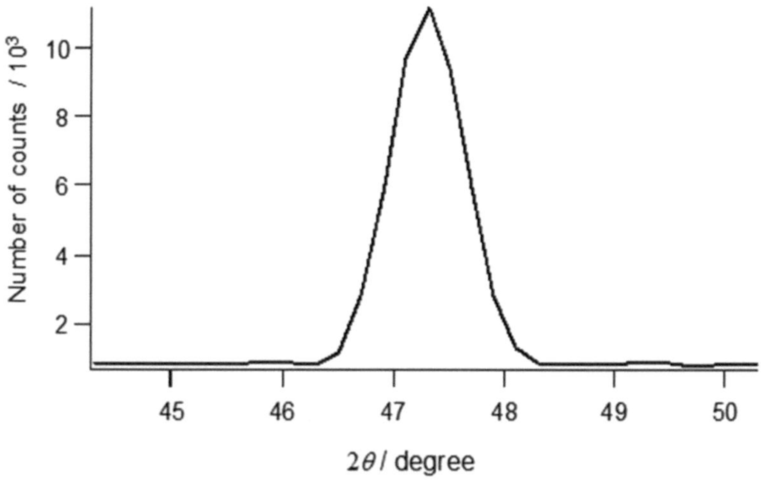

Fig. 16. Diffraction profile for the (040) plane. Baseline is not subtracted. (Reprinted from (F. Kimura et al., 2010a): Cryst. Growth Des. Vol.10(2010) pp. 48-51, No. 1, DOI: 10.1021/cg90132h. Copyright 2010 American Chemical Society)

After subtraction of the baseline, the intensity of the peak is calculated by integration and corrected for Lorentz factor; no absorption correction was performed. The structure refinement was performed with Shelxl-97 using the crystallographic data of neutron diffraction of L-alanine reported in the literature (Wilson et al., 2005), where the atom coordinates, the temperature factors, and the anisotropic temperature factors were fixed. The results are summarized in Table 6. The values of $R1$ and $wR2$ are 0.184 and 0.368, respectively. In the present preliminary study, the integration was performed only along the 2θ direction, and the integration with respect to χ and φ directions was not performed. In addition, absorption correction was not made. Hence, the intensity data could include some error. However, the experimental and calculated results appear to agree satisfactorily, indicating that the MOMA can provide diffraction data that will lead to a successful structure analysis.

7. Possibility of MOMS for application of structure analysis (Matsumoto et al., 2011)

Figure 17 shows the diffraction patterns obtained for $\omega = 0$ rpm (static) and 20 rpm. At $\omega = 0$ rpm, the {002} diffractions did not appear, and the {120} diffractions appeared on the equator. This suggests that the reciprocal vectors **G**{002} are aligned in the direction of the applied magnetic field ($\parallel x$) and the reciprocal vectors **G**{120} are distributed randomly on the yz plane (see Fig. 5). Because the crystal has three mutually orthogonal 2-fold axes (a, b, and c axes), the diffractions belonging to {120} are not distinguishable in a MOMA, neither are the diffractions {002}. With an increase in rotation speed, the {120} diffractions on the equator disappeared and moved to locations around ±45° and ±135° when $\omega = 20$ rpm. At the same time, the {002} diffractions appeared on the meridian at this rotation speed. This indicates that the b^* axis was directed to the z-axis, and there was rotational symmetry about the b^* axis. The alignment of the χ_3 axis ($\parallel b^*$) in the direction of the rotation axis ($\parallel z$) is

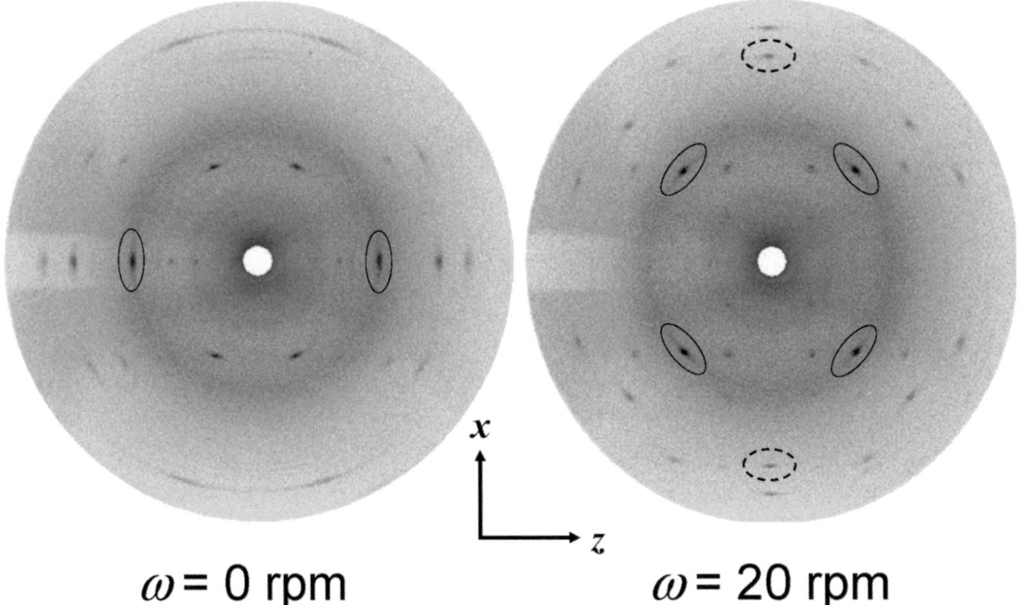

Fig. 17. X-ray diffraction images of a magnetically oriented microcrystal suspension (MOMS) obtained at $\omega = 0$ and 20 rpm. Solid and broken circles indicate the {120} and {002} diffraction spots, respectively. (Reprint from (Matsumoto et al., 2011): Cryst. Growth Des. Vol.11(2011) pp. 945-948, No. 4, DOI: 10.1021/cg200090u. Copyright 2011 American Chemical Society)

based on the behavior of a magnetically uniaxial particle (T. Kimura, 2009; Yamaguchi et al., 2010). We successfully obtained two fiber X-ray diffraction patterns from a microcrystal suspension (crystal size of ca. 2–20 μm) oriented in a magnetic field of 1 T generated by permanent magnets. Diffraction spots were sharp (ca. 2-3° in half width) and well separated. The results suggest a potential use of MOMS for crystal structure analysis of solid materials that do not grow into large crystals but are obtained in the form of microcrystal suspensions.

To summarize, we have introduced a new method, 3D-MOMA (three-dimensional magnetically oriented microcrystal array) method. We use a 3D-MOMA that is a composite in which microcrystals are three-dimensionally aligned. With a 3D-MOMA, we can obtain diffraction equivalent to that from a real single crystal, successfully determining the molecular structure. With this method we can convert a suspension of biaxial microcrystals with sizes of 10 μm or larger to a 3D-MOMA of centimeter sizes by using magnetic fields of 2 T or more. Successful examples of single crystal X-ray analyses of several crystals, including inorganic, organic, and protein crystals, by using 3D-MOMA are presented. A preliminary result of neutron diffraction of 3D-MOMA is also reported. Though the application to the neutron diffraction is preliminary, the MOMA method has high potential for the neutron diffraction method because a large size 3D-MOMA can easily be prepared. The problems to be solved for the future development of this method in the area of neutron single crystal diffraction analyses are discussed.

hkl	F_c^2	F_o^2
2 0 0	12.86	9.50
4 0 0	10.42	8.69
1 1 0	8.26	7.89
2 1 0	1.02	1.06
3 1 0	0.44	1.43
0 2 0	3.55	5.42
1 2 0	77.06	82.74
2 2 0	3.96	3.58
3 2 0	5.95	3.62
1 3 0	0.26	0.37
2 3 0	1.15	0.89
3 3 0	0.00	-0.43
0 4 0	95.88	167.23
1 4 0	19.92	22.84
2 4 0	0.00	-0.10
1 5 0	8.57	10.76
2 5 0	1.99	0.97
0 6 0	10.09	22.55
1 6 0	8.99	12.59
2 6 0	56.46	33.00
0 8 0	0.42	1.51
3 0 1	13.51	11.27
4 0 1	4.90	2.57
0 0 2	23.59	18.36
1 0 2	23.75	21.53
2 0 2	1.88	2.84
1 0 3	0.06	-0.81
2 0 3	2.53	4.15
3 0 3	2.48	-0.81
0 0 4	14.04	16.37
1 0 4	8.01	7.80

Table 6. Calculated and measured F^2 values for diffractions from (*hkl*) planes for L-alanine MOMA.

8. References

Asai, S.; Fujiwara, M.; Kimura, T.; Liang, Z.; Uyeda, C.; Wang, B.; Yamamoto, I.; Zhang, C. (2006). Magnetic Orientation, In Magneto-Science, M. Yamaguchi, Y. Tanimoto, (Eds), ch. 5. pp. 191-247, Tokyo: Kodansha Springer. ISSN 0933-033X

Ataka, M.; Tanaka, S. (1986). The growth of large single crystals of lysozyme. Biopolymers, Vol. 25, No. 2, (February 1986), pp. 337–350.

Blake, C. C.; Koenig, D. F.; Mair, G. A.; North, A. C.; Phillips, D. C.; Sarma, V. R. (1965). Structure of hen egg-white lysozyme. A three-dimensional Fourier synthesis at 2 Angstrom resolution, Nature, Vol. 206, (22 May 1965), pp. 757 - 761, doi:10.1038/206757a0.

Blundell, T. L.; Johnson, L. N. Protein Crystallography; Academic Press: 1976.

Finke, W. (1909), Magnetische Messungen an Platinmetallen und monoklinen Kristallen, insbesondere der Eisen-, Kobult- und Nickelsalze; von Wilhelm Fink., *Ann. Phys.*, Vol. 336, Issue 1, (6. November 1909), PP. 149–168.

Gielen, J. C.; Wolffs, M.; Portale, G.; Bras, W.; Henze, O.; Kilbinger, A. F. M.; Feast, W. J.; Maan, J. C.; Schenning, A. P. H. J.; Christianen, P. C. M. (2009). Molecular Organization of Cylindrical Sexithiophene Aggregates Measured by X-ray Scattering and Magnetic Alignment, *Langmuir* Vol. 25, No. 3, (January 9, 2009), PP. 1272-1276.

Glucksman, M. J.; Hay, R. D.; Makowski, L. (1986). X-ray diffraction from magnetically oriented solutions of macromolecular assemblies, *Science*, Vol. 231, no. 4743, (14 March 1986), pp. 1273-1276.

Graves, b. J.; Crownther, R. L.; Chandran, C.; Rumberger, J. M.; Li, S. H.; Huang, K.-S.;Presky, D. H.; Familletti, P. C.; Wolitzky, B. A.; Burns, D. K. Nature 1994, 367, 537.

Hanson, J. C., Sieker, L. C. & Jensen, L. H. (1973). Sucrose: X-ray refinement and comparison with neutron refinement, *Acta Cryst. Section B*, Vol. 29, Issue 4, (April 1973), pp. 797–808.

Hariss, K. D. M.; Cheung, E. Y. (2004), How to determine structures when single crystals cannot be grown: opportunities for structure determination of molecular materials using powder diffraction data, Chem. Soc. Rev., Vol. 33, No. 8, (22 Sep 2004) pp. 526-538

Herrmann, T.; Güntert, P.; Wüthrich, K. (2002). Protein NMR Structure Determination with Automated NOE Assignment Using the New Software CANDID and the Torsion Angle Dynamics Algorithm DYANA, J. Mol. Biol. Vol. 319, No. 1, (24 May 2002), pp. 209-227.

Kachalova, G. S.; Morozov, V. N.; Morozova, T. Ya.; Myachin, E. T.; Vagin, A. A.; Strokopytov, B. V.; Nekrasov, Yu. V. (1991). Comparison of structures of dry and wet hen egg-white lysozyme molecule at 1.8 Å resolution, *FEBS Lett.*, Vol. 284, Issue 1, (17 June 1991), pp. 91-94.

Kendrew, J. C.; Bodo, G.; Dintzis, H. M.; Parrish, R. G.; Wyckoff, H.; Phillips, D. C. (1958). X-ray crystallography – the first image of myoglobin, Nature, Vol. 181, (8 March, 1058), pp. 662-666.

Kimura, F.; Kimura, T.; Matsumoto, K.; Metoki, N. (2010a). Single-Crystal Neutron Diffraction Study of Pseudo Single Crystal Prepared from Microcrystalline Powder, *Crystal growth & Design*, Vol. 10, No. 1, pp. 48-51.

Kimura, F.; Kimura, T.; Oshima, W., Maeyama, W.; Aburaya, K. (2010b). X-ray diffraction study of a pseudo single crystal prepared from a crystal belonging to point group 2, *J. Appl. Crystallogr.*, Vol. *43*, pp. 151-153.

Kimura, F.; Mizutani, K.; Mikami, B.; Kimura, T. (2011). Single-Crystal X-ray Diffraction Study of a Magnetically Oriented Microcrystal Array of Lysozyme, *Cryst. Growth. Des.*, Vol. *11, No. 1*, pp. 12-15.

Kimura, T.; Yamato, M.; Koshimizu, W.; Koike, M.; Kawai, T. (2000). Magnetic Orientation of Polymer Fibers in Suspension, Langmuir, Vol. 16, No. 2, PP. 858-861.

Kimura, T.; Yoshino, M. (2005). Three-Dimensional Crystal Alignment Using a Time-Dependent Elliptic Magnetic Field, *Langmuir,* Vol. 21, No. 11, (April 30, 2005), pp. 4805-4808.

Kimura, T.; Kimura, F.; Yoshino, M. (2006). Magnetic Alteration of Crystallite Alignment Converting Powder to a Pseudo Single Crystal , *Langmuir*, Vol. 22, No. 8, (March 9, 2006) pp. 3464-3466.

Kimura, T. (2009a). Orientation of Feeble Magnetic Particles in Dynamic Magnetic Fields, *Jpn. J. Appl. Phys.*, Vol. 48, No. 2, (Feb. 2009), pp. 020217(1-3).

Kimura, T.; Chang, C.; Kimura, F.; Maeyama, M. (2009b). The pseudo-single-crystal method: a third approach to crystal structure determination, *J. Appl. Crystallogr.*, Vol. 42, pp. 535–537. ISSN 0021-8898

Kohama, T.; Takeuchi, H.; Usui, M.; Akiyama, J.; Sung, M.-G.; Iwai, K.; Shigeo, A. (2007). In-Situ Observation of Crystal Alignment under a Magnetic Field Using X-ray Diffraction, *Mater. Trans.*, vol. 48, No. 11, pp. 2867-2871. ISSN 1345-9678.

Maret, G. & Dransfeld, K. (1985). Biomolecules and Polymers in High Steady Magnetic Field, in Topics in Applied Physics, Vol. 57, F. Herlach, (Ed), ch. 4. pp. 143-204, Berlin: Springer Verlag

Margiolaki, I.; Wright, J. P. (2008). Powder crystallography on macromolecules, Acta Crystallogr. Section A, Vol. 64, Part 1, (January 2008), pp. 169-180.

Matsumoto, K.; Kimura, F.; Tsukui, S., Kimura, T. (2011). X-ray Diffraction of a Magnetically Oriented Microcrystal Suspension of L-Alanine, *Crystal growth & Design*, Vol. 11, (March 08, 2011), pp. 945-948.

Nagendra, H. G.; Sukumar, N.; Vijayan, M. (1998). Role of water in plasticity, stability, and action of proteins: The crystal structures of lysozyme at very low levels of hydration, *Proteins*, Vol. 32, Issue 2, (1 August 1998), pp. 229-240.

Niimura, N. (1999). Neutrons expand the field of structural biology, Curr. Opin. Struct. Biol., Vol. 9, Issue 5, (1 October 1999), pp. 602-608.

Nye, J. F. (1985). Paramagnetic and diamagnetic susceptibility, in Physical Properties of Crystals, ch. 3, pp. 53-67. Oxford University Press, ISBN0-19-851165-5, New York

Samulski, E. T. (1986). Magnetically Oriented Solutions, *Science*, Vol. 234, No. 4782, (12 December 198), p. 1424.

Sato, T.; Yamada, Y.; Saijo, S.; Hori, T.; Hirose, R.; Tanaka, N.; Sazaki, G.; Nakajima, K.; Igarashi, N.; Tanaka, M.; Matsuura, Y. (2000). Enhancement in the perfection of orthorhombic lysozyme crystals grown in a high magnetic field(10 T), *Acta Crystallogr. Section D*, vol. 56, Issue 8, (August 2000) pp. 1079-1083. ISSN 0907-4449

Sousa, M. C.; Trame, C. B.; Tsuruta, H.; Wilbanks, S. M.; Reddy, V. S.; McKay, D. B. (2000). Crystal and Solution Structures of an HslUV Protease–Chaperone Complex, Vol. 103, Issue 4, (10 November 2000), pp. 633-643

Terada, N.; Suzuki, H. S.; Suzuki, T. S.; Kitazawa, H.; Sakka, Y.; Kaneko, K.; Metoki, N. (2008). *In situ* neutron diffraction study of aligning of crystal orientation in diamagnetic ceramics under magnetic fields, *Appl. Phys. Lett.*, Vol. 92, Issue 11, (17 Mar 2008), pp. 112507(1-3).

van der Beek, D.; Petukhov, A. V.; Davidson, P.; J. Ferr_e, J.; Jamet, J. P.; Wensink, H. H.; Vroege, G. J.; Bras, W.; Lekkerkerker, H. N. W. (2006). Magnetic-field-induced orientational order in the isotropic phase of hard colloidal platelets, *Phys. Rev. E*, Vol. 73, Issue 4, (5 April 2006), pp. 041402(1-10).

Wessels, T., Baerlocher, C. & McCusker, L. B. (1999). Single-Crystal-Like Diffraction Data from Polycrystalline Materials, Science, *Vol. 284 no. 5413, (16 April 1999),* pp. 477–479.

Wilson, C. C.; Myles, D.; Ghosh, M.; Johnson, L. N.; Wang, W. (2005). Neutron diffraction investigations of L- and D-alanine at different temperatures: the search for structural evidence for parity violation, *New J. Chem.*, Vol. 29, (07 Sep 2005), pp. 1318-1322.

Yamaguchi, M.; Ozawa, S.; Yamamoto, I. (2010). Dynamic Behavior of Magnetic Alignment in Rotating Field for Magnetically Weak Particles, Jpn. J. Appl. Phys. Vol. 49, (5 August, 2010), pp. 080213(1-3)

10

Temperature Evolution of the Double Umbrella Magnetic Structure in Terbium Iron Garnet

Mahieddine Lahoubi

Badji Mokhtar-Annaba University, Faculty of Sciences, Department of Physics, Annaba Algeria

1. Introduction

The rare earth iron garnets $RE_3Fe_5O_{12}$ (REIG hereafter) have been discovered at Grenoble (France) (Bertaut & Forrat, 1956; Bertaut et al., 1956) then independently at Murray Hill (USA) (Geller & Gilleo, 1957a, 1957b). These most studied ferrimagnetic materials have a general formula $\{RE^{3+}{}_3\}[Fe^{3+}{}_2](Fe^{3+}{}_3)O_{12}$ where RE^{3+} can be any trivalent rare earth ion or the Yttrium Y^{3+}. The crystal structure is described by the cubic space group Ia$\overline{3}$d-(O_h^{10}) No. 230. Three type of brackets are used to indicate the different coordinations of the cations with respect to the oxygen O^{2-} ions situated in the general positions x, y, z of the sites 96h(1) (Fig. 1). The RE^{3+} ions are located in the dodecahedral sites {24c}(222) whereas the two Fe^{3+} ions are distributed in the octahedral [16a]($\overline{3}$) and tetrahedral (24d)($\overline{4}$) sites.

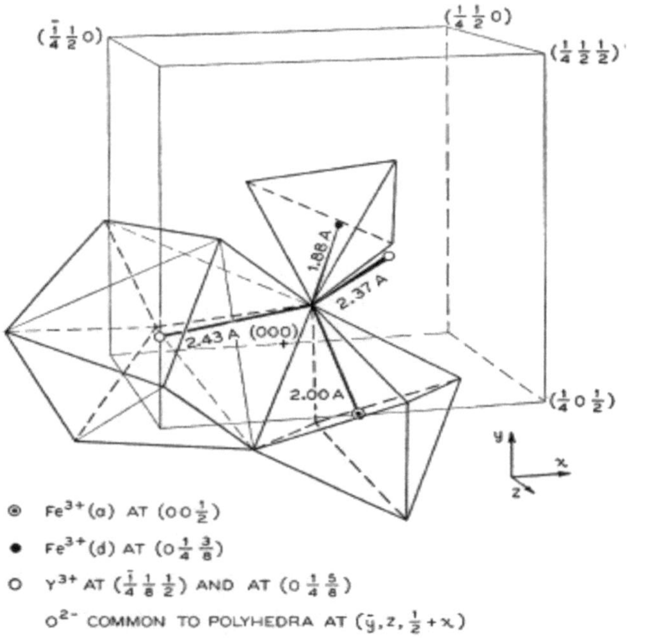

Fig. 1. Crystallographic sites of YIG in the space group Ia$\overline{3}$d (Geller & Gilleo, 1957b)

It is a rather loose structure with a volume of 236.9 Å^3 per formula unit, which has the great technical advantage that it is possible to accommodate a very large variety of cations in the garnet structure. Thus, it is feasible to achieve an enormous range of control of the magnetic properties in the garnet structure system. The largest of REIG that can be formed is SmIG with a lattice parameter of 12.529 Å and the smallest is LuIG with a lattice parameter of 12.283 Å. The REIG have became these famous magnetic compounds by illustrating the Néel theory of ferrimagnetism (Néel, 1948). The strongest superexchange interactions between the two iron sublattices Fe^{3+}[a]–Fe^{3+}(d) are antiferromagnetic and the magnetic moment vectors $\boldsymbol{m}_a$ and $\boldsymbol{m}_d$ are antiparallel. They make YIG, an ideal ferrimagnet with the Néel temperature (T_N) equal to 560 K (Pauthenet, 1958a, 1958b). However, substitution of a magnetic rare earth ion for the diamagnetic Y^{3+} in YIG introduces a third sublattice in the crystallographic site {c} in which $\boldsymbol{m}_{RE}$ are the magnetic moment vectors. In this three sublattices model only weaker and negative antiferromagnetic interactions RE^{3+}\{c\}–Fe^{3+}(d) exist. If M_a, M_d and M_{RE} are the magnetizations of each sublattice, the total bulk ferrite magnetization of REIG is given by the following equation

$$M(\text{REIG}) = |M_{\text{RE}} - (M_{\text{d}} - M_{\text{a}})| \quad (1)$$

For the heavier RE^{3+} ions (Eu^{3+}, ..., Yb^{3+}), M_{RE} is antiparallel to the net resultant of the iron magnetizations $M_{Fe} = M_d - M_a$. We can consider that if we have $M_{Fe} \approx M$(YIG) then the interactions between the rare earth ions are negligible and the equation (1) becomes in a first approximation

$$M(\text{REIG}) = |M_{\text{RE}} - M_{\text{Fe}}| \approx |M_{\text{RE}} - M_{\text{YIG}}| \quad (2)$$

The magnetizations M_a and M_d are still given by the N.M.R values found in YIG (Gonano et al., 1967). Below T_N which is nearly the same for all REIG compounds (554 ± 6) K (Pauthenet, 1958a, 1958b) the magnetization of the rare earth ions M_{RE} can dominate the magnetization M_{Fe}. If the temperature is decreasing, a rapid increasing of M_{RE} is observed because of the large magnetic moment m_{RE}. In heavy rare earth iron garnets, there exists a compensation temperature (T_{comp}) or inversion temperature (T_I) (Herpin, 1968) at which the bulk ferrite magnetization vanishes. For TbIG, T_{comp} is equal to (243.5 ± 0.5 K) and (249.0 ± 0.5 K) for the single crystal and powder samples respectively (Lahoubi et al., 1985; Lahoubi, 1986). In the vicinity of T_{comp}, the magnetic behavior is equivalent to that observed in the antiferromagnet compounds with the existence of the so-called field induced phase transitions which have been studied previously theoretically and experimentally (Zvezdin, 1995). In the Néel model, the RE^{3+} magnetic behavior is described by the pure free ion Brillouin function assuming that the superexchange interactions are represented by the isotropic Weiss molecular field coefficients.

The optical and magneto-optical (MO) properties of REIG and their substituted compounds have also received a substantial interest due to their strong Faraday and Kerr effects. The REIG had their first industrial use in bubble memories more than twenty years ago. Today, these MO materials are the key elements of several technical applications. There are used in Faraday rotators, optical isolators, holographic storage and magnetic field sensors. These applications can be enhanced by using photonic crystals with REIG and such research has yielded promising results. Recently, an additional interest to the REIG has been caused by the prospects for developing materials based on these ferrimagnets for hardware components in the next-generation spintronic devices. The fundamental properties of these REIG and their applications have been enormously reviewed previously and only a few authors are reported here (Dahlbäck, 2006; Geller, 1977; Guillot, 1995; Kazei et al., 1991). At low temperature when both crystal field acting on the RE^{3+} ions and the $RE^{3+}\{c\}$–$Fe^{3+}(d)$ magnetic exchange anisotropies become important (Nekvasil & Veltrusky, 1990), the Néel model must be replaced by a new spin configuration in which the RE^{3+} magnetic moments cease to be antiparallel to the Fe^{3+} magnetization M_{Fe}. So, neutron diffraction experiments have been performed previously to study, the non collinear magnetic structures of the RE^{3+} moments (RE = Dy, Er, Ho, Tb, Tm and Yb) which appear at liquid-helium temperatures (Bertaut et al., 1970; Guillot et al., 1982; Herpin et al., 1960; Hock et al., 1990, 1991; Lahoubi et al., 1984; Lahoubi, 1986, 2012; Pickart et al., 1970; Tchéou et al., 1970a, 1970b). Neutron diffractions experiments have been also made to follow the temperature dependence of these *''umbrella''* magnetic structures in HoIG (Guillot et al., 1983, 1984), ErIG (Hock et al., 1991), DyIG (Lahoubi et al., 2009, 2010).

In the present chapter, we will present the temperature evolution of the Tb^{3+} magnetic ordering in TbIG using powder neutron diffraction experiments combined with magnetic field magnetization measurements on single crystal (Lahoubi et al., 1984, 1985; Lahoubi, 1986; Lahoubi et al., 1997; Lahoubi, 2012). The experimental techniques are described in the section 2. The principle of the non-polarized neutrons diffraction with the preliminarily experiment at 614 K are introduced in the main section 3. The neutron diffraction results obtained at high and low temperatures are discussed using the predictions of the symmetry analysis and compared with the data of magnetization measurements in the sections 4 and 5 respectively. The *"Representation Analysis"* of Bertaut (Bertaut, 1968, 1971, 1972) is applied to the paramagnetic space group $Ia\bar{3}d$ for determining all possible *''umbrella''* magnetic structures in this *"cubic description"*. The method of the so-called *''symmetry lowering device''* (Bertaut, 1981) is required in the treatment for the determination of the best subgroups of $Ia\bar{3}d$ when the temperature is decreasing below T_N until liquid-helium temperatures. The *"basis vectors of irreductible representations"* of the distorted space group $R\bar{3}c$ are chosen in the *"rhombohedral description"*. The thermal variations of the parameters of the *"double umbrella"* magnetic structure constitute the section 6 which will be followed by a conclusion in the section 7.

2. Experimental techniques

The neutron diffraction experiments were measured on polycrystalline sample of TbIG owing to some severe extinctions which appear when a high quality single crystal is used (Bonnet et al., 1979). The first set of patterns have been recorded previously at the Centre d'Etudes Nucleaires de Grenoble (CENG) CEA Grenoble, France (Lahoubi et al., 1984; Lahoubi, 1986; Lahoubi et al., 1997) using the famous "Position Sensitive Detector" (PSD) detector (Convert et al., 1983; Roudaut et al., 1983). For the study, ten temperatures are chosen in the cryostat: 4.2, 20, 54, 68, 80, 109 ± 2, 127 ± 5, 160, 208 ± 2 and around T_{comp} at 244 ± 10 K. In the furnace, the temperatures are: above T_N (T = 614 K) and below T_N (T = 283, 400 and 453 K). The time of counting for some temperatures has been in the order of ten hours. The patterns were recorded with a wavelength λ equal to 2.49 Å and filters to avoid $\lambda/2$ contaminations are used. The second set of patterns has been collected recently on the high flux diffractometer D1B at the Institut Laue–Langevin Grenoble, France. The value of the wavelength λ is equal to 2.52 Å and four temperatures 5, 13, 20 and 160 K have been chosen below room temperature. The resolution of the multidetector is 0.2°. For the magnetic study, the magnetization of a flux-grown single crystal of TbIG was measured in dc magnetic fields produced either by a superconducting coil up to 80 kOe or a Bitter coil up to 150 and 200 kOe. The first series of experiments was performed at the Louis Néel Laboratory of Grenoble, France (currently Néel Institut) and the second series at the Service National des Champs Intenses (SNCI) of Grenoble (currently LNCMI). In the 4.2–300 K temperature range, the external magnetic field was applied parallel to the crystallographic directions <111>, <110> and <100> successively. The spherical sample with 5.5 mm of diameter and 0.4313 g of weight is oriented along these three main crystallographic directions by the X-ray Laüe technique with an error less than 1°. The isothermal magnetizations $M_T(H)$ as a function of internal magnetic field H (the external magnetic field H_{ex} minus the demagnetizing magnetic field of the sphere H_d) are analyzed by the least-squares refinement technique. The measured spontaneous magnetizations $M_S^{mes}(T)$ are reported in (μ_B/mol) where one mole is equal to 2(TbIG) formula units.

3. Principle of neutron diffraction

We present here only the case of the diffraction of the non-polarized (or unpolarized) neutrons. This method was used firstly in the determination of the magnetic structure in MnO (Shull & Smart, 1949). A multitude of others followed after: approximately thousands of magnetic structures have now been solved. The use of polarized neutrons ten years after has been the next progress in magnetic neutron scattering (Nathans et al., 1959). In this second method, the incident neutron beam is polarized either up or down, and the neutron intensities scattered by the sample are compared for the two possible states of the incident polarization. Compared to the non-polarized neutrons experiments, an interference term between the nuclear and the magnetic amplitudes adds or subtracts to the intensities depending on the direction of the polarization. For small magnetic amplitudes, the enhancement of sensitivity is remarkable and we know in the case of ferromagnetic and ferrimagnetics to determine the form factor of the magnetic atoms and to reconstruct the spin (or magnetization) density within the cell. Such investigations are now very common: several hundred spin density investigations have already been performed. In YIG (Bonnet et al., 1979), the polarized neutrons were used in addition to the study of the covalency effects. A book edited recently (Chatterji, 2006) is mostly devoted to the application of polarized neutron scattering from magnetic materials.

3.1 Non-polarized neutrons diffraction: Determination of the diffracted intensities

The scattering of neutron by a magnetic atom is composed essentially of two terms: a nuclear neutron scattering and a magnetic neutron scattering. The first term is due to a "*neutron-nucleus*" nuclear interaction giving the nuclear diffraction which yields information on the spatial arrangement of the nuclei of the atoms in crystal. In the second term, the neutron has a magnetic moment which can interact with the unpaired electrons of the magnetic atoms through the "*dipole-dipole*" interaction conducting to the magnetic diffraction (Bacon, 1975). The nuclear and the magnetic neutron scattering are incoherent in our non-polarized neutrons diffraction. For a magnetic material in the paramagnetic state obtained at $T > T_N$, there is no magnetic contribution and only the nuclear diffraction exists with the diffracted nuclear intensity (I_N). The total diffracted intensity $I(T > T_N)$ is equal to

$$I(T > T_N) = I_N \quad (3)$$

In the ordered magnetic state ($T < T_N$) a magnetic intensity (I_M) is added to the intensity I_N. We can measure finally the total diffracted intensity $I(T < T_N)$ with the following equation

$$I(T < T_N) = I_N + I_M \tag{4}$$

The diffracted magnetic intensity I_M is then obtained by the difference

$$I_M(T < T_N) = I - I_N \tag{5}$$

This method is applied if we have a good counting statistics and also if I_M is not too lower then I_N. We will recall briefly the following useful expressions of the nuclear and magnetic diffracted intensities for the case of a polycrystalline sample.

3.1.1 Nuclear intensity I_N

In the paramagnetic state, the nuclear intensity $I_N(\boldsymbol{H})$ can be calculated through the nuclear structure factor $F_N(\boldsymbol{H})$ with the following expressions

$$I_N(\boldsymbol{H}) = P(\boldsymbol{H})|F_N(\boldsymbol{H})|^2 \tag{6}$$

$$F_N(\boldsymbol{H}) = \sum_{j=1}^{N} b_j \exp(2\pi i\boldsymbol{H}\cdot\boldsymbol{r}_j)\exp(-B_j|\boldsymbol{H}|^2/2) \tag{7}$$

with $P(\boldsymbol{H})$, the multiplicity of the reflector plane, and $\boldsymbol{H}$, the corresponding scattering vector which is concentrated in the Bragg peaks (hkl) of the reciprocal lattice, b_j and B_j being respectively the neutron scattering length and the Debye-Waller factor of the jth atom in the jth position vector $\boldsymbol{r}_j$ among N, the total number of atoms in the crystallographic unit cell. The nuclear structure factor appears as a scalar factor.

3.1.2 Magnetic intensity I_M

When the materials are magnetically ordered, the magnetic diffracted intensity $I_M(\boldsymbol{H})$ has the same form that found before for $I_N(\boldsymbol{H})$ but we consider only the perpendicular component to $\boldsymbol{H}$ of the magnetic structure factor $\boldsymbol{F}_M(\boldsymbol{H})$. We can write these equations

$$I_M(\boldsymbol{H}) = \left||\boldsymbol{F}_M(\boldsymbol{H})|^2 - |\boldsymbol{H}\cdot\boldsymbol{F}_M(\boldsymbol{H})|^2/|\boldsymbol{H}|^2\right| \tag{8}$$

$$\boldsymbol{F}_M(\boldsymbol{H}) = \eta\sum_{j=1}^{N} \boldsymbol{S}_j f_j(\boldsymbol{H})\exp(2\pi i\boldsymbol{H}\cdot\boldsymbol{r}_j)\exp(-B_j|\boldsymbol{H}|^2/2) \tag{9}$$

with $f_j(\boldsymbol{H})$, the magnetic form factor of the jth spin $\boldsymbol{S}_j$, and $\eta = |\gamma|e^2/2m_0c = +0.02696$ cm, the magnetic diffusion length ($\gamma = -1.91348$). The magnetic structure factor has a vector form.

3.1.3 Observed I_{obs} and calculated I_{cal} intensities

The observed diffracted intensity I_{obs} is integrated on number of counts. The value (in barns) is corrected by the Lorentz factor $L(\boldsymbol{H})$ and normalized by the scale factor K where the definitions are expressed respectively by the following equations

$$L(\boldsymbol{H}) = 1/\sin\theta \sin(2\theta) ; K = I_{cal} / I_{obs} \quad (10)$$

The value of K is obtained by the refinement of the nuclear structure. For each I_{obs}, an absolute error ΔI_{obs} is found. She is associated to the sum of a statistical error with the appreciated one on the continu fund noise. The corresponding observed relative error is noted $\Delta I_{obs}/I_{obs}$. Using the least square method, we can define the reliability factor R

$$R = \sum|I_{obs} - I_{cal}| / \sum|I_{obs}| \quad (11)$$

For each calculated diffracted intensity I_{cal}, a calculated relative error $\Delta I_{cal}/I_{cal}$ is attributed. The non-polarized neutron diffraction is then based in the comparison between I_{cal} and I_{obs}.

3.2 Preliminarily neutron diffraction at 614 K

In the pattern (not show) recorded at T = 614 K higher than T_N (568 ± 2 K) (Pauthenet, 1958a, 1958b), only the nuclear contribution exists and the reflections (hkl) are indexed with the general extinction rule of the cubic space group Ia$\overline{3}$d, h + k + l = 2n. Attention was paid to the thermal agitation of the jth atom in the different sites by introducing the corresponding isotropic Debye-Waller factors B_j. For the determination of the calculated nuclear intensities $(I_N)_{cal}$, the previous scattering lengths (Bacon, 1972) are used: b(O) = 0.580; b(Fe) = 0.95; b(Tb) = 0.76 (in units of 10^{-12} cm/atom). We can determine the number of refinement cycles and to choose the parameters with a sufficient number of iterations. The previous best values of the parameters x = -0.0279, y = 0.0555, z = 0.1505 found previously at T = 693 K (Tchéou et al., 1970c) for the general positions (96h) of the oxygen atoms are used as constant parameters and the refinement is made only on the scale factor K and after the corresponding isotropic Debye-Waller factors B_j. A good agreement with a reliability factor of order 10 % is found for the lattice parameter a = 12.470 ± 0.004 Å and K = 0.42 ± 0.02 with B_h(O) = 0.88; B_d(Fe) = B_a(Fe) = 0.82 and B_c(Tb) = 0.81. We observe that the observed intensities of the reflections (400) and (420) are lower than the corresponding calculated intensities and the refinement of the temperature parameters B_j has a little influence on the values of the observed intensities (for example when B_c(Tb) change from 0.80 to 0.30).

4. Neutron diffraction study at high temperature

The neutron diffraction patterns below T_N are reported in Fig. 2 for 453 and 400 K. Both patterns at (283 K) and at 5 K (D1B) are presented for a useful comparison in Fig. 3. A magnetic intensity I_M(hkl) is superimposed to the nuclear intensity I_N(hkl): we have then, I(hkl) = I_N(hkl) + I_M(hkl). In these patterns, we have indexed all the reflections (hkl) in the chemical cell with the same extinction rule characterized by a wave vector $\boldsymbol{k}$ = 0. In these temperatures, the magnetic structure factors $\boldsymbol{F}_M^{(a)}$(hkl), $\boldsymbol{F}_M^{(d)}$(hkl) and $\boldsymbol{F}_M^{(c)}$(hkl) associated to each magnetic sublattice of TbIG are used to describe the collinear ferrimagnetic state of the Néel model along the easy axis [111] found by magnetization measurements (Lahoubi et al., 1985 and refs. herein). Two types of reflections (hkl) appear in the patterns: (h = odd, k = odd, l = even) and h = even, k = even, l = even). For the related reflections (hkl) with (h = even, k = odd, l = odd) and (h = odd, k = even, l = odd) and cyclic permutations (c.p.) of h, k, l must be done in the expressions of the magnetic structure factors. A complete description can be found in the previous paper on the neutron diffraction of HoIG (Guillot et al., 1984).

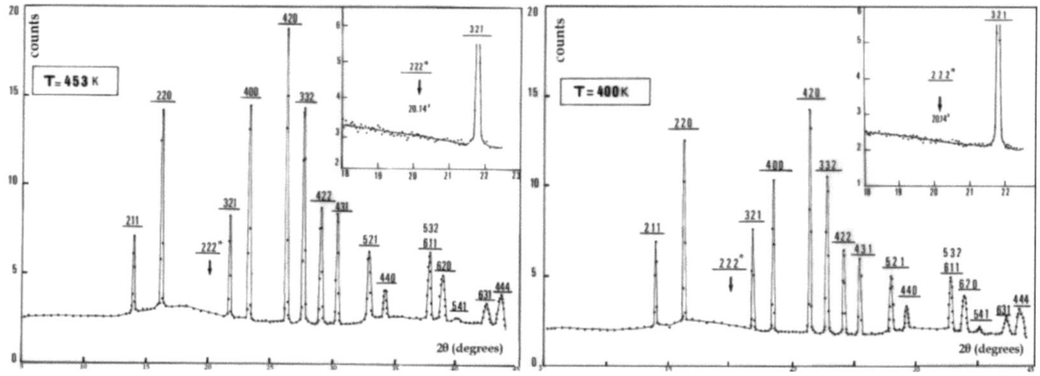

Fig. 2. Neutron diffraction patterns at 453 and 400 K

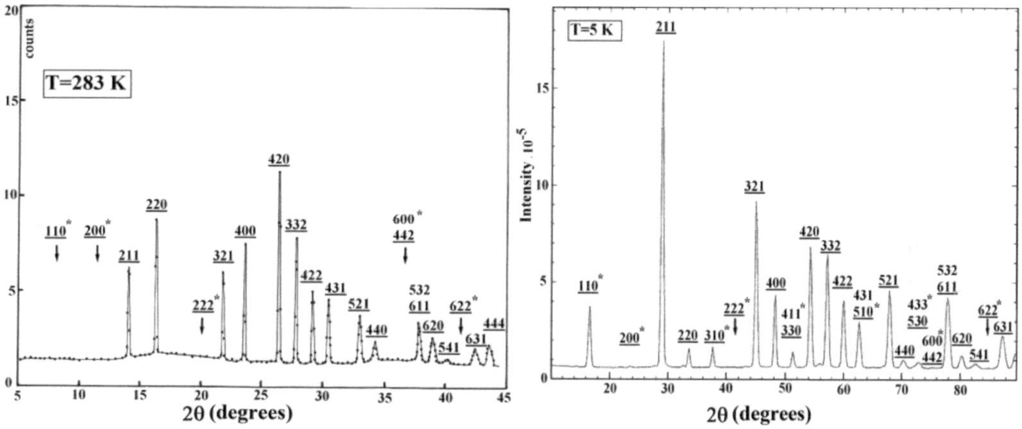

Fig. 3. Neutron diffraction patterns at 283 and 5 K (D1B)

4.1 Results and discussion

We have observed during the refinement that the calculated parameters m_{Tb}, m_a and m_d of the three magnetic sublattices are depending highly of the choice of the magnetic form factors, $f_{Tb}(hkl)$ and particularly those related to the two iron sublattices $f_a(hkl)$ and $f_d(hkl)$. We choose at first, the theoretical magnetic form factors determined in the Hartree-Fock description based on the free ion model for the Tb^{3+} ion, $f_{Tb}(hkl)$ (Blume, et al., 1962) and those calculated (Watson & Fremann, 1961) $f_a(hkl)$ and $f_d(hkl)$ with the equality $f_a(hkl) = f_d(hkl)$ for the iron sublattices. Secondly, two previous experimental magnetic form factors $f_a(hkl)$ and $f_d(hkl)$ are also tentatively used. The first values (Bonnet, 1976; Bonnet et al., 1979) obtained by polarized neutrons experiments on YIG single crystal indicate that $f_a(hkl)$ and $f_d(hkl)$ are different that the free ion value and $f_a(hkl) > f_d(hkl)$. The second values of $f_a(hkl)$ and $f_d(hkl)$ with the relation $f_a(hkl) < f_d(hkl)$ have been found previously by powder neutron diffraction experiments (Guillot et al., 1983). In the study of the *"umbrella structure"* at low temperature on HoIG sample prepared by grinding of single crystals (Guillot et al., 1984), an evaluation of m_a and m_d using the Bonnet' $f_a(hkl)$, $f_d(hkl)$ determinations was made. At 4.2 K, these moments were found equal respectively to 4,01 and 4,26 μ_B. These values are smaller than the theoretical ground state $^6S_{5/2}$ saturated magnetic moment (5 μ_B). The observed reduction of the moments is explained by covalent bonding for YIG (Bonnet, 1976; Fuess et al., 1976) or topological frustration for FeF_3 (Ferey et al., 1986). When the proposed values (Guillot., et al 1983) are chosen in the refinement of the neutron diagrams at 453, 400 and 283 K, we observe that the calculated intensities of the reflections (211) and (220) which have a high magnetic contribution are lower than the observed intensities. Consequently, we shall consider in this work for the Fe^{3+} ions only the theoretical values of f_a(hkl) and f_d(hkl) (Watson & Fremann, 1961). In this condition, m_a and m_d are not considered as fitting parameters in the first cycle of the refinement and the N.M.R values are used (Gonano et al., 1967). Working in this hypothesis leads to the best values of m_{Tb} at each temperature. Finally, in the second cycle of the refinement based on twenty reflections, the parameters m_a and m_d are fitted by the self-consistent calculation of m_{Tb}. At 453 and 400 K, we obtain respectively for m_{Tb} the refined values 0.50 ± 0.10 μ_B and 0.60 ± 0.10 μ_B with a reliability factor R varying in the range of 11.8–9.3 and 11.6–8.6 % if the refinement is makes only on the twelve first reflections (Lahoubi, 1986). The results lead to a good agreement between the bulk calculated magnetizations M_S^{cal}(TbIG) and the observed spontaneous magnetization M_S^{obs}(TbIG) (Pauthenet, 1958a, 1958b). The result at 283 K (1.15 μ_B with R = 8.5 %) (Lahoubi, 1986) is similar to that found previously (Bonnet, 1976; Fuess et al., 1976) but with a poor agreement for m_a and m_d which have been found lower by comparison with those determined by N.M.R (Gonano et al., 1967).

5. Neutron diffraction study at low temperature and symmetry analysis

The neutron patterns recorded at 5 (D1B), 13 (D1B), 20, 54, 68, 80, 109, 127, 160, 208 and 244 K are presented from Fig. 3 up to Fig. 9.

5.1 Results and discussion

At T = 5 K (Fig. 3), two types of reflection appear. In addition to the earlier reflections (hkl) observed previously, pure superstructure lines (hkl)* forbidden by the nuclear space group Ia$\overline{3}$d are present and we have $I = I_M$(hkl)*: (110)*, (310)*, (411, 330)*, (433, 530)* and (510)*.

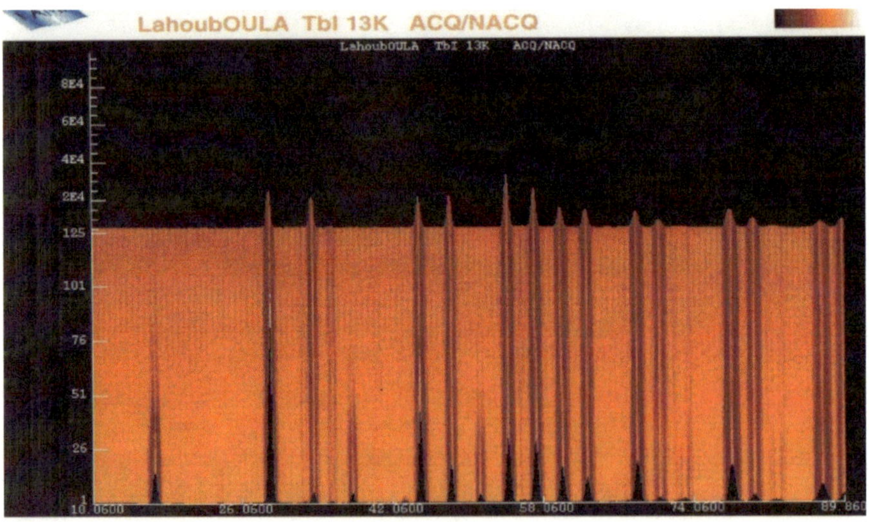

Fig. 4. Neutron diffraction pattern at 13 K (D1B)

In contrast to previous findings at 1.5 K (Bertaut et al., 1970; Tchéou al., 1970a) and 4.2 K (Lahoubi et al., 1984), the small superstructure lines (200)* and (600, 442)* have been observed recently and confirmed only at 5 K (D1B) (Lahoubi, 2012) with a sensibility equal to 0.5 and 1% respectively, the line (110)* being chosen as a reference. The same order of magnitude (1/276) by comparison with the previous result (Hock et al., 1990) was found for the line (200)*. Above 5 K, they are not observed in the neutron diffraction pattern at 13 K (Fig. 4).

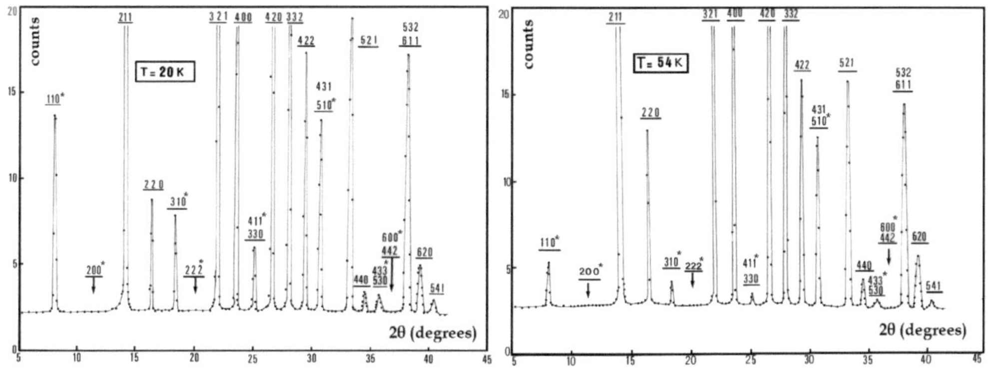

Fig. 5. Neutron diffraction patterns at 20 and 54 K

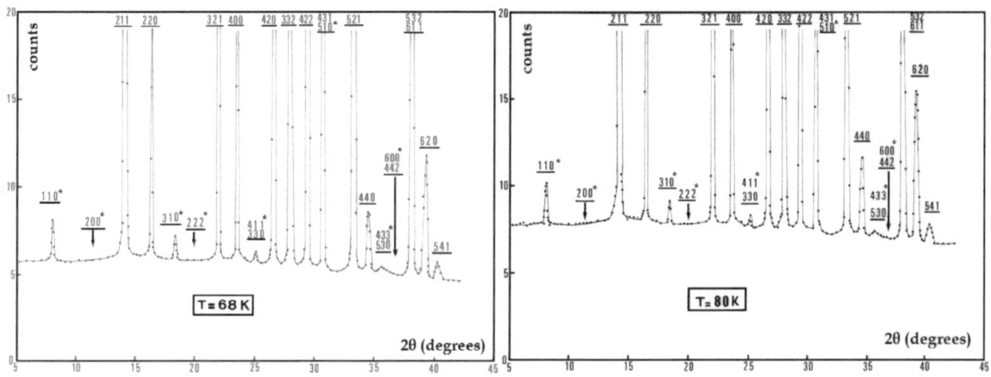

Fig. 6. Neutron diffraction patterns at 68 and 80 K

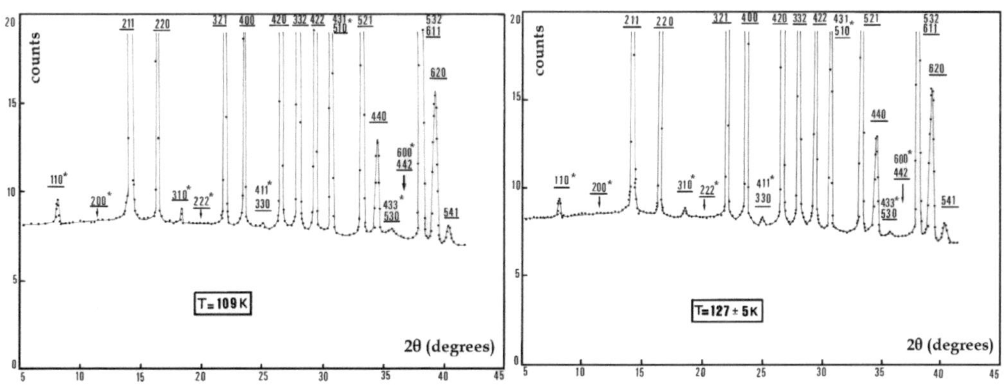

Fig. 7. Neutron diffraction patterns at 109 and 127 ± 5 K

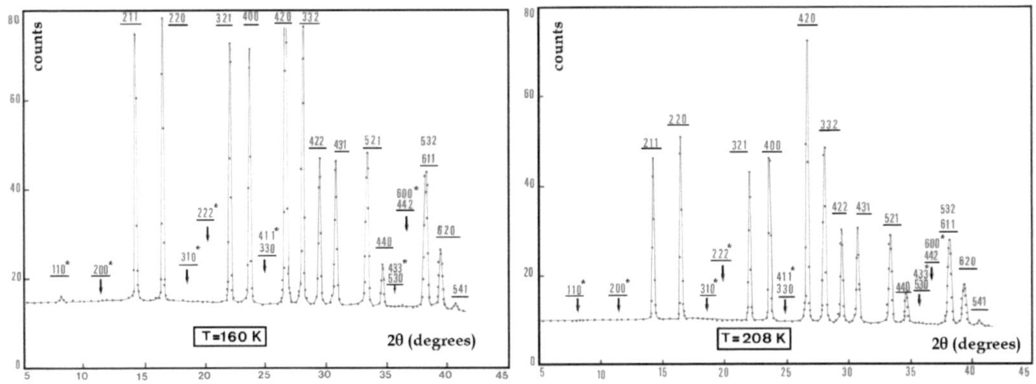

Fig. 8. Neutron diffraction patterns at 160 and 208 K

All the superstructure lines appear without any ambiguity from 5 K up to 127 ± 5 K. They are indexed with the same extinction rule ($h^* + k^* + l^* = 2n$) and imply the signature of a change of the collinear magnetic structure of the RE^{3+}\{24c\} ions of the Néel model to a non collinear magnetic structure with the wave vector $\boldsymbol{k} = 0$. At 160 and 208 K, the superstructure line $(110)^*$ is resolved with the best sensibility (0.3 and 1% respectively) which is equal to the ratio of the peak to background normalized to the intensity of the line (211) (see the details for 2θ up to 15° in the left of Fig. 9).

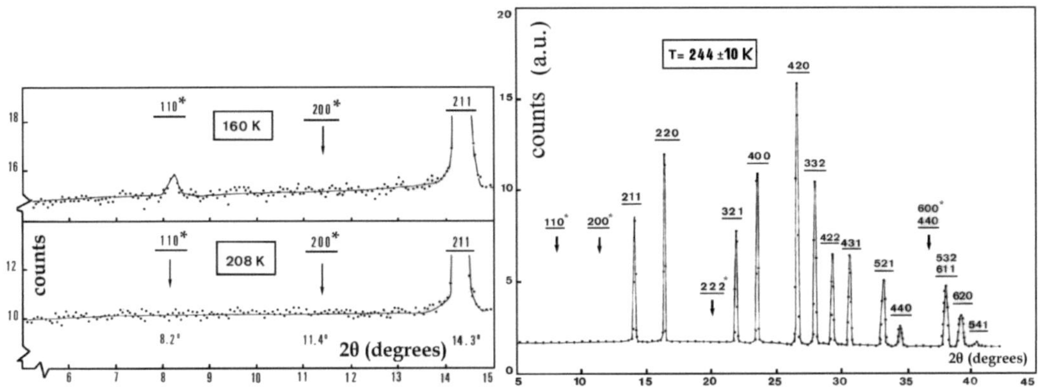

Fig. 9. Details at 160 and 208 K (left) and neutron diffraction pattern at 244 ± 10 K (right)

The chemical cell is equal to the magnetic cell, thus the primitive translation noted (1|1/2,1/2,1/2) of the crystallographic lattice (I) (Hahn, 1983) is a primitive translation of the magnetic lattice (I). Based on the numbered positions gathered on Table 1 we can write for the RE^{3+}\{24c\} and Fe^{3+}(24d) sublattices that $S_j = S_j + S_{j+12}$ (j = 1-3); for Fe^{3+}[16a] sublattice, we have $S_p = S_p + S_{p+8}$ (p = 1-4) and $S_{p'} = S_{p'} + S_{p'+8}$ (p' = 5-8). It means that two spin vectors S_j, S_{j+12}, S_p, S_{p+8} and $S_{p'}$, $S_{p'+8}$ are coupled ferromagnetically. To discuss the corresponding magnetic structure factors $\boldsymbol{F}_M^{(a)}(hkl)^*$, $\boldsymbol{F}_M^{(d)}(hkl)^*$ and $\boldsymbol{F}_M^{(c)}(hkl)^*$ associated to each magnetic sublattice in TbIG, the four earlier linear combinations of four spin vectors introduced by Bertaut (Bertaut, 1963) labeled ***F***, ***G***, ***C*** and ***A*** are used. These four magnetic modes (one ferromagnetic and three antiferromagnetics) form the "*basis of irreductible representations*".

Notations C_j , C'_j (j = 1-3) and {i = 1-6} (Wolf et al., 1962; Wolf, 1964 and Pearson et al., 1965) Local axes in $D_2(222)$ symmetry ***U*** g_z ***V*** g_x (g_y) ***W*** g_y (g_x) Numbered positions of RE^{3+} ions in site {24c}(222)			Notations D_j, D'_j (j = 1-3) Numbered positions of Fe^{3+} ions in site (24d)($\bar{4}$)		Unprimed and primed (') notations respectively for p = 1-4 and p' = 5-8 with c.p. Numbered positions of Fe^{3+} ions in site [16a]($\bar{3}$)	
C_1 1 3	100 $01\bar{1}$ 011	**(1)** 1/8,0,1/4 **(7)** 7/8,0,3/4 **(13)** 5/8,1/2,3/4 **(19)** 3/8,1/2,1/4	D_1	**(1)** 3/8,0,1/4 **(7)** 5/8,0,3/4 **(13)** 7/8,1/2,3/4 **(19)** 1/8,1/2,1/4		**(1)** 0,0,0 **(9)** 1/2,1/2,1/2
C'_1 2 4	100 011 $0\bar{1}1$	**(4)** 3/8,0,3/4 **(10)** 5/8,0,1/4 **(16)** 7/8,1/2,1/4 **(22)** 1/8,1/2,3/4	D'_1	**(4)** 1/8,0,3/4 **(10)** 7/8,0,1/4 **(16)** 5/8,1/2,1/4 **(22)** 3/8,1/2,3/4		**(2)** 0,1/2,1/2 **(10)** 1/2,0, 0
C_2 3 5	010 $\bar{1}01$ 101	**(2)** 1/4,1/8,0 **(8)** 3/4,7/8,0 **(14)** 3/4,5/8,1/2 **(20)** 1/4,3/8,1/2	D_2	**(2)** 1/4,3/8,0 **(8)** 3/4,5/8,0 **(14)** 3/4,7/8,1/2 **(20)** 1/4,1/8,1/2		**(3)** 1/2,0,1/2 **(11)** 0,1/2,0
C'_2 4 6	010 101 $10\bar{1}$	**(5)** 3/4,3/8,0 **(11)** 1/4,5/8,0 **(17)** 1/4,7/8,1/2 **(23)** 3/4,1/8,1/2	D'_2	**(5)** 3/4,1/8,0 **(11)** 1/4,7/8,0 **(17)** 1/4,5/8,1/2 **(23)** 3/4,3/8,1/2		**(4)** 1/2,1/2,0 **(12)** 0,0,1/2
C_3 5 1	001 $\bar{1}10$ 110	**(3)** 0,1/4,1/8 **(9)** 0,3/4,7/8 **(15)** 1/2,3/4,5/8 **(21)** 1/2,1/4,3/8	D_3	**(3)** 0,3/4,3/8 **(9)** 0,3/4,5/8 **(15)** 1/2,3/4,7/8 **(21)** 1/2,3/4,3/8	(')	**(5)** 1/4,1/4,1/4 **(13)** 3/4,3/4,3/4
C'_3 6 2	001 110 $\bar{1}10$	**(6)** 0,3/4,3/8 **(12)** 0,1/4,5/8 **(18)** 1/2,1/4,7/8 **(24)** 1/2,3/4,1/8	D'_3	**(6)** 0,3/4,1/8 **(12)** 0,1/4,7/8 **(18)** 1/2,1/4,5/8 **(24)** 1/2,3/4,3/8	(')	**(6)** 1/4,3/4,3/4 **(14)** 3/4,1/4,1/4
					(')	**(7)** 3/4,1/4,3/4 **(15)** 1/4,3/4,1/4
					(')	**(8)** 3/4,3/4,1/4 **(16)** 1/4,1/4,3/4

Table 1. Notations, numbers and positions in the unit cell of the three magnetic sublattices. ***U***, ***V*** and ***W*** are the local axes of the RE^{3+} in D2(222) symmetry

The adapted magnetic modes for the RE^{3+}\{24c\} and Fe^{3+}(24d) sublattices with j = 1-3 are

$$\boldsymbol{F}_j = \boldsymbol{S}_j + \boldsymbol{S}_{j+3} + \boldsymbol{S}_{j+6} + \boldsymbol{S}_{j+9}; \boldsymbol{G}_j = \boldsymbol{S}_j - \boldsymbol{S}_{j+3} + \boldsymbol{S}_{j+6} - \boldsymbol{S}_{j+9};$$
$$\boldsymbol{C}_j = \boldsymbol{S}_j + \boldsymbol{S}_{j+3} - \boldsymbol{S}_{j+6} - \boldsymbol{S}_{j+9}; \boldsymbol{A}_j = \boldsymbol{S}_j - \boldsymbol{S}_{j+3} - \boldsymbol{S}_{j+6} + \boldsymbol{S}_{j+9} \quad (12)$$

For the Fe^{3+}[16a] sublattice, it is necessary to consider two distinct magnetic modes. The first chosen notation is the unprimed magnetic modes for the numbered spins $S_p = S_p + S_{p+8}$ (p = 1-4)

$$\boldsymbol{F} = \boldsymbol{S}_1 + \boldsymbol{S}_2 + \boldsymbol{S}_3 + \boldsymbol{S}_4; \boldsymbol{G} = \boldsymbol{S}_1 - \boldsymbol{S}_2 + \boldsymbol{S}_3 - \boldsymbol{S}_4; \boldsymbol{C} = \boldsymbol{S}_1 + \boldsymbol{S}_2 - \boldsymbol{S}_3 - \boldsymbol{S}_4; \boldsymbol{A} = \boldsymbol{S}_1 - \boldsymbol{S}_2 - \boldsymbol{S}_3 + \boldsymbol{S}_4 \quad (13)$$

The second chosen notation of the magnetic modes is the primed notation (') for the numbered spins $S_{p'} = S_{p'} + S_{p'+8}$ (p' = 5-8)

$$\boldsymbol{F}' = \boldsymbol{S}_5 + \boldsymbol{S}_6 + \boldsymbol{S}_7 + \boldsymbol{S}_8; \boldsymbol{G}' = \boldsymbol{S}_5 - \boldsymbol{S}_6 + \boldsymbol{S}_7 - \boldsymbol{S}_8; \boldsymbol{C}' = \boldsymbol{S}_5 + \boldsymbol{S}_6 - \boldsymbol{S}_7 - \boldsymbol{S}_8; \boldsymbol{A}' = \boldsymbol{S}_5 - \boldsymbol{S}_6 - \boldsymbol{S}_7 + \boldsymbol{S}_8 \quad (14)$$

For the RE^{3+} ions in the Wyckoff site \{24c\}, the local axes (***U,V,W***) in the D_2(222) symmetry are identified to the parameters g_α ($\alpha = x,y,z$) of the magnetic tensor $\tilde{g}$ in the hypothesis of the effective spin Hamiltonian model (Wolf et al., 1962, Wolf, 1964). The particular superstructure lines (222)* and (622)* are not observed in the whole temperature range below T_N. For example, the associated magnetic structure factors are all equal to zero

$$\boldsymbol{F}_M^{(a)}(222)^* = \boldsymbol{F} - \boldsymbol{F}' = 0; \boldsymbol{F}_M^{(d)}(222)^* = +i(\boldsymbol{A}_1 + \boldsymbol{A}_2 + \boldsymbol{A}_3) = 0 \quad (15)$$

$$\boldsymbol{F}_M^{(c)}(222)^* = -i(\boldsymbol{A}_1 + \boldsymbol{A}_2 + \boldsymbol{A}_3) = 0; \boldsymbol{F}_M^{(c)}(622)^* = +i(\boldsymbol{A}_1 - \boldsymbol{A}_2 - \boldsymbol{A}_3) = 0 \quad (16)$$

In these conditions, the spins vectors $\boldsymbol{S}_j$ and $\boldsymbol{S}_{j+6}$ (j = 1-3) of Tb^{3+}\{24c\} and Fe^{3+}(24d) ions are coupled ferromagnetically and the symmetry operation ($\bar{1}$|0,0,0) is an inversion center. These significant properties related to magnetic symmetry involve that all the magnetic modes $\boldsymbol{A}_j$ are absent. This absence will be accompanied by the elimination of the modes $\boldsymbol{C}_j$. Consequently, the above crystallographic sites split into six magnetically inequivalent sublattices C_j, C'_j and D_j, D'_j with (j = 1-3) respectively as it is indicated on Table 1. In each sublattice we have four ions which are equivalent under the symmetry operations: ($\bar{1}$|0,0,0) and (1| 1/2,1/2,1/2). The sublattices C_2 and C_3 are related to C_1 by a rotation of 120 and 240° around the 3-fold symmetry [111] axis (also for C'_j); the same remarks can be made for the sublattices D_j, D'_j (j = 1-3). Contrary to previous spin rotation observed at T_{comp} (260 ± 5 K) by mössbauer spectroscopy (Hong et al., 2004), no deviation from the colinearity along the easy axis [111] for the spins of the Fe^{3+}[16a] ions is evidenced around T_{comp} (244 ± 10 K) (right of Fig. 9). In this *T*-region, this sublattice is described by one ferromagnetic configuration of the cubic magnetic modes ***F*** and ***F'*** with the equality of the left of the equation (15). The magnetic structure factors $\boldsymbol{F}_M^{(c)}$(hkl)* calculated for all the observed superstructure lines imply to know if we need both magnetic modes $\boldsymbol{F}_j^{(c)}$ and $\boldsymbol{G}_j^{(c)}$ in the description of the non collinear magnetic structures at low temperature

$$\boldsymbol{F}_M^{(c)}(100)^* = \sqrt{2}/2(\boldsymbol{G}_1 - \boldsymbol{G}_2) + i[\sqrt{2}/2(\boldsymbol{C}_1 + \boldsymbol{C}_2) + \boldsymbol{A}_3] \quad (17)$$

$$F_M^{(c)}(200)^* = -F_2 + F_3 + iA_1 \tag{18}$$

$$F_M^{(c)}(310)^* = \sqrt{2}/2(-G_1 + G_2) + i[\sqrt{2}/2(C_1 - C_2) + A_3] \tag{19}$$

$$F_M^{(c)}(411)^* = \sqrt{2}/2(G_2 - G_3) + i[\sqrt{2}/2(C_2 + C_3) - A_1] \tag{20}$$

$$F_M^{(c)}(330)^* = \sqrt{2}/2(-G_1 + G_2) + i[\sqrt{2}/2(C_1 + C_2) - A_3] \tag{21}$$

$$F_M^{(c)}(510)^* = \sqrt{2}/2(-G_1 - G_2) + i[\sqrt{2}/2(-C_1 + C_2) + A_3] \tag{22}$$

$$F_M^{(c)}(433)^* = \sqrt{2}/2(-G_2 + G_3) + i[\sqrt{2}/2(C_2 + C_3) + A_1] \tag{23}$$

$$F_M^{(c)}(530)^* = \sqrt{2}/2(-G_1 - G_2) + i[\sqrt{2}/2(-C_1 - C_2) - A_3] \tag{24}$$

$$F_M^{(c)}(600)^* = -F_2 + F_3 - iA_1 \tag{25}$$

$$F_M^{(c)}(442)^* = -F_2 + F_1 + iA_3 \tag{26}$$

From these expressions, one can observe that both magnetic modes $F_j^{(c)}$ and $G_j^{(c)}$ are necessary in the description of the non collinear structures of the Tb^{3+} ions. The absence of the small superstructure lines (200)* and (600, 442)* above 13 K gives rise to the equality between the magnetic modes $F_j^{(c)}$ (j = 1-3). It must be noted that the superstructure line (510)* appears at the same Bragg peak of the pure nuclear reflection (431) (θ = 30.7°). A magnetic contribution of the RE^{3+} ions exists for (431). We observe also a magnetic contribution for another reflection (541). It is very difficult to isolate only the magnetic contribution of these two reflections which are represented by their magnetic structure factors

$$F_M^{(c)}(431) = \sqrt{2}/2(-G_2 + G_3) + i[\sqrt{2}/2(C_2 - C_3) - A_1] \tag{27}$$

$$F_M^{(c)}(541) = \sqrt{2}/2(G_1 + G_3) + i[\sqrt{2}/2(-C_1 + C_3) - A_2] \tag{28}$$

5.2 Representation analysis of magnetic structures

Bertaut (Bertaut, 1963, 1968, 1971, 1972) has created a group theory method called "*Representation Analysis*" which has been widely used in the last four decades by Bertaut himself and by other researchers. The essential role is plaid by the "*Basis Vectors of Irreductible Representations*" of the paramagnetic space group Ia$\overline{3}$d of TbIG and its highest subgroups.

5.2.1 Representation analysis of Ia$\overline{3}$d

The representation analysis of the cubic space group Ia$\overline{3}$d was applied in the past (Bertaut et al., 1970; Tchéou et al., 1970a) in order to determine the spin configurations of the Tb^{3+}

ions in TbIG. In the point group O_h ten possible irreductible representations Γ_{ig} and Γ_{iu} (i = 1-5) are present and listed usually on quantum mechanics text books (Flury Jr, 1980; Kahan, 1972). The subscripts g and u refer to irreductible representations which are even (gerade) and odd (ungerade) respectively under the inversion I = ($\overline{1}$|0,0,0). In their original works, two representations ($\Gamma_{4g} = T_{1g}$) and ($\Gamma_{5g} = T_{2g}$) were used. However, the coupling between the RE^{3+} in sites {24} and the two iron ions Fe^{3+} in sites [16a] and (24d) has not been taken account in their study. The representation analysis of Ia$\overline{3}$d has been developed later completely by the author (Lahoubi, 1986) in order to choose the common irreductible representation which could be able to represent both the Néel model at high temperature below T_N and the non collinear magnetic structures observed below 160 K. For both RE^{3+}{24c} and Fe^{3+}(24d) sublattices, we need the even Γ_{ig} (i = 1-5) which appear only for Fe^{3+}[16a] sublattice. Using the precedent linear combinations of the spins of equations (12), (13) and (14), the sets of magnetic basis vectors of the three sublattices belonging to Γ_{ig} (i = 1-5) which are formed by the functions Ψ_{lm} (l = 1-d_{ig}, d_{ig} the dimension of Γ_{ig} and m fixed) are listed on Tables 2, 3 and 4. The letters A and E are assigned to one and two dimensional representations where $(\Psi_{11})^*$ belongs to the complex conjugate representation with ϵ = {exp(2 πi/3)}. Due to the equation (13), it can be show that ($\Gamma_{5g} = T_{2g}$) is excluded and only ($\Gamma_{4g} = T_{1g}$) may be used to describe formally in a first approximation all magnetic structures.

$\Gamma_{1g} = A_{1g}$			
$\Gamma_{2g} = A_{2g}$	$\Psi_{11} = G_{1x} + G_{2y} + G_{3z}$		
$\Gamma_{3g} = E_g$	$\Psi_{11} = G_{1x} + \epsilon G_{2y} + \epsilon^2 G_{3z}$ $\Psi_{21} = -\epsilon(\Psi_{11})^*$		
$\Gamma_{4g} = T_{1g}$	$\Psi_{11} = F_{1x}$ $\Psi_{21} = F_{2y}$ $\Psi_{31} = F_{3z}$	$\Psi_{12} = F_{2x} + F_{3x}$ $\Psi_{22} = F_{1y} + F_{3y}$ $\Psi_{32} = F_{1z} + F_{2z}$	$\Psi_{13} = G_{2z} + G_{3y}$ $\Psi_{23} = G_{3x} + G_{1z}$ $\Psi_{33} = G_{1y} + G_{2x}$
$\Gamma_{5g} = T_{2g}$	$\Psi_{11} = F_{3x} - F_{2x}$ $\Psi_{21} = F_{1y} - F_{3y}$ $\Psi_{31} = F_{2z} - F_{1z}$		$\Psi_{13} = G_{3y} - G_{2z}$ $\Psi_{23} = G_{1z} - G_{3x}$ $\Psi_{33} = G_{2x} - G_{1y}$

Table 2. Basis vectors of the RE^{3+} in site {24c} of Ia$\overline{3}$d

$\Gamma_{1g} = A_{1g}$	$\Psi_{11} = G_{1x} + G_{2y} + G_{3z}$		
$\Gamma_{2g} = A_{2g}$			
$\Gamma_{3g} = E_g$	$\Psi_{11} = G_{1x} + \epsilon G_{2y} + \epsilon^2 G_{3z}$ $\Psi_{21} = +\epsilon(\Psi_{11})^*$		
$\Gamma_{4g} = T_{1g}$	$\Psi_{11} = F_{1x}$ $\Psi_{21} = F_{2y}$ $\Psi_{31} = F_{3z}$	$\Psi_{12} = F_{2x} + F_{3x}$ $\Psi_{22} = F_{1y} + F_{3y}$ $\Psi_{32} = F_{1z} + F_{2z}$	$\Psi_{13} = G_{3y} - G_{2z}$ $\Psi_{23} = G_{1z} - G_{3x}$ $\Psi_{33} = G_{2x} - G_{1y}$
$\Gamma_{5g} = T_{2g}$	$\Psi_{11} = F_{3x} - F_{2x}$ $\Psi_{21} = F_{1y} - F_{3y}$ $\Psi_{31} = F_{2z} - F_{1z}$	$\Psi_{12} = G_{3y} + G_{2z}$ $\Psi_{22} = G_{1z} + G_{3x}$ $\Psi_{32} = G_{2x} + G_{1y}$	

Table 3. Basis vectors of the Fe^{3+} in site [24d] of Ia$\overline{3}$d

$\Gamma_{1g} = A_{1g}$	$\Psi_{11} = (A_x + C_y + G_z) - (G'_x + A'_y + C'_z)$		
$\Gamma_{2g} = A_{2g}$	$\Psi_{11} = (A_x + C_y + G_z) + (G'_x + A'_y + C'_z)$		
$\Gamma_{3g} = E_g$	$\Psi_{11} = A_x + \epsilon C_y + \epsilon^2 G_z$	$\Psi_{12} = -(A'_y + \epsilon^2 G'_x + \epsilon C'_z)$	
	$\Psi_{21} = -(A'_y + \epsilon G'_x + \epsilon^2 C'_z)$	$\Psi_{22} = A_x + \epsilon^2 C_y + \epsilon G_z$	
$\Gamma_{4g} = T_{1g}$	$\Psi_{11} = F_x + F'_x$	$\Psi_{12} = C_z + C'_y$	$\Psi_{13} = G_y + A'_z$
	$\Psi_{21} = F_y + F'_y$	$\Psi_{22} = G_x + G'_z$	$\Psi_{23} = A_z + C'_x$
	$\Psi_{31} = F_z + F'_z$	$\Psi_{32} = A_y + A'_x$	$\Psi_{33} = C_x + G'_y$
$\Gamma_{5g} = T_{2g}$	$\Psi_{11} = F_x - F'_x$	$\Psi_{12} = C_z - C'_y$	$\Psi_{13} = G_y - A'_z$
	$\Psi_{21} = F_y - F'_y$	$\Psi_{22} = G_x - G'_z$	$\Psi_{23} = A_z - C'_x$
	$\Psi_{31} = F_z - F'_z$	$\Psi_{32} = A_y - A'_x$	$\Psi_{33} = C_x - G'_y$

Table 4. Basis vectors of the Fe^{3+} in site (16a) of Ia$\overline{3}$d

In the high temperature region, the observed spectra which are well interpreted within the ferrimagnetic model of Néel are easily identified to the magnetic modes $\boldsymbol{F}_j$, $\boldsymbol{F}$ and $\boldsymbol{F'}$. At T_{comp} = 243.5 K the mean exchange field acting on the Tb^{3+} ions by the iron sublattices is too strong (~ 174 kOe) by comparison with the Tb^{3+}–Tb^{3+} exchange field (~ 8 kOe) (Lahoubi, 1986); this last coupling beetwen the Tb3+ ions will be not able to decouple at low temperature the two sublattices. This remarkable property excludes permanently the intervention of the three-dimensional irreductible representation (Γ_{5g} = T_{2g}) where the magnetic modes of the rare earth sublattice are along the <0$\overline{1}$1> directions. Using the basis vectors of ($\Gamma_{4g} = T_{1g}$), we present in Fig. 10 the four cubic models of "*double umbrella*" of type II (Lahoubi, 1986) in the irreductible representation ($\Gamma_{4g} = T_{1g}$) of Ia$\overline{3}$d

$$\text{"\textit{Double umbrella of type} II"}: S_1 = +S_7 = (\text{f, F+G, F+G}); S_4 = +S_{10} = (\text{f, F-G, F-G});\ \text{c.p.} \quad (29)$$

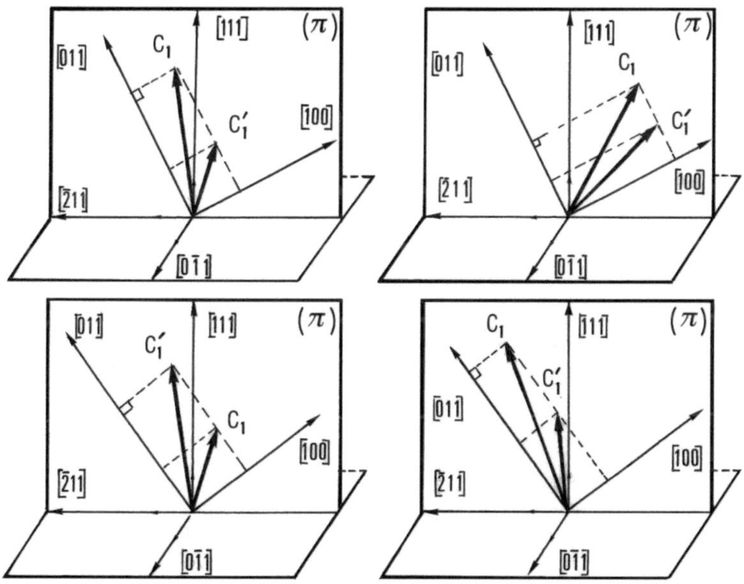

Fig. 10. Four cubic "*double umbrella*" models of type II for ($\Gamma_{4g} = T_{1g}$) of Ia$\overline{3}$d.

In these four models, the angles and the modulus of the moments $\boldsymbol{m}_1$ and $\boldsymbol{m}'_1$ are different. They must to respect some requirements: the components of the moments $\boldsymbol{m}_1$, $\boldsymbol{m}'_1$ along the cubic axis [100] are necessary equals; also for the components of the moments $\boldsymbol{m}_2$, $\boldsymbol{m}'_2$ and $\boldsymbol{m}_3$, $\boldsymbol{m}'_3$ respectively along [010] and [001]. The components of the moment $\boldsymbol{m}_1$ along the cubic axes [010] and [001] must be equals; also for $\boldsymbol{m}_2$ and $\boldsymbol{m}'_2$ along [100] and [001]; also for $\boldsymbol{m}_3$ and $\boldsymbol{m}'_3$ along [100] and [010]. The first refinements at 4.2 K (with $m_a = m_d = 5\ \mu_B$) lead to values of the moment m'_1 (~ 10 μ_B), i.e., above the free Tb^{3+} ion value (9 μ_B). The reason of this discrepancy is associated to the deviation of the cubic description which imply a rhombohedral distortion observed on powder sample by X-ray diffraction at 6 K (Bertaut et al., 1970, Sayetat, 1974, 1986 and refs. herein) and neutron diffractions at 5 K (Hock et al., 1990) where two subgroups of Ia$\overline{3}$d have been proposed respectively: R$\overline{3}$c and R$\overline{3}$.

5.2.2 Representation analysis of R$\overline{3}$c

A detailed description of the representation analysis of the subgroup R$\overline{3}$c is presented here. According to the earlier precise X-ray diffraction measurements of the rhombohedral distortions carried out on single crystals of the terbium-yttrium iron garnet system $Tb_xY_{3-x}Fe_5O_{12}$ with $0 \le x \le 3$ (hereafter Tb_xY_{3-x}IG) (Levitin et al., 1983), the choice of the subgroup R$\overline{3}$c seems more appropriate. This choice will be confirmed later by our high field magnetization measurements and the use of the method of "*the symmetry lowering device*" (Bertaut, 1981) which is connected with representation analysis of Bertaut (Bertaut, 1968, 1971, 1972). From the International Tables (Hahn, 1983), the Bravais lattice of the crystallographic space group R$\overline{3}$c-(D_{3d}^6) No. 167 is defined in the system of the rhombohedral directions [$\overline{1}$11], [1$\overline{1}$1], [11$\overline{1}$] with the three fundamental vectors: $\boldsymbol{A}_1$, $\boldsymbol{A}_2$, $\boldsymbol{A}_3$ with the same parameter A_{Rh} and an angle α_{Rh} # 90° and < 120° respectively along the unit vectors g_j, j = 1-3). There are related to the cubic axes by unit vectors {$\boldsymbol{i}, \boldsymbol{j}, \boldsymbol{k}$}

$$g_1 = \frac{1}{\sqrt{3}}(-\boldsymbol{i}+\boldsymbol{j}+\boldsymbol{k});g_2 = \frac{1}{\sqrt{3}}(+\boldsymbol{i}-\boldsymbol{j}+\boldsymbol{k});g_3 = \frac{1}{\sqrt{3}}(+\boldsymbol{i}+\boldsymbol{j}-\boldsymbol{k}) \tag{30}$$

$$X = y + z;\ Y = x + z;\ Z = x + y \tag{31}$$

The correspondence between the ion positions in the two space groups Ia$\overline{3}$d and R$\overline{3}$c is reported on Table 5. For TbIG, the values of the parameters found at 6.75 K (Sayetat, 1974, 1986) are: A_{Rh} = 10.7430 Å; α_{Rh} = 109°24′40″. The rhombohedral unit cell contains only the half atoms of the cubic unit cell. In order to use the representation analysis to determine the magnetic basis vectors, we choose the following generators of the space group R$\overline{3}$c: the identity E = (1| 0,0,0), the inversion I = ($\overline{1}$ | 0,0,0), a ternary axis 3 = (3| 0,0,0) and a diagonal binary axis 2_d = ($2_{X\overline{Y}}$ | 1/2,1/2,1/2). This axis is perpendicular to the glide plane c = I. 2_d of the symbol R$\overline{3}$c. The wave vector being $\boldsymbol{k}$ = 0, the six irreductible representations of the space group R$\overline{3}$c are those of the point group D_{3d}, Γ_{jg} and Γ_{ju} with j = 1-3. The previous three-dimensional irreductible representation ($\Gamma_{4g} = T_{1g}$) of O_h is reduced to: $\Gamma_{2g} = A_{2g} + E_g$. Only the one-dimensional irreductible representation will be chosen in our study (the two-dimensional irreductible representation E_g being complex, she is not considered). It appears that the previous linear combinations of the spin vectors ***F, G, C*** and ***A*** are not in reality

	$Ia\bar{3}d$			$R\bar{3}c$		
Ions	Sites Symmetry		Positions	Sites Symmetry		Positions
RE^{3+}	24c	222	±(1/8,0,1/4); c.p. ±(3/8,0,3/4); c.p.	6e 6e′	2 2	±(X,1/2-X,1/4) X ≈ 3/8; c.p. ±(X,1/2-X,1/4) X ≈ 7/8; c.p.
Fe^{3+}	24d	$\bar{4}$	±(3/8,0,1/4); c.p. ±(1/8,0,3/4); c.p.	12f	1	±(X,Y,Z); c.p. ±(Y+1/2,X+1/2,Z+1/2); c.p. X≈5/8,Y≈3/8,Z≈1/4
Fe^{3+}	16a	$\bar{3}$	0,0,0; 1/4,1/4,1/4 0,1/2,1/2; c.p. 1/4,3/4,3/4; c.p.	2b 6d	$\bar{3}$ $\bar{1}$	0,0,0; 1/2,1/2,1/2 0,1/2,1/2; c.p. 1/2,0,0; c.p.
O^{2-}	96h	1		12f	1	O_I O_{II} O_{III} O_{IV}

Table 5. Correspondence between the positions of the ions in $Ia\bar{3}d$ and $R\bar{3}c$

adapted to describe the magnetic structures of the three magnetic ions. For the RE^{3+} ions, the preceding magnetically inequivalent sublattices C_j and C'_j (j = 1-3) become crystallographic inequivalent sites 6e and 6e′; they are described in the rhombohedral axis X, Y, Z by a new linear combination for the ferromagnetic mode $\boldsymbol{f}_j = S_j + S_{j+6}$ (j = 1-3). Concerning the iron ions sublattices D_j and D'_j of the 24d site, they will be associated to a new 12f site where the basis vectors are described with the ferromagnetic mode $\boldsymbol{V}_j = S_j + S_{j+6}$ (j = 1-3). The associated basis vectors of ($\Gamma_{2g} = A_{2g}$) are presented here on Table 6. A new combination of these basis vectors is proposed and four modified double umbrella models are presented for the six sublattices C_j and C'_j on Table 7. In this description, the first part ($f\boldsymbol{n}$) of the moment $\boldsymbol{m}_j$ represents the collinear ferromagnetic mode of the component along the [111] direction.

		$R\bar{3}c$
Ions	Sites	Basis vectors
RE^{3+}	6e	$\Psi_{11}(I) = f_{1X} + f_{2Y} + f_{3Z}$ $\Psi_{11}(II) = f_{1Y} + f_{1Z} + f_{2Z} + f_{2X} + f_{3X} + f_{3Y}$
	6e′	$\Psi_{11}(IV) = f_{4X} + f_{5Y} + f_{6Z}$ $\Psi_{11}(V) = f_{4Y} + f_{4Z} + f_{5Z} + f_{5X} + f_{6X} + f_{6Y}$
Fe^{3+}	12f	$\Psi_{11}(I) = V_{1X} + V_{2Y} + V_{3Z} + V_{4X} + V_{5Y} + V_{6Z}$ $\Psi_{11}(II) = V_{2X} + V_{3Y} + V_{1Z} + V_{6X} + V_{4Y} + V_{5Z}$ $\Psi_{11}(III) = V_{3X} + V_{1Y} + V_{2Z} + V_{5X} + V_{6Y} + V_{4Z}$
Fe^{3+}	2b	$\Psi_{11}(I) = (S_1 + S_5)_X + (S_1 + S_5)_Y + (S_1 + S_5)_Z$
	6d	$\Psi_{11}(II) = (S_2 + S_6)_X + (S_3 + S_7)_Y + (S_4 + S_8)_Z$ $\Psi_{11}(III) = (S_3 + S_8)_X + (S_4 + S_6)_Y + (S_2 + S_7)_Z$ $\Psi_{11}(IV) = (S_4 + S_7)_X + (S_2 + S_8)_Y + (S_3 + S_6)_Z$

Table 6. Basis vectors of ions in the irreductible representation A_{2g} of $R\bar{3}c$

Four A_{2g} models in $R\bar{3}c$ with $f \neq f'$ and $a \neq a'$	New model for TbIG with $f \neq f'$ and $a' \approx a - \varepsilon$ ($\varepsilon \to 0$ at $T > 5$ K)
(6e): $C_j : S_j = S_{j+6} = f\boldsymbol{n} \pm a\boldsymbol{g}_j$	$C_1 : \boldsymbol{m}_1 = S_1 = S_7 = (f - a/3)\boldsymbol{n} - (2\sqrt{2}/3)a\boldsymbol{p}_1$
(6e'): $C'_j : S'_j = S'_{j+6} = f'\boldsymbol{n} \pm a'\boldsymbol{g}_j$	$C'_1 : \boldsymbol{m}'_1 = S_4 = S_{10} = (f' + a'/3)\boldsymbol{n} + (2\sqrt{2}/3)a'\boldsymbol{p}_1$

Table 7. Four rhombohedral models in A_{2g} of $R\bar{3}c$ and new model for TbIG (Lahoubi, 2012)

The second part ($\pm a\boldsymbol{g}_j$) represents the non collinear antiferromagnetic modes of the components of the moment $\boldsymbol{m}_j$ along the three rhombohedral axes {[$\bar{1}$11], [1$\bar{1}$1], [11$\bar{1}$]}. Some requirements from the previous "*cubic description*" are now ignored in the "*rhombohedral description*": the axis [100] and equivalent directions cease to be principal axes. In the four models of non collinear arrangements for the RE^{3+} ions around the ternary axis [111], the sublattices C_j, C'_j are situated in the three glide planes c, c.3 and c.3^2 of the subgroup $R\bar{3}c$: each plane containing [111] (unit vector $\boldsymbol{n}$) and one of the rhombohedral directions {[$\bar{1}$11], [1$\bar{1}$1], [11$\bar{1}$]} represented by the unit vectors $\boldsymbol{g}_j$, j = 1-3. In this condition, we have: c = ($\boldsymbol{n}$, $\boldsymbol{g}_3$); c.3 = ($\boldsymbol{n}$, $\boldsymbol{g}_1$) and c.3^2 = ($\boldsymbol{n}$, $\boldsymbol{g}_2$) (Fig. 11a). Furthermore, the projection on the plane (111) of the rhombohedral direction [$\bar{1}$11] ($\boldsymbol{g}_1$) for example, is equivalent to the low symmetry axis [$\bar{2}$11] (with the unit vector $\boldsymbol{p}_1$); the local axis [0$\bar{1}$1] ($\boldsymbol{W}$) is also chosen. If one takes into account the smallness superstructure lines (200)* and (600,442)* at 5 K (D1B), new parameters are found for C_1 (m_1, θ_1, ϕ_1) and C'_1 (m'_1, θ'_1, ϕ'_1) sites (Lahoubi, 2012) (Fig. 11b). The moments of C_1 and C'_1 are drawed in the (π) plane which corresponds to glide plane c.3: this plane remains a principal plane of the magnetic tensor $\tilde{g}$ of the earlier "*cubic description*". A good reliability factor R of the order of 6.7 % is found for a refinement based on all the reflections and the method of validation of the magnetic structures (Wills, 2007 and refs. herein):

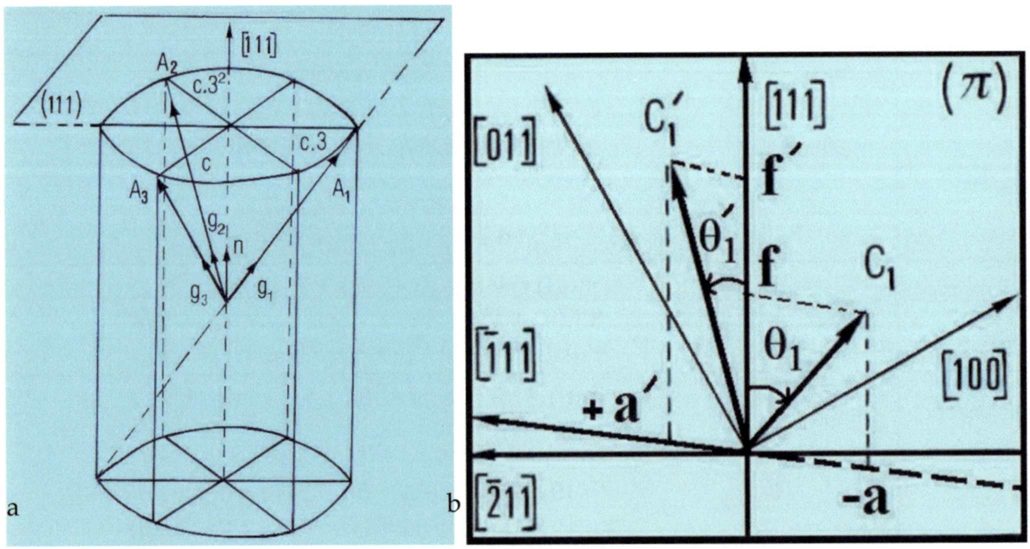

Fig. 11. a. The three glide planes of $R\bar{3}c$, b. New model for C_1 and C'_1 at 5 K (D1B)

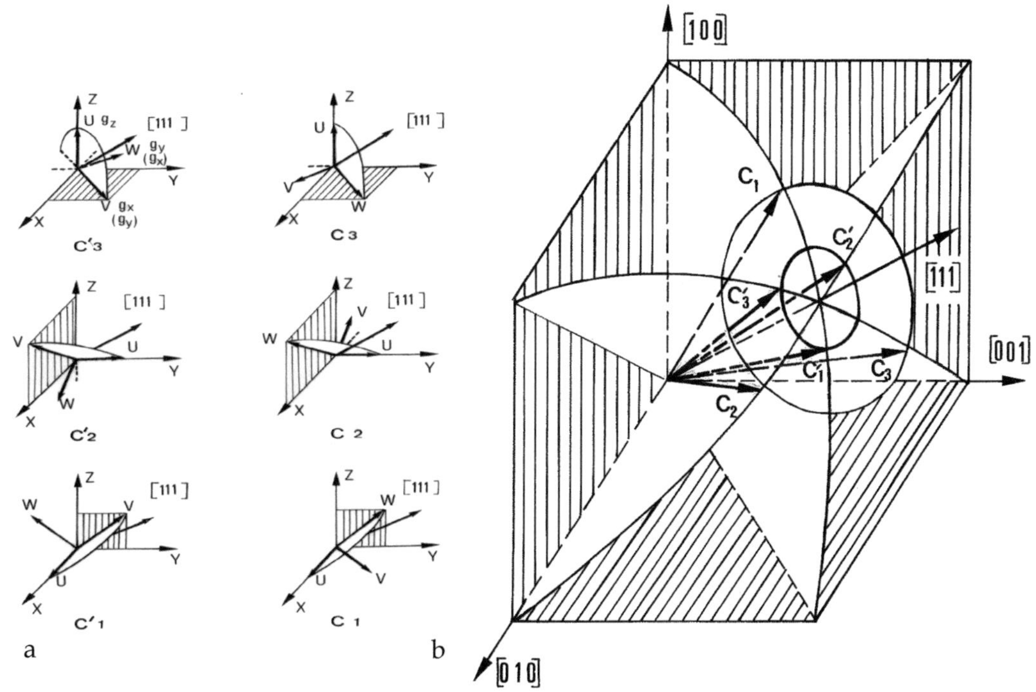

Fig. 12. a. Local axes of the D2(222) symmetry, b. Novel "*double umbrella*"at 5 K

$$C_1: m_1 = 8.07\ \mu_B\ ;\ \theta_1 = 32°\ ;\ \phi_1 = 180° \tag{32}$$

$$C'_1: m'_1 = 8.90\ \mu_B\ ;\ \theta'_1 = 27°\ ;\ \phi'_1 = 0° \tag{33}$$

where, ϕ_1 and ϕ'_1 are the angles from $\boldsymbol{p}_1$ in the (111) plane and $\varepsilon = 0.20\ \mu_B$; with $\delta(M_S)$<111> = $M_S^{cal} - M_S^{mes} = 0.03\ \mu_B$/mol. For a better presentation one can show the novel "*double umbrella* "(Fig. 12b) in the local axes of the D2(222) symmetry (Fig. 12a). These results are in good agreement with our recent high magnetic field magnetizations performed at 4.2 K (Lahoubi, 2012) where a third low critical field H_{c0} (Fig. 13) is observed along the <100> direction and added to the previous H_{c1} and H_{c2} (Lahoubi et al., 1984). They confirm unambiguously the presence of the three magnetic glide planes c' of the symbol R$\bar{3}$c'. The earlier results at 5 K described in the subgroup R$\bar{3}$ (Hock et al., 1990) lead to values of the components m_{1x} and m'_{1z} above the value (9 μ_B) of the free Tb^{3+} ion.

6. Temperature evolution of the double umbrella structure

The parameters for the sites C_1 (m_1, θ_1, $\phi_1 = 180°$) and C'_1 (m'_1, θ'_1, $\phi'_1 = 0°$) are refined with the same model found at 5 K with the condition $\varepsilon \to 0$ ($m_1{}^{\perp} = -m'_1{}^{\perp}$) due to the absence of the superstructure lines (200)* and (600,442)* above 13 K (D1B) (Fig. 4). The magnetic moments m_a and m_d of the iron sublattices were not concerned by the refinements for all reflections of the patterns and the observed values found by N.M.R. experiments (Gonano et al., 1967) are used. Good agreement is obtained with the reliability factors R varying between 6 and 10 % in the 4.2–283 K temperature range.

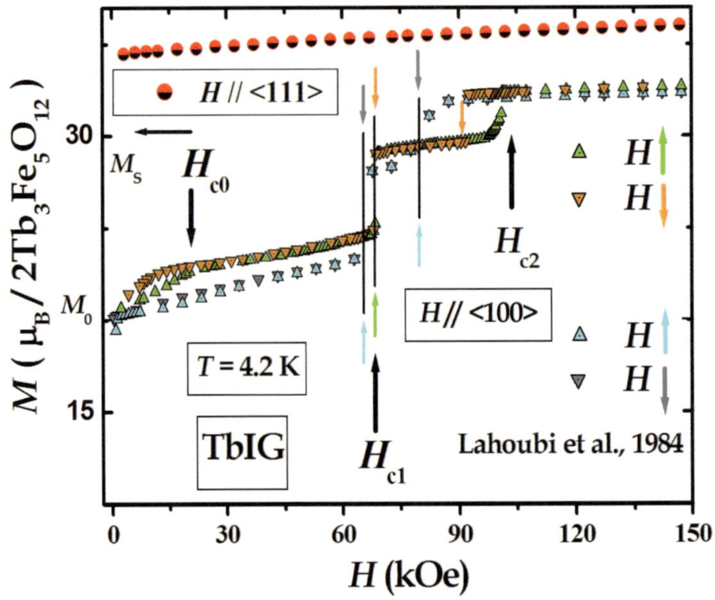

Fig. 13. $M_T(H)$ versus H at 4.2 K along the <111> and <100> directions (Lahoubi, 2012)

The equation (5) in the section 3, permit us to deduce the magnetic intensities I_M by assuming negligible the T-variations of I_N with a rather good precision for the reflections (hkl) which have a great magnetic contribution such as (211), (321), (521) and (532, 611) where the ratio $(I_M/I_N)_{cal}$ is equal respectively to: 200, 20, 4 and 2. Consequently, the high magnetically reflections with a small nuclear contribution are only (211) and (321). The thermal variations of these two magnetic reflections which are responsible of the collinear ferrimagnetic ordering are reported in Fig. 14. They appear at T_N and present a first increase below 160 K with a second rapid increase at 68 K. Different and complex temperature dependences are observed for the reflections (220) and (440). At first, the values of $(I_M)_{cal}$ are higher than those observed for $(I_M)_{obs}$. These two reflections increase simultaneously between 4.2 and 160 K and present an inflexion point near 68 K. Above 160 K, they tend to a plateau until room temperature after which they decrease progressively and reach zero at T_N. The possible explanation of this characteristic behavior seems to be related to the magnetic contribution of the irons in the octahedral site [16a] to the total intensity I, which is not the case for the reflections (211) and (321). Another good agreement between $(I_M)_{obs}$ and $(I_M)_{cal}$ is found in the thermal variations of the superstructure lines (310)*, (110)*, (411, 330)* and (530, 433)* plotted in Fig. 15 where a rapid variation is observed at 68 K. Two distinct magnetic behaviors separated by the specific temperature 160 K are clearly evidenced.

The refined values of the parameters (m_j, m'_j) and (θ_j, θ'_j) listed on Table 8 in the 4.2–283 K temperature range lead to a good agreement between M_S^{cal} and M_S^{mes}. During the refinement at 109 ± 2 K, two different results related to the set {a, b} are found and lead to identical values of M_S^{cal} and reliability factor R. The same feature is observed for the set {c, d} at 127 ± 5 K. We observe in the thermal variations of the parameters (m_j, m'_j) and (θ_j, θ'_j) plotted in Fig. 16 a broad variation between 54 K and 80 K which disappears beyond 160 K.

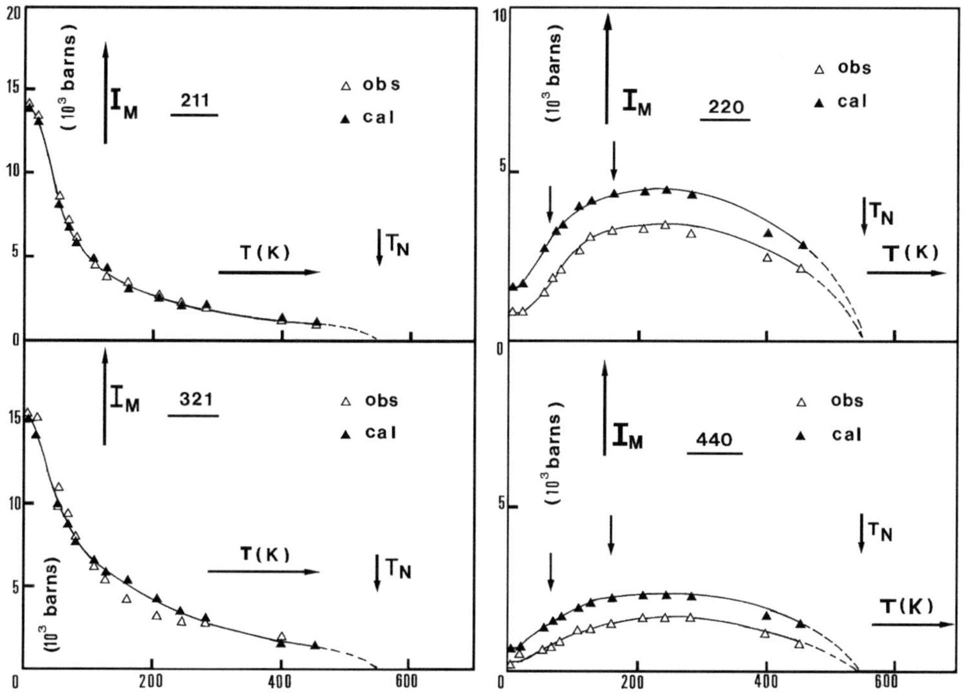

Fig. 14. Thermal variations of $(I_M)_{obs}$ and $(I_M)_{cal}$ for the reflections (211), (321), (220) and (440)

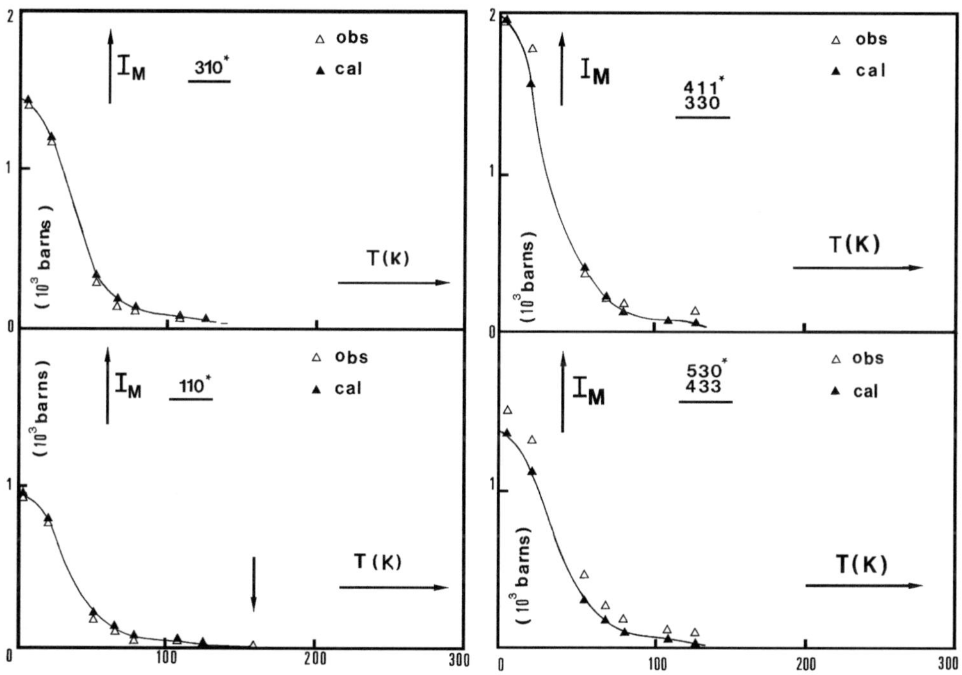

Fig. 15. Thermal variations of $(I_M)_{obs}$ and $(I_M)_{cal}$ for the superstructure lines $(310)^*$, $(110)^*$, $(411, 330)^*$ and $(530, 433)^*$

	Sites C_1 ; C'_1		M_S^{cal}			M_S^{mes} (μ_B/mol)		
T(K)	$m_1(\mu_B)$ $m'_1(\mu_B)$	$\theta_1(°)$ $\theta'_1(°)$	[111]	[110]	[100]	[111]	[110]	[100]
4.2	8.18 8.90	30.79 28.07	34.64	28.28	20.00	34.53	28.35	20.17
5	8.07 8.99	32.00 27.00	34.56	28.22	19.95	"	"	"
20	7.49 8.77	30.11 25.39	33.21	27.12	19.18	33.34	27.21	19.56
54	4.79 6.27	22.57 17.03	21.32	17.41	12.31	21.24	17.24	12.80
68	4.23 5.21	19.57 15.76	17.10	13.97	9.87	17.01	13.44	10.05
80	3.57 4.72	16.72 12.56	14.25	11.63	8.23	13.97	11.70	8.38
109 ± 2	2.76 [a] 3.79 2.78 [b] 3.78	15.04 10.86 15.23 11.20	9.48	7.74	5.47	8.78	7.20	5.26
127 ± 5	2.59 [c] 3.12 2.71 [d] 3.00	14.94 12.30 14.52 13.08	7.13	5.82	4.12	6.52	5.40	3.88
160	2.30 2.31	0.11 0.00	4.65	3.80	2.68	4.61	3.76	2.66
208 ± 2	1.74(3) 1.75(7)	0.05 0.00	1.92	1.56	1.11	1.97	1.36	1.12
244 ± 10	1.34(4) 1.35(8)	0.08 0.00	0	0	0	0	0	0
283	1.15 1.15	0 0	0.66	0.54	0.38	0.79	0.59	0.63

Table 8. Values of the parameters m_j, m'_j, θ_j and θ'_j in the 4.2–283 K temperature range with a comparison between the calculated M_S^{cal} and measured M_S^{mes} magnetizations.

The thermal variations of the parallel (m_j//, m'_j//) and perpendicular ($m_j{\perp}$, $m'_j{\perp}$) components are also reported in Fig. 17. The double umbrella magnetic structure appears to close slowly around the <111> direction in the three magnetic glide planes c' near 160 K with an abrupt increase between 54 and 68 K. Previous temperature dependences of the calculated non collinear magnetic structure in TbIG (Druzhinina & Shkarubskii, 1988) and the recent neutron scattering on TbIG single crystal, (Louca et al., 2009) are not consistent with these thermal variations. In another recent study of the magnetic and magneto-optical properties of the Tb^{3+} ions in TbIG and in the mixed system of terbium-yttrium ferrites garnets $Tb_xY_{3-x}IG$ (x < 1) (Zhang et al., 2009) the differences between the two non collinear magnetic structures which exist at low temperature were not taken account in their calculations.

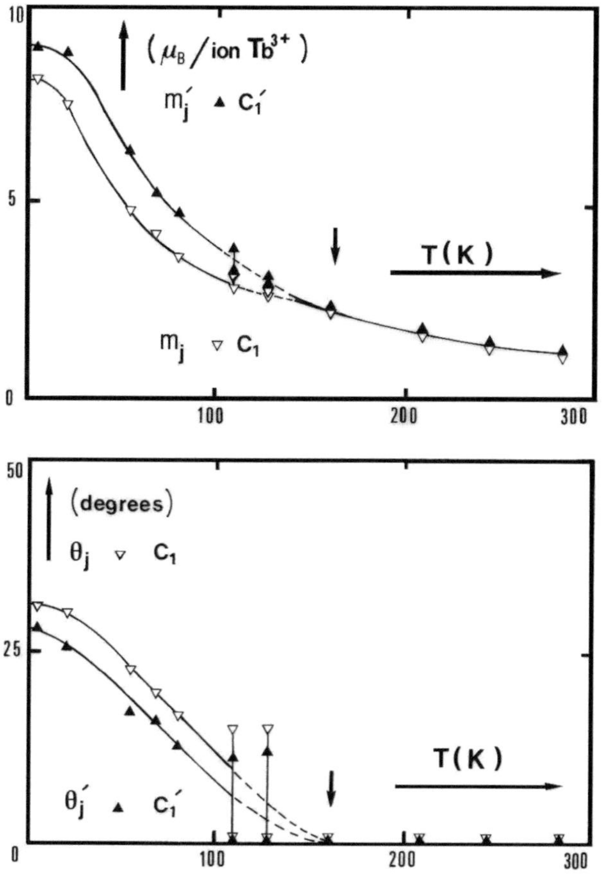

Fig. 16. Thermal variations of the parameters (m_j, m'_j) and (θ_j, θ'_j)

The <111> direction ceases to be the easy axis of magnetization and changes below 140 K to the <100> direction up to 4.2 K in $Tb_{0.37}Y_{2.63}IG$ for example with the appearance of the low symmetry phases <uuw> (Lahoubi et al, 2000) in the spontaneous spin reorientation phase transitions.

These results are in good agreement with the previous observed rhombohedral distortion below 190 K (Rodić & Guillot, 1990) and 200 K (Sayetat, 1974, 1986). They are also in good agreement with the anomalous behaviors observed previously below 200 K on TbIG single crystals without applied magnetic fields, in the acoustic properties (Kvashnina et al., 1984; Smokotin et al., 1985) and in the elastic constant measurements (Alberts et al., 1988) along the [100], [110] and [111] crystallographic directions, precisely in the temperature ranges, 60–140 K and 50–165 K respectively.

It seems that the behavior around 160 K has a relation with the previous predicted momentum angular compensation point T_J (Nelson & Mayer, 1971) localized at 150 and 190 K by assuming the free and quenched ion value respectively.

The large magnetodielectric (MD) effects which have been recently revealed on TbIG single crystal at low temperature as well when a very small external magnetic field ($H_{ex} < 0.2$ T) is

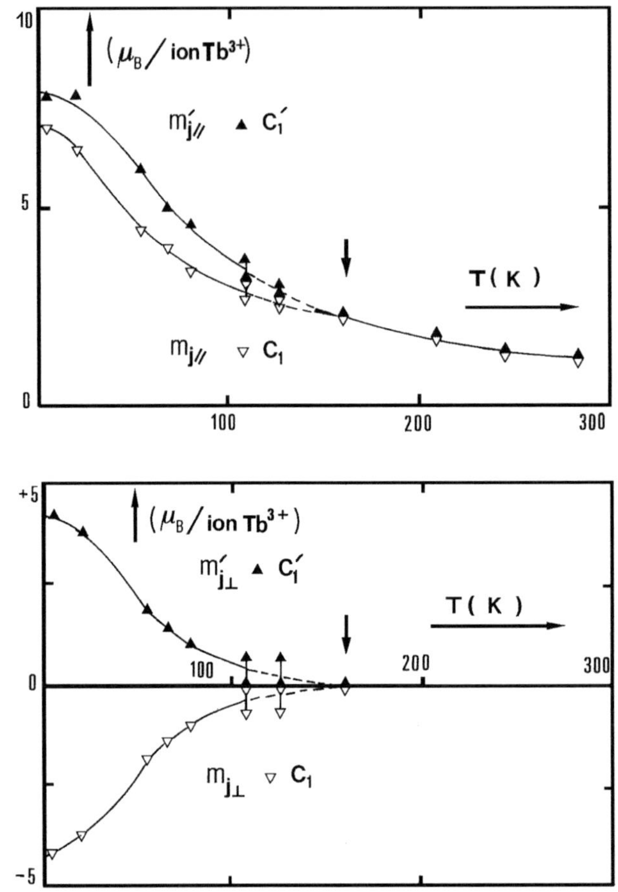

Fig. 17. Thermal variations of the parameters ($m_{j}//$, $m_j\perp$) and ($m'_{j}//$, $m'_j\perp$)

applied (Hur et al., 2005) could be combined with the previous huge spontaneous magnetostriction measurements (Sayetat, 1974, 1986; Guillot et al., 1980) where a peak near 70 K has been observed in the thermal variations of $\lambda^{\varepsilon,2}(Tb^{3+})_{exp}/\lambda^{\varepsilon,2}(Tb^{3+})_{cal}$, the ratio of the experimental values of the magnetostriction constant to the theoretical values derived from the one ion model. It correspond to the abrupt change in the long-range magnetic order in the Tb^{3+} sublattice near the previous predicted low-temperature point T_B = 58 K (Belov, 1996 and refs. herein) which is situated between 54 and 68 K in this study. More recently, some magnetoelectric (ME) and MD effects in weak and high external magnetic fields (H_{ex} up to 2 T and H_{ex} up to 10 T respectively) have been reported (Kang et al., 2010). A possible coupling between the magnetic exchange and the ligand-field excitations which occurs at a specific temperature situated between 60 and 80 K has been discovered without external magnetic field (H_{ex} = 0) with two distinct behaviors above and below another characteristic temperature (~ 150 K). All results confirm that Landau's theory of second order phase transitions does not apply to TbIG in the 5 K–T_N temperature range without applied external magnetic field and the magnetic space group is $R\bar{3}c'$ (Bertaut, 1997; Lahoubi et al., 1997).

7. Conclusion

In this chapter, the temperature evolution of the magnetic structure in TbIG is studied by neutron diffraction experiments below T_N (568 K). The "double umbrella" structure observed at 5 K appears below a specific temperature (~ 160 K) which could be related to the previous predicted T_J-point situated in the 150–190 K temperature range. The rapid variation of the Tb^{3+} moments is observed between 54 and 68 K where the predicted value (58 K) of the T_B-point is located. The magnetic symmetry doesn't change with the temperature and the magnetic space group is $R\bar{3}c'$. The author is convinced that the symmetry considerations of the Representation Analysis of Bertaut presented in this chapter have demonstrated their usefulness in the determination of the thermal variation of the double umbrella magnetic structure in TbIG below T_N. It is hoped that these results which are in good agreement with the magnetization measurements will facilitate a better understanding of the possible correlations between the magnetic properties via the double umbrella magnetic structure and the recent ME and MD effects found in this ferrite garnet.

8. Acknowledgements

This work is dedicated to the Academician Dr. E F Lewy Bertaut, the father *"on group theoretical techniques in magnetic structure analysis"*, who died in 2003. Some of unpublished results presented here constitute a part of my thesis. I am indebted to Dr. M Guillot (LNCMI, Grenoble) for his experimental contribution as supervisor and for providing the samples. We thank Dr. F Tchéou (UJF, Grenoble) for his helpful discussions during the refinement of the neutron diffraction patterns. Many acknowledgements are also addressed to Dr. B Ouladdiaf (ILL, Grenoble) for his assistance during the recent D1B experiments.

9. References

Alberts, H. L.; Palmer, S. B. & Patterson, C. (1988). *J. Phys. C: Solid State Phys.*, Vol. 21, pp. 271-275

Bacon, G. E. (1972). *Acta. Cryst. A*, Vol. 28, pp. 357-358

Bacon, G. E. (1975). *Neutron Diffraction*, (Third Edition), Clarendon Press, Oxford, UK

Belov, K. P. (1996). *Phys. Usp.*, Vol. 39, No. 6, pp. 623-634

Bertaut, E. F. & Forrat, F. (1956a). *C. R. Acad. Sci., Paris*, Vol. 242, pp. 382-384

Bertaut, E. F.; Forrat, F.; Herpin, A. & Mériel, P. (1956b). *C. R. Acad. Sci., Paris*, Vol. 243, pp. 898-901

Bertaut, E. F. (1963). Spin Configurations of Ionic Structures: Theory and Practice, In: *Treatise on Magnetism*, Suhl & Rado, Vol. III, Chap. 4, pp. 149-209, Acad. Press, New York

Bertaut, E. F. (1968). *Acta Cryst. A*, Vol. 24, No.1, pp. 217-231

Bertaut, E. F.; Sayetat, F. & Tchéou, F. (1970). *Solid State Commun.*, Vol. 8, No.4, pp. 239-245

Bertaut, E. F. (1971). *J. Phys., Colloque C1*, Suppl. No. 2-3, t. 32, pp. C1. 462-470

Bertaut, E. F. (1972). *Ann. Phys.*, t. 7, pp. 203-232

Bertaut, E. F. (1981). *J. Magn. Magn. Mater.*,Vol. 24, pp. 267-278

Bertaut, E. F. (1997). *J. Phys. IV France* 7, *Colloque C1*, Suppl. J. Phys III, pp. C1.11-26

Blume, M.; Freeman, A. J. & Watson, R. E. (1962). *J. Chem. Phys.*, Vol. 37, pp. 1245-1253

Bonnet, M. (1976). *Thesis, Doct. of Sciences in Physics*, Grenoble University, France

Bonnet, M.; Delapalme, A.; Fuess, H. & Becker, P. (1979). *J. Phys. Chem. Solids*, Vol. 40, (No. 11), pp. 863-876

Chatterji, T. (2006). Editor, *Neutron Scattering from Magnetic Materials*, Elsevier B.V., pp. 1-559

Convert, P.; Fruchart, D.; Roudaut, E. & Wolfers, P. (1983). 12 Years of Life with Bananas (Curved One-Dimensional Neutron PSDs), In: *Position-Sensitive Detection & Thermal Neutrons*, pp. 302-309, Academic Press Inc., ISBN: 0-12-186180-5, London, UK

Druzhinina, R. F. & Shkarubskii, V.V. (1988). *Sov. Phys. Solid State,* Vol. 30, (No. 2), pp. 342-343

Ferey, G.; De Pape, R.; Le Blanc, M. & Pannetier, J. (1986). *Rev. Chim. Miner.,* Vol. 23, pp. 474-484

Flury Jr, R. L. (1980). *Symmetry Groups: Theory and Chemical Applications*, Prentice-Hall, Inc., Englewood Cliffs, New-Jersey 07632

Fuess, H.; Bassi, G.; Bonnet, M. & Delapalme, A. (1976). *Solid State Commun.*, Vol. 18, (No. 5) pp. 557-562

Geller, S. (1978). Crystal and Static Magnetic Properties of Garnets, In: *Proc. of the International School of Physics "Enrico Fermi", 1977, Course LXX, Physics of magnetic Garnets*, A. Paoletti (Ed.), pp. 1-55, North-Holland Publishing Co, Amsterdam

Geller, S. & Gilleo, M. A. (1957a). *Acta. Cryst.*, Vol. 10, p. 239

Geller, S. & Gilleo, M. A. (1957b). *J. Phys. Chem. Solids*, Vol. 3, (No.1-2), pp. 30-36

Gonano, R.; Hunt, E. & Meyer, H. (1967). *Phys. Rev.*, Vol. 156, pp. 521-533

Guillot, M. & du Tremolet de Lacheisserie, E. (1980). *Z. Phys. B- Condensed Matter,* Vol. 39, pp. 109-114

Guillot, M.; Marchand, A.; Tchéou, F & Feldmann, P (1982). *J. Appl. Phys.,* Vol. 53, (No.3), pp. 2719-2721

Guillot, M.; Tchéou, F.; Marchand, A. & Feldmann, P. (1983).*J. Magn. Magn. Mater.,*Vol. 31-34, pp. 631-632

Guillot, M.; Tchéou, F.; Marchand, A. & Feldmann, P. (1984). *Z. Phys. B - Condensed Matter,* Vol. 56, pp. 29-39

Guillot, M. (1994). Magnetic Properties of Ferrites, In: *Materials Science and Technology: A Comprehensive Treatment,* Cahn, Haasen & Kramer, (editors), Vol. 3B, Electronic and Magnetic Properties of Metals and Ceramics, Part II, VCH Publishers Inc., Chapter 8, K. H. J. Buschow (Ed.), pp. 7–92, VCH Verlagsgesellschaft mbH, VCH Publishers Inc. Weinheim, New York, USA

Hahn, Th. (1983). (Ed.) Space Group Symmetry, Vol. A, In: *International Tables for Crystallography,* Published by the IUCr, Dordrecht Reidel Boston, London, kluwer academic publisher's edition

Herpin, A. (1968). *Théorie du Magnétisme,* Puf, Paris, France

Herpin, A.; Koehler, W. & Meriel, P. (1960). *C. R. Acad. Sci.,* Vol. 251, p. 1359

Hock, R.; Fuess, H.; Vogt, T & Bonnet, M. (1990). *J. of Solid State Chem.,*Vol. 84, (No. 1) pp.39-51

Hock, R.; Fuess, H.; Vogt, T & Bonnet, M. (1991). *Z. Phys. B-Cond. Matter,* Vol. 82, pp. 283-294

Hong, Y. J.; Kum, J.S.; Shim, I. B. & Kim, C. S. (2004). *I.E.E.E. Trans. on Magn.*, Vol. 40, (No. 4), pp. 2808-2810

Hur, N.; Park, S.; Guha, S.; Borissov, A.; Kiryukhin, V. & Cheong, S.-W. (2005). *Appl. Phys. Lett.,* Vol. 87, (No.4), 042901(3 pages)

Kahan, T. (1972). *Théorie des Groupes en Physique Classique et Quantique*, tome 3, Dunod, Paris

Kang, T. D.; Standard, E.; Ahn, K. H.; Sirenko, A. A.; Carr, G. L.; Park, S.; Choi, Y. J.; Ramazanoglu, M.; Kiryukhin, V. & Cheong, S.-W. (2010). *Phys. Rev. B,* Vol. 82, (No.1), 014414 (7 pages)

Kazei, Z. A.; Kolmakova, N.P.; Novak, P. & Sokolov, V. I. (1991). Magnetic Properties of Non-Metallic Inorganic Compounds Based on Transition Elements, In: *Landolt-Börnstein Group III,* H. P. J. Wijn (Ed.), Vol. 27, ISBN 3-540-53963-8, Springer Verlag edition, Berlin Heidelberg

Kvashnini, O. P.; Kapitonov, A. M.; Smokotin, É. M. & Titova, A. G. (1984). *Sov. Phys. Solid State,* Vol. 26, (No.8), pp. 1458-1459

Lahoubi, M.; Guillot, M.; Marchand, A.; Tchéou, F. & Roudaut, E. (1984). *I.E.E.E. Trans. on Magn.,* Vol. MAG-20, (No. 5), pp. 1518-1520

Lahoubi, M.; Guillot, M.; Marchand, A.; Tchéou, F. & Le Gall, H. (1985). High Magnetic-Field Magnetization in Terbium Iron Garnet (TbIG), In: *Advances in Ceramics, Proceedings of the Fourth Inter. Conf. on Ferrites (ICF4),* Vol. 15, Part I, Wang F. F. Y. (Ed.), pp. 275-282, ISSN 0730-9546, San Francisco, USA, Oct.31-Nov.2, 1984

Lahoubi, M. (1986). *Thesis, Doct. of Sciences in Physics,* Grenoble University, France, pp.1-245

Lahoubi, M.; Fillion, G. & Tchéou, F. (1997). *J. Phys. IV France 7, Colloque C1,* Suppl. J. Phys III, pp. C1. 291-292

Lahoubi, M.; Kihal, A. & Fillion, G. (2000). *Physica B,* Vol. 284-288, pp. 1503-1504

Lahoubi, M.; Younsi, W.; Soltani, M.-L.; Voiron, J. & Schmitt, D. (2009). *J. Phys.: Conf. Ser.,* Vol. 150, 042108(4 pages)

Lahoubi, M.; Younsi, W.; Soltani, M.-L. & Ouladdiaf, B. (2010). *J. Phys.: Conf. Ser.,* Vol. 200, 082018(4 pages)

Lahoubi, M. (2012). Symmetry Analysis of the Magnetic Structures in TbIG and Tb:YIG at Low Temperature, In: *J. Phys.: Conf. Ser.,* Vol. 340, 012068(10 pages) *Proceedings of the 5th European Conf. on Neutron Scattering,* Prague, Czech Republic, July 17-22, 2011

Levitin, R. Z.; Markosyan, A. S. & Orlov, V. N. (1983). *Sov. Phys. Solid State,* Vol. 25, (No. 6), pp. 1074-1075

Louca, D.; Kamazawa, K & Proffen, T. (2009). *Phys. Rev. B,* Vol. 80, 214406(6 pages)

Nathans, R.; Shull, C. G.; Shirane, G. & Andresen, A. (1959). *J. Phys. Chem. Solids,* Vol. 10, (No. 2-3), pp. 138-146

Néel, L. (1948). *Ann. Phys.* Paris, Vol. 3, pp. 137-198

Nelson T. J. & Mayer D.C. (1971). *I.E.E.E. Trans. on Magn.,* Vol. 7, (No. 3), pp. 616-617

Nekvasil, V. & Veltrusky, I. (1990). *J. Magn. Magn. Mater.,* Vol. 86, pp. 315-325

Pauthenet, R. (1958a). *Ann. Phys.,* Paris, Vol. 3, pp. 424

Pauthenet, R. (1958b). *Thesis Doct. of Sciences in Physics,* Grenoble University, France, pp.1-39

Pearson, R. F. (1965). *Proc. Phys. Soc.,*Vol. 86, pp. 1055-1066

Pickart, S. J.; Halperin, H. A. & Clark, A. E. (1970). *J. Appl. Phys.,* Vol. 41, pp.1192-1193

Rodić, D. & Guillot, M. (1990). *J. Magn. Magn. Mater.,* Vol. 86, pp. 7-12

Roudaut, E. (1983). Evolution of Position-Sensitive Detectors for Neutron Diffraction Experiments From 1966 to 1982 in the Nuclear Centre of Grenoble, In: *Position-Sensitive Detection & Thermal Neutrons,* pp. 294-301, Academic Press Inc., ISBN: 0-12-186180-5, London, UK

Sayetat, F. (1974). *Thesis, Doct. of Sciences in Physics,* Grenoble University, France

Sayetat, F. (1986). *J. Magn. Magn. Mater.,* Vol. 58, pp. 334-346

Shull, C. G. & Smart, J. S. (1949). *Phys. Rev.* Vol. 76, pp. 1256-1257

Smokotin, É. M., Kvashnini O. P. & Kapitonov, A. M. (1985). *Phys. Status Solidi (a)* Vol. 87, K pp. 53-56

Tchéou, F.; Bertaut, E. F.; Delapalme, A.; Sayetat, F. & Fuess, H. (1970a). *Colloque International C.N.R.S. "Les Eléments des Terre Rares"* 180, Vol. II, pp. 313-332

Tchéou, F.; Bertaut, E. F. & Fuess, H. (1970b). *Solid State Commun.*, Vol. 8, pp. 1751-1758

Tchéou, F.; Fuess, H. & Bertaut, E. F. (1970c). *Solid State Commun.*, Vol. 8, pp. 1745-1749

Watson, R. E. & Fremann, A. J. (1961). *Acta. Cryst.*, Vol. 14 pp. 27-37

Wills, A. S. (2007). *Zeitschrift f ür Kristallographie,* Suppl., Vol. 26, pp. 53-58

Wolf, W. P.; Ball, M.; Hutchings, M. T.; Leask, M. J. M. & Wyatt, A. F. G. (1962). *J. Phys. Soc. Japan,* Vol. 17, Suppl., B-I, pp. 443-448

Wolf, W. P. (1964). Local Anisotropy in Rare earth Garnets, In: *Proceedings of the International Conference on Magnetism* (ICM) Nottingham, pp. 555-560

Zhang, G.-y.; Wei, M.; Xia, W.-s. & Yang. G. (2009). *J. Magn. Magn. Mater.,*Vol. 321, pp. 3077-3079

Zvezdin, A. K. (1995). Field Induced Phase Transitions in Ferrimagnets, In: *Handbook of Magnetic Materials,* K. H. J. Buschow (Ed.), 4, Vol. 9, pp. 405-543, Elsevier Science, Amsterdam

11

Superspace Group Approach to the Crystal Structure of Thermoelectric Higher Manganese Silicides $MnSi_\gamma$

Yuzuru Miyazaki
*Department of Applied Physics, Graduate School of Engineering, Tohoku University
Japan*

1. Introduction

Thermoelectric (TE) materials directly convert waste heat into electricity based on the Seebeck effect. This process itself yields no extra gas, noise, or vibration, and it is thus recognized as a clean power generator for next decades. Currently, large amounts of waste heat, ranging from <100 °C (PCs, TVs, etc.) to ~ 1000 °C (power stations, incinerators, etc.), are emitted into the environment, but some of which can be recovered as electricity by simply placing TE materials on the waste heat sources. Of the waste heat, exhaust gases from automobiles account for a total energy of 460 Pcal (4.6×10^{17} cal) per year in Japan (Terasaki, 2003) and if we can recover 10% of this energy, the total generated electricity would equate to that of a typical thermal power station.

TE generators usually consist in series of $\sim$100-pair p-n junctions of TE materials. The performance of TE materials is commonly evaluated by the "figure-of-merit" using the Seebeck coefficient S, electrical conductivity σ, and thermal conductivity κ as $Z = S^2\sigma/\kappa$. We also use the dimensionless figure-of-merit ZT (T the absolute temperature) and the "power factor" given by $S^2\sigma$. For good characteristics in a TE material, a large S and σ as well as a small κ are necessary although all three parameters are dependent on carrier concentration and hence are correlated. The best TE materials are to be found in doped-semiconductors as the Seebeck coefficient is significantly smaller in conventional metals. A ZT value larger than unity is regarded as a measure of practical application because it roughly corresponds to a thermal-to-electric conversion efficiency of $\eta \sim 10$ %. However, $\eta > 10$ % can be achieved at higher operating temperatures above 600 K even if $ZT < 1$.

Since the 1960's, higher manganese silicides (HMSs) have been extensively studied as potential p-type thermoelectric materials both in Russia and Japan (Nikitin et al., 1969; Nishida, 1972). The compounds exhibit ZT = 0.3-0.7 at around 800 K (Fedorov & Zaitsev, 2006), but different structure formulae, e.g., Mn_4Si_7 (Gottlieb et al., 2003), $Mn_{11}Si_{19}$ (Schwomma et al., 1963; 1964), $Mn_{15}Si_{26}$ (Flieher et al., 1967; Knott et al., 1967), $Mn_{27}Si_{47}$(Zwilling & Nowotny, 1973), were proposed as HMS phases. Figure 1 shows the crystal structures (Momma & Izumi, 2008) of the first three of these; all three have great resemblance apart from the c-axis length. Until recently, controversy existed as to whether the compounds were an identical phase or a series of phases with different structures. The existence of Mn_7Si_{12}, $Mn_{19}Si_{33}$ and $Mn_{39}Si_{68}$ was also reported but these phases were only recognized in a microscopic domain observed using transmission microscopy (Ye & Amelinckx, 1986). In contrast to the existence of several phases, there is only one line compound, $Mn_{11}Si_{19}$, in the Mn-Si binary phase diagram near the corresponding composition (Okamoto, 1991).

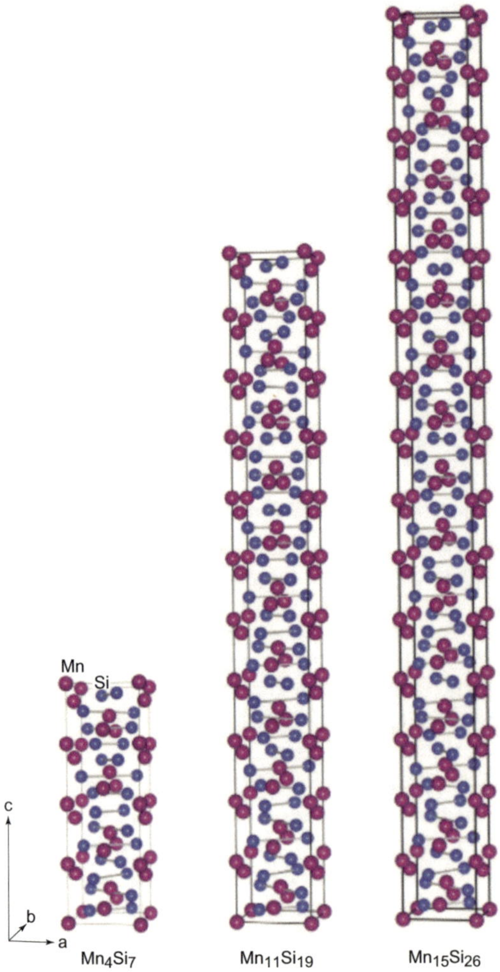

Fig. 1. Crystal structure of commensurate HMSs; Mn_4Si_7 ($c \sim$ 1.75 nm), $Mn_{11}Si_{19}$ ($c \sim$ 4.81 nm) and $Mn_{15}Si_{26}$ ($c \sim$ 6.53 nm).

Based on the observed electron diffraction patterns, Ye and Amelinckx (Ye & Amelinckx, 1986) proposed that the HMS phases are composed of two tetragonal basic units of Mn and Si, with an identical *a*-axis but different *c*-axes. With this idea, a variety of complicated superlattice reflections can be well indexed based on the two basic units distinguished by *c*-axis lengths, c_{Mn} and c_{Si}. Yamamoto (Yamamoto, 1993) employed the concept of superspace group (De Wolff, 1974) to appropriately describe the detailed crystal structure of HMSs. He proposed that the HMSs belong to a family of "composite crystals", consisting of two tetragonal subsystems of [Mn] and [Si]. By means of neutron diffraction data, Miyazaki et al (Miyazaki, 2008) succeeded in determining the detailed modulated structure using this superspace group approach. In this chapter, we describe the method by which the structure of such a composite crystal is determined based on the superspace group formalism.

2. Superspace group approach

2.1 Backgrounds

Suppose we have a crystal consisting of two tetragonal basic unit cells (i.e., subsystems) of [Mn] and [Si] with a common a-axis but different c-axes, c_{Mn} and c_{Si} as shown in Fig. 2. If the ratio c_{Mn}:c_{Si} can be expressed as simple integers, such as 2:1, 3:2, etc., the whole crystal structure can be represented by a three dimensional (3D) unit cell with $c = m \times c_{Mn} = n \times c_{Si}$ (m and n are integers) and we no longer need to assume two basic units. The three HMSs shown in Fig. 1 are examples of such commensurate structures of Mn_mSi_n, although rather complicated ones. In contrast, if the c_{Mn}:c_{Si} ratio is irrational, we then have to assume a unit cell with an infinite length along the c-axis. In such cases, the number of atomic sites grows too large and structural analysis becomes practically impossible. The concept of superspace group is the best way to accurately describe such a structure (De Wolff, 1974; Janner & Janssen, 1980).

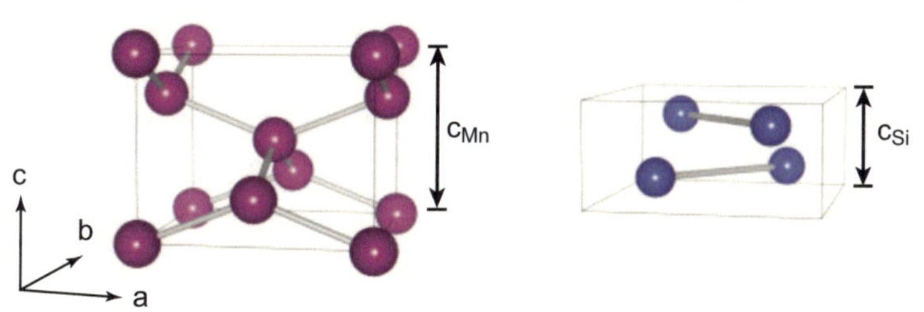

Fig. 2. Two tetragonal subsystems of [Mn] (left) and [Si] (right) with a common a-axis but different c-axis lengths, c_{Mn} and c_{Si}.

2.2 General description of composite crystals

To apply the superspace group approach, we need to introduce a (3+1)-dimensional unit vectors, $\mathbf{a}_{s1}$, $\mathbf{a}_{s2}$, $\mathbf{a}_{s3}$, and $\mathbf{a}_{s4}$. Using the 3D unit bases and another $\mathbf{d}$, perpendicular to this 3D space, the relationship between these basis sets can be defined as,

$$\begin{pmatrix} \mathbf{a}_{s1} \\ \mathbf{a}_{s2} \\ \mathbf{a}_{s3} \\ \mathbf{a}_{s4} \end{pmatrix} = \begin{pmatrix} 1 & 0 & 0 & -\alpha \\ 0 & 1 & 0 & -\beta \\ 0 & 0 & 1 & -\gamma \\ 0 & 0 & 0 & 1 \end{pmatrix} \begin{pmatrix} \mathbf{a} \\ \mathbf{b} \\ \mathbf{c} \\ \mathbf{d} \end{pmatrix} \tag{1}$$

where $\mathbf{k} = (\alpha\beta\gamma)$ is a modulation vector to index the electron diffraction patterns. For the HMSs, the modulation exists only in the c-axis, i.e., $\mathbf{k} = (00\gamma)$ with $\gamma = c_{Mn}/c_{Si}$. From equation 1, both the $\mathbf{a}_{s1}$- and $\mathbf{a}_{s2}$-axes lie in the 3D physical space $\mathbf{R}_3$ and the $\mathbf{a}_{s4}$-axis is perpendicular to $\mathbf{R}_3$, as shown in Fig. 3. Similar to the unit bases, fractional coordinates in the (3+1)D space,

$\overline{x}_{s1}$, $\overline{x}_{s2}$, $\overline{x}_{s3}$ and $\overline{x}_{s4}$ can be defined as,

$$\begin{pmatrix} \overline{x}_{S1} \\ \overline{x}_{S2} \\ \overline{x}_{S3} \\ \overline{x}_{S4} \end{pmatrix} = \begin{pmatrix} 1 & 0 & 0 & 0 \\ 0 & 1 & 0 & 0 \\ 0 & 0 & 1 & 0 \\ \alpha & \beta & \gamma & 1 \end{pmatrix} \begin{pmatrix} \overline{x} \\ \overline{y} \\ \overline{z} \\ t \end{pmatrix} \tag{2}$$

where the parameter t, called as an internal coordinate, represents the distance from $\mathbf{R_3}$ and is related to the fourth coordinate $\overline{x}_{s4}$ by $\overline{x}_{s4} = \gamma\overline{z} + t$.

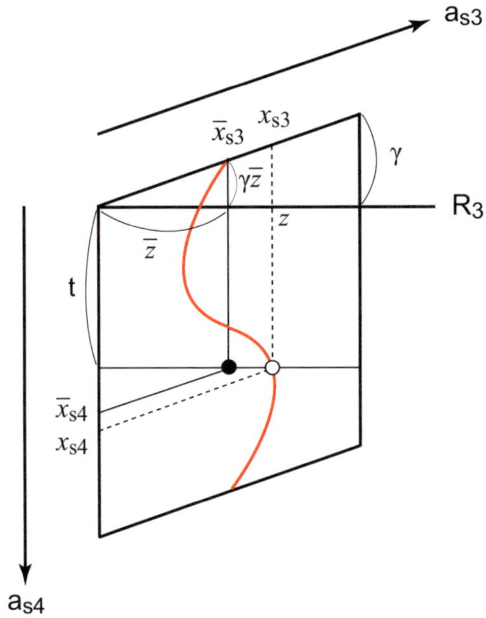

Fig. 3. A unit cell in the (3+1)-dimensional superspace and the relationship between the coordinates without modulation, $\overline{x}_{si}$, with modulation x_{si} (i = 1-4), and internal coordinate t. The red curve represents the positional modulation of an atom in subsystem 1. $\mathbf{R_3}$ denotes the 3D real space.

Due to the periodic difference along a particular direction, the composite crystals show positional (*displacive*) modulation of atomic sites. The positional modulation is periodic with an interval of $0 \leq \overline{x}_{s4} \leq 1$ as shown with the red curve in Fig. 3. The modulated position of an atom is then mathematically represented with a Fourier series as:

$$x_{si} = \overline{x}_{si} + \sum_{n=0}^{m} [A_n \cos(2\pi n \overline{x}_{s4}) + B_n \sin(2\pi n \overline{x}_{s4})], \tag{3}$$

where i = 1,2,3.

2.3 Symmetry of HMSs

To specify the superspace group to the present HMSs, we need to consider an additional axis c_{Si} parallel to c_{Mn}. The first subsystem [Mn] has a β-Sn type arrangement of Mn atoms classified under the 3D space group of $I4_1/amd$. For the second subsystem [Si], originally having dimensions of $a/\sqrt{2} \times a/\sqrt{2} \times c_{Si}$, the a-axis length is adjusted to that of [Mn]. The Si atoms are at the origin and the equivalent positions generated by the $P4/nnc$ space group. According to Yamamoto (Yamamoto, 1993), the most appropriate (3+1)D superspace group is designated as $P : I4_1/amd : 1\bar{1}ss$ (Mn) $W : P4/nnc : q\bar{1}q1$ (Si), or based on the modulation vector $\mathbf{k} = (00\gamma)$ more conveniently as $I4_1/amd(00\gamma)00ss$.

Based on the adopted superspace group, the translation parts for the [Mn] subsystem are expressed as:
+(0, 0, 0, 0); +(1/2, 1/2, 1/2, 0),
with symmetry operations represented as:
(i) x, y, z, t;
(ii) $-y, x+1/2, z+1/4, t$;
(iii) $-x, y, z, t+1/2$;
(iv) $-x, -y, z, t$;
(v) $-y, -x+1/2, z+1/4, t+1/2$;
(vi) $y, -x+1/2, z+1/4, t$;
(vii) $x, -y, z, t+1/2$;
(viii) $y, x+1/2, z+1/4, t+1/2$;
(ix) $-x, -y+1/2, -z+1/4, -t$;
(x) $y, -x, -z, -t$;
(xi) $x, -y+1/2, -z+1/4, -t+1/2$;
(xii) $x, y+1/2, -z+1/4, -t$;
(xiii) $y, x, -z, -t+1/2$;
(xiv) $-y, x, -z, -t$;
(xv) $-x, y+1/2, -z+1/4, -t+1/2$;
(xvi) $-y, -x, -z, -t+1/2$.

The translation parts and the symmetry operation for the [Si] subsystem can be obtained by simply interchanging the third and the fourth components.

2.4 Structural modulation

Polycrystalline samples were prepared in a tetra-arc-type furnace under an Ar atmosphere using tungsten electrodes and a water-cooled copper hearth. Appropriate amounts of Mn (99.9%) and Si (99.99%) powders were mixed in an alumina mortar and pressed into pellets. The pellets were melted four times, and turned over each time to obtain full homogeneity. As the satellite reflections were observed up to eighth order in the electron diffraction patterns, positional modulation of the atomic sites was introduced taken to the eighth order of cosine and sine components of the Fourier terms, A_n and B_n (n= 0-8), of equation 3.

Figure 4 shows the observed, calculated, and difference profiles of the neutron diffraction (ND) data for $MnSi_\gamma$ at 295 K. The ND data were collected using a Kinken powder diffractometer for high efficiency and high resolution measurements, HERMES (Petricek et al., 2000), at the Institute for Materials Research (IMR), Tohoku University, installed at the JRR-3M reactor of the Japan Atomic Energy Agency (JAEA) at Tokai. A monochromatized incident neutron beam at λ = 1.8265 Å was used. The ND data were analyzed with the JANA2000 software package (Petricek et al., 2000). The bound coherent scattering lengths used for the refinement were -3.730 fm (Mn) and 4.149 fm (Si).

The short vertical lines below the patterns indicate the peak positions of possible Bragg reflections. Small peaks at $2\theta\sim$ 32.5°, 40.5°, and 53.0°, derived from the secondary phase MnSi, were excluded in the refinement cycles. The final R factors were R_{wp} = 9.8% and R_p = 6.9%, and the lattice parameters were refined to a = 5.5242(3) Å, c_{Mn} = 4.3665(3) Å, and c_{Si} = 2.5202(3) Å. The numbers in parentheses represent the estimated standard deviation of the last significant digit. The resulting c-axis ratio was γ = 1.7326(1), different from that of any commensurate HMSs (Flieher et al., 1967; Gottlieb et al., 2003; Knott et al., 1967; Schwomma et al., 1963; 1964; Zwilling & Nowotny, 1973) and that for the sample annealed at 1273 K for 168 h from the same batch of the present sample of γ = 1.7361(1) (Miyazaki, 2008).

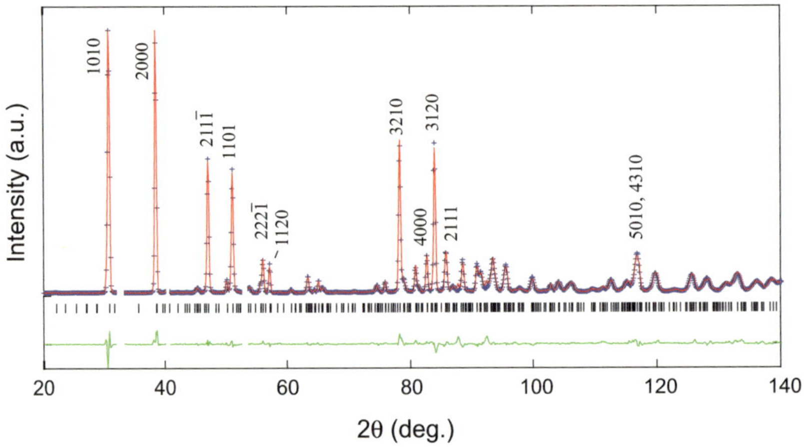

Fig. 4. Observed, calculated, and difference patterns of powder neutron diffraction data for $MnSi_\gamma$ measured at 295 K. Short vertical lines below the patterns indicate positions of Bragg reflections. The difference between the observed and calculated intensities is shown below the vertical lines. The peaks, denoted as $hkl0$ and $hk0m$, are the fundamental reflections derived from the [Mn] and [Si] subsystems, respectively, while the $hklm$ peaks are satellite reflections.

Table 1 summarizes the atomic coordinates and equivalent isotropic displacement parameters, U_{eq}, for the fundamental structure of $MnSi_\gamma$ at 295 K.

Subsystem 1: [Mn]	x	y	z	U_{eq} (Å^2)
	0	0	0	0.0057(12)
Subsystem 2: [Si]	x	y	z	U_{eq} (Å^2)
	1/4	1/4	1/4	0.0139(12)

Table 1. Atomic coordinates and equivalent isotropic atomic displacement parameters, U_{eq}, for the fundamental structure of $MnSi_\gamma$ at 295 K.

2.5 Details of modulated structure

Table 2 summarizes the refined Fourier amplitudes for the positional parameters together with the anisotropic displacement parameters of each atom. Due to the superspace group symmetry, the number of refinable parameters of the Fourier terms is limited. Only even terms of the sine wave along the z-direction, i.e., B_{2z}, B_{4z}, B_{6z}, and B_{8z}, are allowed for the Mn atoms. In contrast, odd terms of the sine and cosine waves are allowed for both the x and y components for the Si atoms, along with B_{4z} and B_{8z} terms. The amplitudes of each cosine wave component in the x-y plane, such as A_{1x} and A_{1y}, are equal, whereas those of each sine wave are equal but their signs are opposite.

Subsystem 1: [Mn]	x	y	z
B_2	0	0	-0.0142(12)
B_4	0	0	0.017(2)
B_6	0	0	0
B_8	0	0	0
$U_{11} = U_{22} = 0.002(2)$ (Å^2)			
$U_{33} = 0.013(2)$ (Å^2)			
Subsystem 2: [Si]	x	y	z
A_1	0.0772(3)	$= A_{1x}$	0
B_1	$= A_{1x}$	$= -A_{1x}$	0
A_3	0.0103(3)	$= A_{3x}$	0
B_3	$= -A_{3x}$	$= A_{3x}$	0
B_4	0	0	-0.0441(19)
A_5	-0.0040(5)	$= A_{5x}$	0
B_5	$= A_{5x}$	$= -A_{5x}$	0
A_7	-0.0034(7)	$= A_{7x}$	0
B_7	$= -A_{7x}$	$= A_{7x}$	0
B_8	0	0	0.017(4)
$U_{11} = U_{22} = 0.009(2)$ (Å^2)			
$U_{33} = 0.023(2)$ (Å^2)			

Table 2. Refined positional modulation wave components and anisotropic displacement parameters, U_{ij}, for MnSi$_\gamma$.

In Fig. 5, we show the revealed positional modulations for the x, y, and z coordinates of each atom plotted against $\overline{x}_{s4}$. The right vertical axes are re-scaled to represent the displacement (in Angstroms) for each atom. All the displacements are periodic in the interval $0 \leq \overline{x}_{s4} \leq 1$. For the Mn atoms, positional modulation is only allowed in the z direction and the maximum displacement (~0.12 Å) from $z = 0$ is recognized at $\overline{x}_{s4} \sim 0.18$, 0.32, 0.68 and 0.82. The displacement of Si atoms along z is comparable to that of the Mn atoms and the maximum displacement of (~0.13 Å) from $z = 1/4$ can be seen at $\overline{x}_{s4} \sim 0.03$, 0.19, 0.28, etc. Based on the large sine and cosine components, the rotational modulation in the x and y directions is significant for the Si atoms. Both the modulation waves in x and y are identical with the phase-shift of $\Delta\overline{x}_{s4} = 1/4$. The maximum displacement of ~0.54 Å from $x = 1/4$, equivalent to $\Delta x \sim 0.1$, is realized at $\overline{x}_{s4} \sim 0.48$ and 0.98, and at $\overline{x}_{s4} \sim 0.23$ and 0.73 from the $y = 1/4$ position. Similar rotational modulations have been reported for related chimney-ladder compounds, such as $(Mo_{1-x}Rh_x)Ge_\gamma$ (Rohrer et al., 2000) and $(Cr_{1-x}Mo_x)Ge_\gamma$ (Rohrer et al., 2001). The deviation from the fundamental position, Δx and Δy, is also ~0.1 in these compounds, although the a-axis lengths of these phases ($a \sim 5.9$ Å) are much larger than that in the present compound.

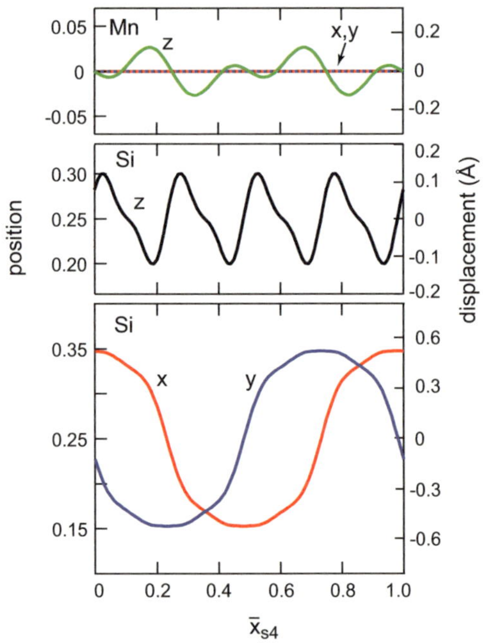

Fig. 5. Revealed positional modulations of the Mn and Si atoms plotted as a function of the fourth superspace coordinate, $\overline{x}_{s4}$. The right vertical axes are re-scaled to represent the displacement for each atom.

Figure 6 shows the revealed modulated structure of $MnSi_\gamma$ at 295 K. The top left figure illustrates the *c*-axis projection to represent the helical arrangement of the Si atoms. The bottom left figure depicts the atoms within a partial unit cell of length $4\times c_{Mn}$. The seven squares on the right represent slices of the 1st to 7th layers of Si atoms from the origin. The 1st layer of Si atoms corresponds to $\overline{x}_{s4} = (1/\gamma)(\overline{z} + t) = (1/\gamma)\times(1/4 - 1/4) = 0$, because the modulation vector component is inverted to γ^{-1} for the [Si] subsystem. As deduced from Fig. 5, the coordinates at $\overline{x}_{s4} = 0$ are around $(x, y) = (0.35, 0.23)$. Since there is a two-fold axis at (1/2, 1/2) parallel to the *c*-axis, the other Si atom in the 1st layer is located at around (0.65, 0.77). Similarly, the 3rd ($\overline{z} = 5/4$) and 5th ($\overline{z} = 9/4$) layers of Si atoms, corresponding to $\overline{x}_{s4} =$ 0.576 and 0.152 ($\equiv$ 1.152), are located at around (0.16, 0.32), (0.84, 0.68), and (0.32, 0.16), (0.68, 0.84), respectively. By symmetry, the coordinates of even numbered layers can be obtained as $(-x, y)$ and $(x, -y)$. The 2nd ($\overline{z} = 3/4$) and 4th ($\overline{z} = 7/4$) layers of Si atoms, corresponding to $\overline{x}_{s4} = 0.288$ and 0.864, are located at around (0.19, 0.85), (0.81, 0.15) and (0.33, 0.67), (0.67, 0.33), respectively. The *z* coordinates of Si and Mn atoms in the modulated structure can also be calculated in a similar way.

Figure 7 shows interatomic distances plotted as a function of *t*. The two periodic curves around the distance of 3.0 Å represent the nearest four Mn-Mn distances, for which each curve is duplicated because of the two equidistant bonds. The nearest Mn-Mn distances, ranging from 2.92 Å to 3.01 Å, are relatively longer than that expected from the atomic radius of Mn, $r_{Mn} = 1.24$ Å (Inoue et al., 2001). However, such long distances have been reported in the structure of α-Mn (Bradley & Thewlis, 1927), wherein the Mn-Mn distances vary from 2.24 Å to 2.96 Å. In the case of the Mn-Si bonds, each Mn atom is coordinated to eight Si atoms, as deduced from Fig. 6. The four curves around 2.4 Å represent the nearest eight Mn-Si distances,

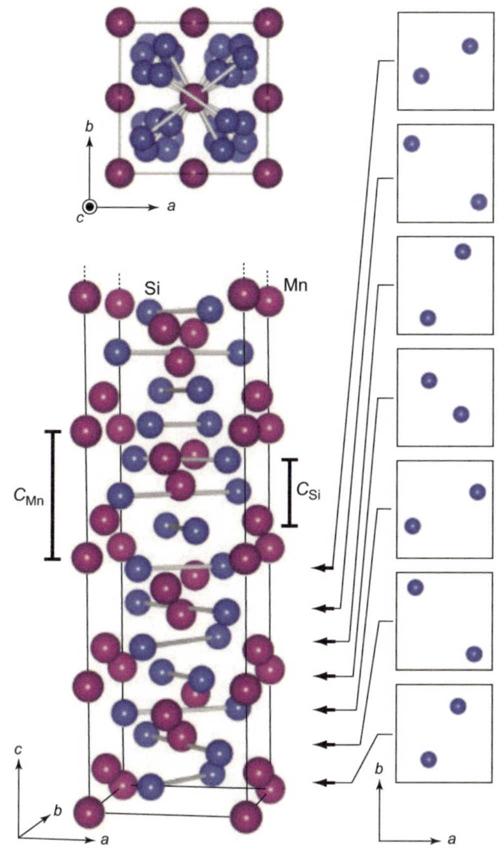

Fig. 6. Revealed modulated composite structure of $MnSi_\gamma$ at 295 K. The upper figure is a c-axis projection to illustrate the rotational arrangement of the Si atoms. The lower figure depicts the atoms within $4\times c_{Mn}$ lengths. The seven squares on the right are slices through the 1st to 7th layers of the Si subsystem from the origin.

with each curve duplicated due to equidistant Mn-Si bond pairs. At $t = 0$, the Mn atom has eight Si neighbors with $d_{Mn-Si}\sim$ 2.4 Å, which is the typical Mn-Si bond distance on the basis of the metallic radii of Mn and Si (r_{Si} = 1.17 Å) (Inoue et al., 2001). With increasing t, the four bonds (two curves) become shorter towards $\sim$2.2 Å, and the remaining four bonds become longer. In this way, the Mn atom is always bounded to eight Si atoms within a distance of $<$ 2.8 Å.

3. Universal treatment of HMSs by means of the superspace description

In principle, the superspace group approach described so far can also be applicable to *all* the reported commensurate structure of HMSs. Towards a uniform treatment of the crystal structure of the present $MnSi_\gamma$ and other HMSs, we will first convert the 3D coordinates of Mn_4Si_7 (Gottlieb et al., 2003) to $\overline{x}_{s4}$. The Mn_4Si_7 phase can be regarded as a case for which γ = $7/4 = 1.75$, wherein $4c_{Mn}$ exactly equals $7c_{Si}$; the 3D unit cell consists of a stacking of four [Mn] subsystems and seven intervening [Si] subsystems, as shown in Fig. 1. Let us consider the [Si] subsystem closest to the origin. The Si atom at Si1 ($8j$) sites with (0.15715, 0.2015, 0.11253) is

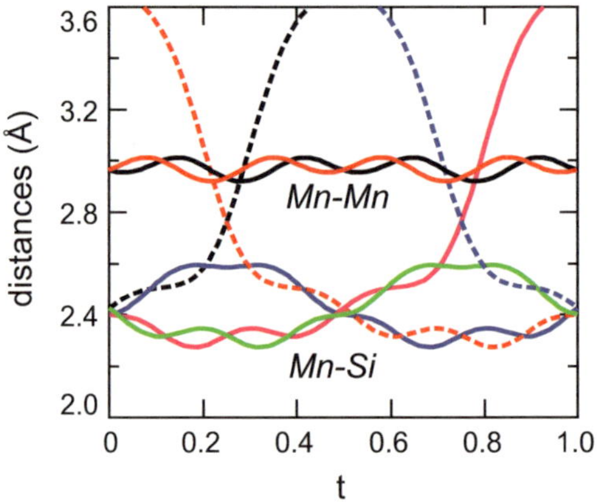

Fig. 7. Interatomic distances as a function of the internal coordinate t for $MnSi_\gamma$ at 295 K.

converted to (0.15715, 0.2015, $\overline{z}+\delta$) in the [Si] subsystem. Since the Si atom is located at the 2nd ($\overline{z}$ = 3/4) layer, we obtain $\overline{x}_{s4} = (1/\gamma)\times(\overline{z} - 1/4) = 0.286$ and $3/4 + \delta = 7\times0.11253 = 0.788$.

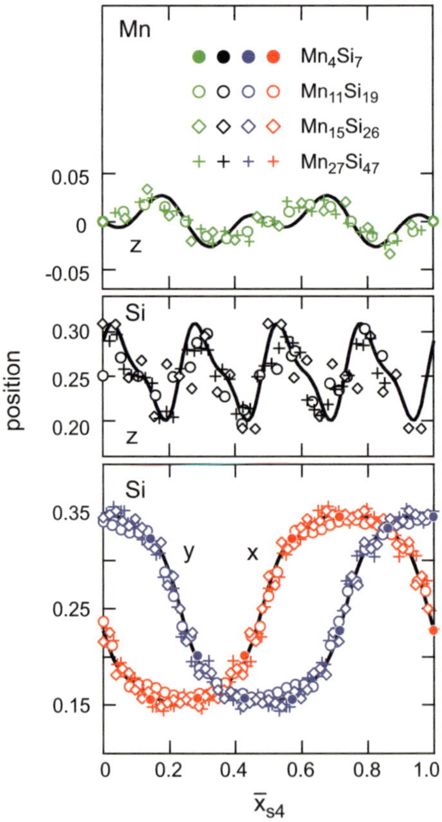

Fig. 8. The converted coordinates of Mn at ~ (0, 0, 0) and Si at ~ (1/4, 1/4, 1/4) in each subsystem of HMSs plotted as a function of the superspace coordinate, $\overline{x}_{s4}$. Solid lines are the data of the present incommensurate $MnSi_\gamma$ (γ = 1.7326(1)).

Similarly, the Si atom at Si4 (8*j*) sites with (0.34518, 0.2274, 0.9620) is converted to (0.34518, 0.2274, 27/4+δ') with $\overline{x}_{s4}$ = 0.714 ($\equiv$ 3.714). In this case, 27/4 + δ' equals 7×0.9620 = 6.734 in the [Si] subsystem, corresponding to z = 0.734 in the 7th [Si] subsystem from the origin. By applying all the symmetry operations to the Mn1-Mn5 and Si1-Si4 sites, the equivalent positions close to Mn (0, 0, 0) and Si (1/4, 1/4, 1/4) in each subsystem can be obtained as filled circles shown in Fig. 8.

Based on this structuring, those equivalent positions as a function of $\overline{x}_{s4}$ of $Mn_{11}Si_{19}$ ($\gamma = 1.7\dot{2}\dot{7}$) (Schwomma et al., 1963; 1964), $Mn_{15}Si_{26}$(γ = $1.7\dot{3}$) (Flieher et al., 1967; Knott et al., 1967) and $Mn_{27}Si_{47}$(γ = $1.7\dot{4}0\dot{7}$) (Zwilling & Nowotny, 1973) are converted into the respective marks as shown in Fig. 8. The solid lines in these panels correspond to those shown in Fig. 5 but the x and y lines are interchanged to preserve the original (x, y) coordinates given in the reports (Flieher et al., 1967; Gottlieb et al., 2003; Knott et al., 1967; Schwomma et al., 1963; 1964; Zwilling & Nowotny, 1973). All the x and y coordinates of the points in the [Si] subsystem are well superposed on the two *universal* lines, suggesting that the helical arrangement of Si atoms is almost identical and independent of the γ values in all HMSs. In contrast, the z coordinates for the Mn and Si atoms deviate from the periodic solid lines of the present sample. It would be, however, reasonable to consider different shape of curves, i.e., different B_z terms, to fit the z coordinates for each HMS because the stacking periodicity might be dependent on γ.

4. Conclusion

A superspace group approach is economical in describing an incommensurate compound by using a much reduced number of parameters. The positional modulation of the atoms can be expressed by means of a Fourier expansion of the modulation functions. For HMSs, both the Mn and Si atoms are at special positions in each subsystem and their modulated positions are excellently described by only *eight* parameters. This is much smaller set compared with other HMSs with a 3D description; the number of refinable positional parameters are 13 (Mn_4Si_7), 41 ($Mn_{11}Si_{19}$), 26 ($Mn_{15}Si_{26}$) and 97 ($Mn_{27}Si_{47}$). These four phases have different 3D symmetries and lattice parameters but by using the superspace group approach, we can treat these phases as an identical compound $MnSi_\gamma$, with slightly different stoichiometric γ values. Moreover, this approach affords a uniform treatment of a compound system which can change its stoichiometry incommensurately upon substitution of the elements. For example, a partial substitution of Cr, Fe, and Co is possible at Mn sites and such a solid solution can also be simply represented as $(Mn_{1-x}M_x)Si_\gamma$ (M = Cr, Fe and Co) (Flieher et al., 1967). Ranging over the various solid solutions, the γ value is known to change from 1.65 through 1.76. The next target should be to optimize the TE properties by tuning the electronic structure of HMSs and we hope that these TE materials can be prepared in the present HMS-based compounds in the near future.

5. References

Bradley, A. J. & Thewlis, J. (1927). The crystal structure of α-manganese. *Proc. Royal Soc. London, Ser. A* 115: 456-471, 0080-4630.

De Wolff, P. M. (1974). The pseudo-symmetry of modulated crystal structures. *Acta Cryst.* A30: 777-785, 0567-7394.

Fedorov, M. I. & Zaitsev, V. K. (2006). Thermoelectrics of transition metal silicides, In: *Thermoelectrics Handbook Macro to Nano*, Rowe, D. M., (Ed.), chap.31, CRC press, Boca Raton, 0-8493-2264-2, London.

Flieher, G.; Vollenkle, H.; Nowotny, H. (1967). Die Kristallstruktur von $Mn_{15}Si_{26}$, *Monatsh. Chem.* 98: 2173-2179, 0026-9247.

Gottlieb, U.; Sulpice, A.; Lambert-Andron, B.; Laborde, O. (2003). Magnetic Properties of Single Crystalline Mn_4Si_7, *J. Alloys Compd.* 361: 13-18, 0925-8388.

Inoue, T.; Chikazumi, S.; Nagasaki, S; Tanuma, S. (2001). *Agne Periodic Table*, AGNE Technology Center, 978-4-901496-37-7,Tokyo.

Janner, A. & Janssen, T. (1980). Symmetry of incommensurate crystal phases. II. Incommensurate basic structure, *Acta Cryst.* A36: 408-415, 0567-7394.

Knott, H. W.; Mueller, M. H.; Heaton, L. (1967). The Crystal Structure of $Mn_{15}Si_{26}$, *Acta Cryst.* 23: 549-555, 0365-110X.

Miyazaki, Y.; Igarashi, D.; Hayashi, K.; Kajitani, T.; Yubuta, K. (20080. Modulated Crystal Structure of Chimney-Ladder Higher Manganese Silicides $MnSi_\gamma$ ($\gamma\sim$ 1.74), *Phys. Rev. B* 78, 214104 (8 pages), 1098-0121.

Momma, K. & Izumi, F. (2008). VESTA: a three-dimensional visualization system for electronic and structural analysis. *J. Appl. Cryst.* 41: 653-658, 0021-8898.

Nikitin, E. N.; Tarasov, V. I.; Tamarin, P. V. (1969). Thermal and electrical properties of the higher manganese silicide from 4.2 to 1300 K and its structure, *Sov. Phys. Solid State* 11: 187-189, 0038-5654.

Nishida, I. (1973). Semiconducting properties of nonstoichiometric manganese silicides. *J. Mater. Sci.* 7: 435-440, 0022-2461.

Ohoyama, K.; Kanouchi, T.; Nemoto, K.; Ohashi, M.; Kajitani, T.; Yamaguchi, Y. (1998). The new neutron powder diffractometer with a multi-detector system for high-efficiency and high-resolution measurements, *Jpn. J. Appl. Phys.* 37: 3319-3326, 0021-4922.

Okamoto, H. (1991). Mn-Si (Manganese-Silicon), *J. Phase Equilibria* 12: 505-507, 1054-9714.

Petricek, V.; Dusek, M.; Palatinus, L. (2000). *JANA2000 The crystallographic computing system.* Institute of Physics, Praha, Czech Republic.

Rohrer, F. E.; Lind, H.; Eriksson, L.; Larsson, A. K.; Lidin, S. (2000). One the question of commensurability - The Nowotny chimney-ladder structures revisited. *Z. Kristallogr.* 215: 650-660, 1433-7266.

Rohrer, F. E.; Lind, H.; Eriksson, L.; Larsson, A. K.; Lidin, S. (2001). Incommensurately modulated Nowotny Chimney-ladder phases: $Cr_{1-x}Mo_xGe_{\sim 1.75}$ with x = 0.65 and 0.84. *Z. Kristallogr.* 216: 190-198, 1433-7266.

Schwomma, O.; Nowotny, H.; Wittmann, A. (1963). Die Kristallarten $RuSi_{1.5}$, $RuGe_{1.5}$ und $MnSi_{\sim 1.7}$, *Monatsh. Chem.* 94: 681-685, 0026-9247.

Schwomma, O.; Preisinger, A.; Nowotny, H.; Wittmann, A. (1964) Die Kristallstruktur von $Mn_{11}Si_{19}$ und deren Zusammenhang mit Disilicid-Typen, *Monatsh. Chem.* 95: 1527-1537, 0026-9247.

Terasaki, I. (2003). Thermoelectric cobalt oxides. *Parity* 64-67 (*in Japanese*).

Yamamoto, A. (1993). Determination of Composite Crystal Structures and Superspace Groups, *Acta Cryst. A* 49: 831-846, 0567-7394.

Ye, H. Q. & Amelinckx, S. (1986). High-Resolution Electron Microscopic Study of Manganese Silicides $MnSi_{2-x}$, *J. Solid State Chem.* 61: 8-39, 0022-4596.

Zwilling, G. & Nowotny, H. (1973). Zur Struktur der Defekt-Mangansilicide, *Monatsh. Chem.* 104: 668-675, 0026-9247.

12

The pH Dependence of Protonation States of Polar Amino Acid Residues Determined by Neutron Diffraction

Nobuo Niimura
Frontier Research Center for Applied Atomic Sciences, Ibaraki University, Shirakata Japan

1. Introduction

The charges of various amino-acid side chains depend on the pH. For example, at a high pH (low acidity conditions), carboxylic acids tend to be negatively charged (deprotonated), and amines tend to be uncharged (unprotonated). At a low pH (high acidity), the opposite is true. The pH at which exactly half of any ionized amino acid is charged in solution is called the pK_a of that amino acid. These pK_a values of such ionizable amino-acid side chains are tabulated in standard textbooks. However, the protonation state of a given amino-acid side chain in a protein cannot be estimated from standard pK_a values measured from isolated amino acids in solution, because inside a protein it may vary significantly depending on the local environment. The electrically charged states of the amino-acid residues are very important in understanding the physiological function of the protein, the interaction between ligands and proteins, molecular recognition, structural stability, and so on.

Neutron Protein Crystallography (NPC) is the unique method to provide the definite protonation states of the amino acid residue in proteins because neutron can identify not only hydrogen atoms but also protons. (N.Niimura et al. 2004,2006, N.Niimura & R. Bau 2008, N.Niimura & A. Podjarny 2011)

Consider one example: Figure 1 shows a typical example of the different protonation states of histidines in met-myoglobin at pH 6.8. To create this figure, the Neutron Protein Crystallography (NPC) of met-myoglobin was carried out at 1.5 Å resolution (A. Ostermann et al. 2005). Met-myoglobin contains 12 histidines. The measured pK_a value of the histidine amino acid in aqueous solution is about 6.0, so the histidines should be neutral at pH 6.8 aqueous solution. However, four kinds of ionization states are observed by NPC in the imidazole ring of histidines as shown in Figure 1. Two histidines are doubly protonated on both $N^{\delta 1}$ and $N^{\varepsilon 2}$ (in red in Figure 1, with an average B-factors of 14.9 $Å^2$), three are singly protonated (neutral) on $N^{\delta 1}$ (in yellow, with an average B-factors of 4.5 $Å^2$), and four are singly protonated (neutral) on $N^{\varepsilon 2}$ (in blue, with an average B-factors of 16.5 $Å^2$). In the three remaining histidines, the protonation state is not clear because of the disordered state of deuterium atoms, the B-factors of which are rather large (in pink in Figure 1, with an average B-factor of 22.9 $Å^2$). Clearly, histidines inside the protein do not always follow the pK_a value in solution.

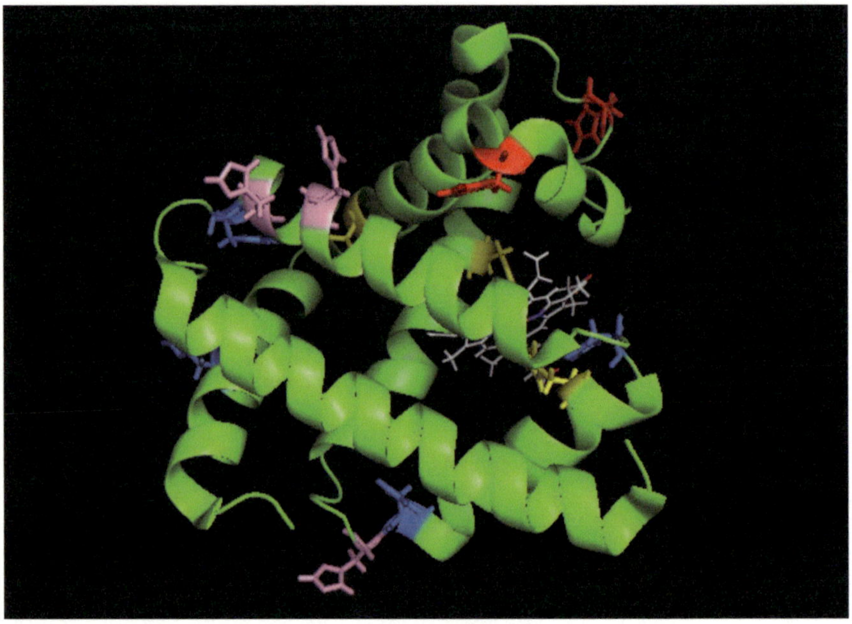

Fig. 1. Protonation states of 12 Histidines in myoglobin.

2. The protonation state of the N^{δ} and N^{ε} atoms of histidine.

2.1 Cubic insulin

Insulin is a polypeptide hormone critical for the metabolism of glucose. The insulin monomer consists of two chains—a 21-residue A-chain and a 30-residue B-chain—linked by a pair of disulfide bonds, A7-B7 and A20-B19, respectively; furthermore, an additional intra-chain disulfide bond links A6 and A11. In a solution free of metal ions, insulin exists as a mixture of monomer, dimer, tetramer, hexamer, and higher aggregates depending on its concentration (T. Blundell et al. 1972). The structure of hexameric insulin, the form in which it is stored in the pancreas β-cells, was reported first in 1969 (Adams et al. 1969); and a comprehensive description of the room-temperature structure at 1.5 Å resolution, as well as the biological implications thereof, was published afterwards (Baker et al. 1988). The structure of the insulin monomer, which is the hormonally active form, was determined by NMR (T. L. Blundell et al. 1971; T. Blundell et al. 1972). Zn^{2+} ions promote the hexamerization of insulin in pancreas, and the other divalent cations, such as Ni^{2+}, Co^{2+}, Cd^{2+} and Cu^{2+}, also have the same effect for the hexamerization in solution (Hill et al. 1991).

Normally, the standard pK_a value of histidine is about 6.0. In insulin, there are two histidine amino acid residues (both on chain B); and it is of interest to know whether at pD 6.6 (pD is the pH in heavy water.) they are protonated or deprotonated, since pD 6.6 is clearly close to the borderline pK_a value. Moreover, it is also important to know which of the two nitrogen atoms (N^{δ}, N^{ε}) in the imidazole ring of histidine are ionized, in order to discuss the physiological function of insulin. (In fact, His *B*10 is a zinc-ion binding site.) (Ishikawa et al. 2008a; Ishikawa et al. 2008b) X-ray structural studies of insulin at various pH values reported that a conformational change of His *B*10 is likely to occur at pH values higher (i.e., more basic) than pH 9, but that of His *B*5 is expected to be unchanged (Diao 2003).

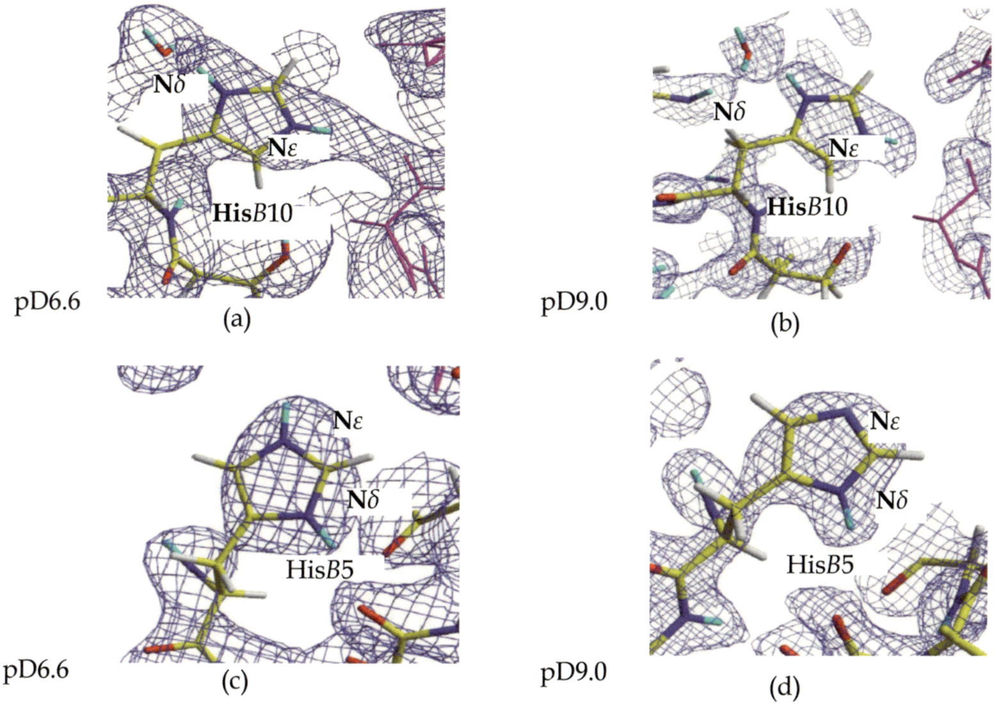

Fig. 2. 2Fo-Fc positive nuclear density maps in insulin of (a) His *B*10 at pD 6.6, (b) His *B*10 at pD 9.0, (c) His *B*5 at pD 6.6, and (d) His *B*5 at pD 9.0.

Porcine insulin can crystallize in a cubic space group without zinc, and neutron diffraction experiments on porcine insulin (crystallized at pD 9.0 and 6.6) were performed at room temperature (Ishikawa et al. 2008a; Maeda et al. 2004). The overall root-mean-square differences of non-hydrogen atoms and all atoms between pD6.6 (Adams et al. 1969) and pD9 are 0.67 Å^2 and 1.04 Å^2, respectively. Porcine insulin has two histidines (HisB5 and HisB10) in the B chain. As for HisB5, both of N^δ and N^ε atom of the imidazole ring are protonated at pD6.6, and only the N^ε atom was protonated at pD9 (Figure 2 (c) and (d)). This fact indicates that the pK_a value of HisB5 has the range between pD6.6 and pD9, which is plausible because of the pK_a value (=6.5) of a simple histidine moiety. However, HisB10 was confirmed to be protonated at both the N^δ and N^ε atoms at both pDs. (See Figure 2 (a) and (b)). Figure 2 (b) shows that HisB10 is doubly protonated and positively charged even at the alkaline pD of 9.0. There are two possible reasons for this double-protonation state; a hydrogen bonding network and an electrostatic potential. In HisB10, two D atoms of the imidazole ring are anchored to the surrounding water molecules and amino acids by hydrogen bonds. At both pD6.6 and pD9, the N^ε-D group was hydrogen bonded to a carbonyl group of the main chain of TyrA14. (The δD-O bond lengths were 2.3 Å and 2.1 Å for pD9 and pD6.6, respectively.) At the same time, the N^δ-D group formed an indirect interaction with a carbonyl group of HisB5 mediated by a water molecule. Moreover, HisB10 is located on the surface of insulin and near a negative electrostatic potential. GluB13 was about 7 Å away from HisB10 and the carboxyl group of this residue was deprotonated, indicating negatively charged states of this functional group. In order to investigate the electrostatic effect for HisB10, molecular surfaces and electrostatic grid potentials were calculated by GRASP2 (Diao 2003). HisB10 was not on a negative potential surface; however, the negative potential derived from GluB13 was in the vicinity of the imidazole ring of HisB10. On the other hand, there was no negative potential near HisB5, which has a normal pK_a value. In the previous X-ray study, it was reported that GluB13 affected a structural change of insulin by pH shift, supporting the hypothesis in the neutron study (Diao 2003). These hydrogen bonds and electrostatic potentials seem to stabilize the double protonation state of HisB10.

The abnormal pK_a value of HisB10 higher than 9.0 indicates the high affinity for positive ions. This property would be effective to the polymerization of insulin depending on various cations. HisB10 captured the D^+ atom instead of the divalent metal cation in the absence of Zn^{2+} ion. This structure would be the active form of insulin at biological pH levels. This mechanism would be achieved by the cation capturing of HisB10; therefore, HisB10 is essential for polymerization of insulin in organisms.

2.2 2Zn Insulin

In the case of the hexamer state, the crystal asymmetric unit contains two insulin monomers (a dimer) related by a pseudo-two-fold axis as shown in; the crystallographic three-fold axis generates the insulin hexamer from three dimers, as shown in Figure 3. (Two zinc ions, on opposite sides of the hexamer, are situated on the crystallographic three-fold axis; each is octahedrally coordinated by the three crystallographically related HisB10 $N^{\varepsilon 2}$ atoms and three water molecules. The coordination of metal ions is intrinsically the problem of protonation/deprotonation of the polar amino acid residues, and only NPC can reveal the protonation states.

To understand the octahedrally coordinated structure of the zinc ions and the coordination mechanism thereof, it is essential to know the protonation and/or deprotonation states of not only HisB10 $N^{\varepsilon 2}$ but also HisB10 $N^{\delta 1}$. In the hexamer insulin crystals, three dimers are assembled around two zinc ions (upper Zn (u) and lower Zn (l)), 16.4 Å apart on the three-fold axis. Parts (a) and (b) of Figure 4 show (2*Fo-Fc*) maps of the areas around upper Zn (u) and lower Zn (l), respectively. The $N^{\varepsilon 2}$ atom of HisB10 is deprotonated and coordinated to Zn, whereas $N^{\delta 1}$ of HisB10 is protonated, making the net charge of HisB10 neutral. The two molecules in the dimer are designated molecule 1 and molecule 2, as shown in Figure 3 (a) and (b). Residues belonging to either molecule 1 or molecule 2 are distinguished by the suffix .1 or .2 after the residue number. Thus, the upper zinc is coordinated to three symmetry-related $N^{\varepsilon 2}$ atoms of HisB10.1 (i.e., from molecule 1) and three water molecules, while the lower Zn is coordinated to the three symmetry-related $N^{\varepsilon 2}$ atoms of HisB10.2 and three water molecules, as shown in Figure 4 (a) and (b), respectively. (Iwai et al. 2009) From the top, three water molecules can be seen coordinating the upper Zn (u) and from the bottom, another three water molecules can be seen coordinating the lower Zn (l). The Former three water molecules are classified as triangular, since the deuterium atoms of these water molecules could be placed. On the other hand, the latter three water molecules are classified as stick-shaped and are presumably rotating about OD axis. These facts are well supported by the B-factors, which are smaller for the triangular molecules (36.5 $Å^2$) than for the ball-shaped ones (74.0 $Å^2$) (Chatake et al. 2003). Differences in the dynamic behavior of the coordinated water molecules, between the upper and lower Zn ions, can be observed in

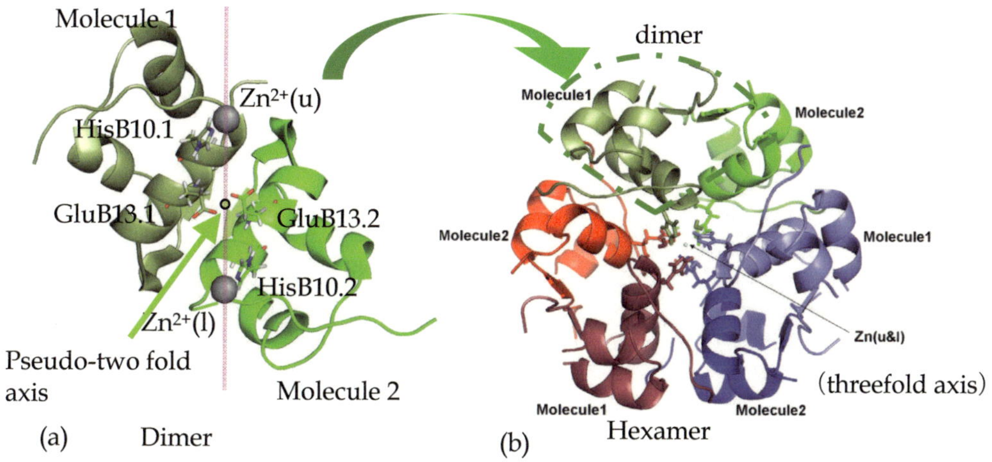

Fig. 3. (a) Model of the crystal asymmetric unit showing two insulin monomers (a dimer) related by a pseudo-two-fold axis. (b) Model of the insulin hexamer from three dimers generated by the crystallographic three-fold axis.

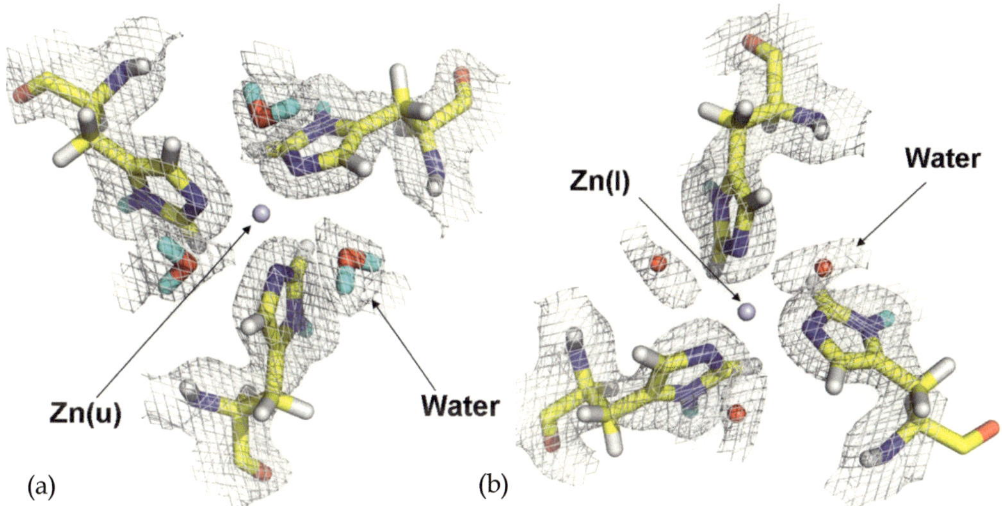

Fig. 4. 2Fo-Fc maps of the areas around upper Zn (u) and lower Zn (l) of insulin.

the structure of hexamer human insulin at 1.0 Å resolution, at 120 K, in which the unambiguous electron density of three water molecules coordinating the upper Zn ion was observed (Smith and Blessing 2003). Continuous electron density for the three water molecules coordinating the lower Zn ion can also be observed, suggesting that it would be possible for a disordered anion—such as a citrate—to occupy this site (Smith et al 2003). Furthermore, in the structure of a dried crystal of hexamer human insulin at 100 K, the upper Zn ion binds three water molecules and adopts an octahedral coordination in the crystal, whereas the lower zinc ion adopts a tetrahedral coordination, with the three water molecules being replaced by a single chloride ion (Smith et al. 2003).

In the neutron crystallography results, a difference in the temperature factor of $N^{\varepsilon 2}$ of HisB10 coordinating with Zn ions was also observed: The temperature factors were 9.3 $Å^2$ for $N^{\varepsilon 2}$ of HisB10.1 coordinating the upper Zn ion and 14.6 $Å^2$ for $N^{\varepsilon 2}$ of HisB10.2 coordinating the lower Zn ion.

2.3 RNase A

Bovine pancreatic ribonuclease A (RNase A) is a major ribonuclease in the bovine pancreas, which cleaves and hydrolyzes RNA exclusively at pyrimidine nucleotide positions. RNase A consists of 124 amino acids, has a mass of about 14 kDa and an isoelectric point of pH 9.6, and is a comparatively stable protein. The structure was initially solved by X-ray diffraction analysis at 5.5 Å resolution (Avey et al. 1967). Subsequently, a high resolution (1.05 Å) X-ray diffraction analysis of RNase A was carried out at six different pH values (Berisio et al. 2002).

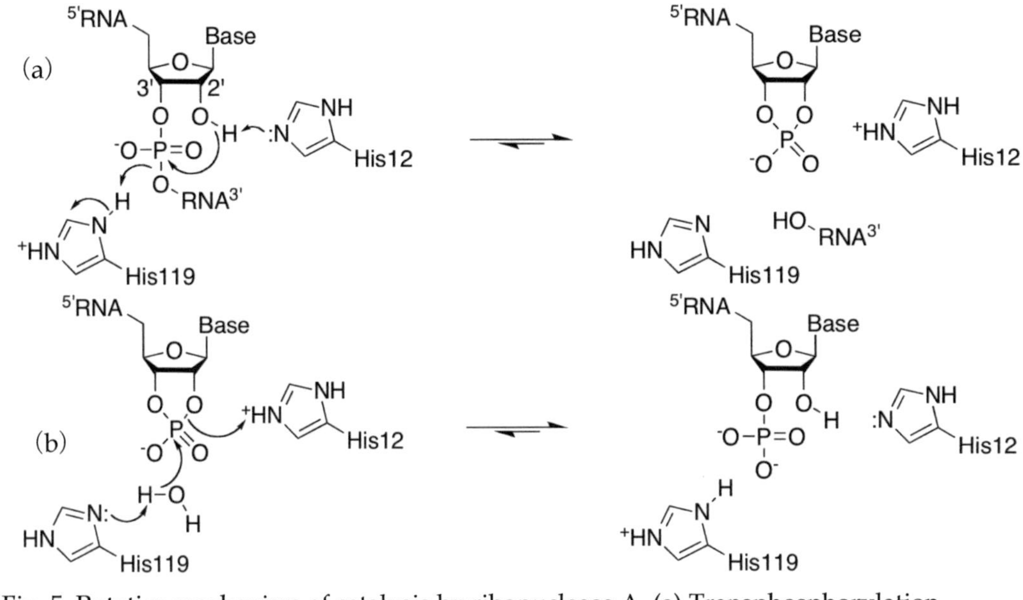

Fig. 5. Putative mechanism of catalysis by ribonuclease A. (a) Transphosphorylation reactions. (b) Hydrolysis reaction. (As taken from ref: Park, C., Schultz, L. W., Raines, R. T. (2001) Contribution of the active site histidine residues of ribonuclease A to nucleic acid binding, Biochemistry 40, 4949-4956.)

Two histidine residues, His12 and His119, play a key role in the cleavage reaction catalyzed by RNase A (Findlay et al. 1962; C. Park et al. 2001; Wlodawer 1980). In this transphosphorylation reaction, His12 acts as a base to abstract a proton from the 2′-hydroxyl group of a ribose ring, while His119 acts as an acid to donate a proton to the 5′-oxygen of the leaving group (Figure 5 (a)). The roles of His12 and His119 as base and acid are switched in the subsequent hydrolysis reaction of 2′,3′-cyclic nucleotide substrates (Figure 5 (b)). It is important to know the protonation states of His12 and His119 to understand the hydrolysis mechanism of ribonuclease A. Neutron protein crystallography can therefore contribute significantly to the understanding of this catalytic mechanism.

The structure of the active site of phosphate-free RNase A was determined by a neutron diffraction experiment (Yagi et al. 2009). In the phosphate-bound model (Figure 6 (a)), O^1 of the phosphate (PO_4) group is hydrogen-bonded to the $D^{\delta 1}$ of His119 and O^2 of the phosphate group is hydrogen-bonded to the $D^{\varepsilon 2}$ of His12. In the phosphate-free model, the PO_4 group is replaced by three water molecules (Wlodawer et al. 1986), DOD41, DOD76, and O89, as shown in Figure 6 (b), and the H-bonds occur now with these water molecules. The oxygen of DOD41 is hydrogen-bonded to the $D^{\varepsilon 2}$ atom of His12. The $D^{\delta 1}$ atom of His119 is hydrogen bonded to the oxygen of O89 (2.3 Å), while the $D^{\varepsilon 2}$ atom of His119 is hydrogen-bonded to the $O^{\delta 1}$ atom of Asp121. It was reported that at pH 6.3, the active site has a sulphate ion, which is hydrogen bonded to the ordered His119, whereas at pH 7.1 the active site releases the sulphate ion and has a disordered His119 (Berisio et al. 2002). Although a soaking solution at pD 6.2 was used in this study, the active site does not have the sulphate group, and His119 is ordered. (DOD37 and DOD39 occupy the place of the alternative conformation of His119.) This difference might come from two different crystallization methods. The oxygen of DOD41 forms possible hydrogen bonds with D^2 of DOD37, D^1 of DOD3, and O of DOD76. DOD3 is hydrogen-bonded to the $D^{\varepsilon 2}$ of Gln11. DOD3 is one of the most deeply buried water molecules in the pocket of the catalytic site and has a B-factor of 17.1 $Å^2$, which is lower than those of other water molecules located at the surface of RNase A. The B-factor of DOD41 is 42.1 $Å^2$, which is higher than those of other water molecules, the average value of which is 29.2 $Å^2$. This fact suggests that DOD41 could be relatively easily replaced by other substrates. The occupancy of DOD41 may also be affected by the protonation states of His12, a hydrogen-bonding partner.

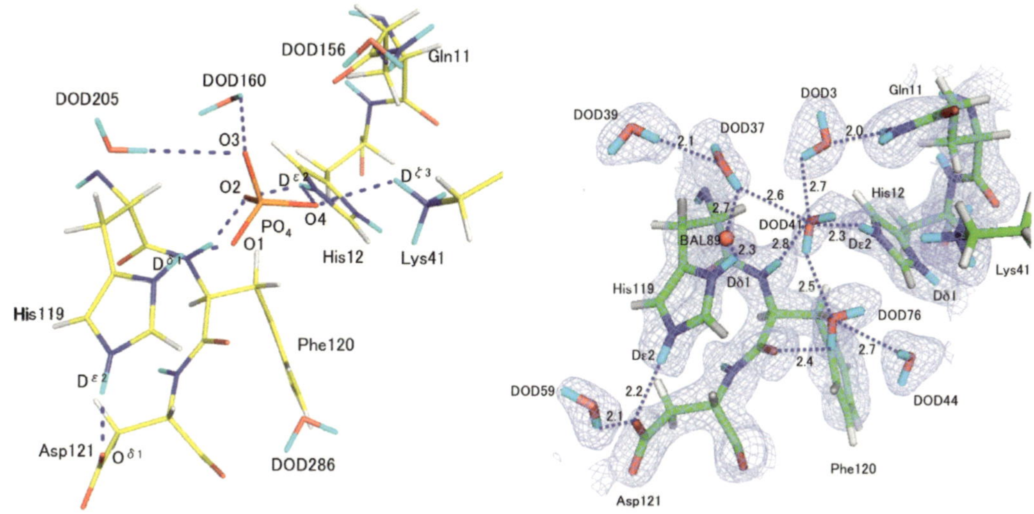

Fig. 6. The structure of the active site of phosphate-free RNase A determined by a neutron diffraction experiment. (a) The phosphate-bound model. (b) The phosphate-free model (blue contours: positive 2Fo-Fc map).

His12 and His119 are located at the catalytic site of RNase A and play important roles in the process of enzymatic activity (Figure 5). His12 is known to act as a general base that extracts a proton from the RNA 2′-OH group and thereby promotes its nucleophilic attack on the adjacent phosphorus atom. On the other hand, His119 is believed to act as general acid, enabling bond scission by donating a proton to the leaving group. To act as a base, His12 should be singly protonated. However, double protonation of His12 at acidic pH has been reported from both X-ray and neutron diffraction analyses (Berisio et al. 2002; Wlodawer et al. 1983; Wlodawer et al. 1986). Neutron diffraction analysis suggested at first glance that His12 is doubly-protonated (positively charged) (Usher et al. 1972). If that would be the case, it would be impossible for His12 to extract a proton from the 2′-OH group of RNA under these conditions. To confirm the protonation state of His12, an $(|Fo|-|Fc|)$ nuclear density map was calculated after first omitting $D^{\delta1}$ and $D^{\varepsilon2}$ of His12, as shown in Figure 7. The nuclear density for $D^{\delta1}$ is higher than that for $D^{\varepsilon2}$ at the 5σ level, implying a lower occupancy for $D^{\varepsilon2}$. This result means that a certain proportion of His12 residues have a singly protonated imidazole ring, allowing the proton from the RNA 2'-OH group to be extracted at pD 6.2. These results support the hypothesis that His12 is singly-protonated part of the time and is capable of acting as a general base in the catalytic mechanism of RNase A.

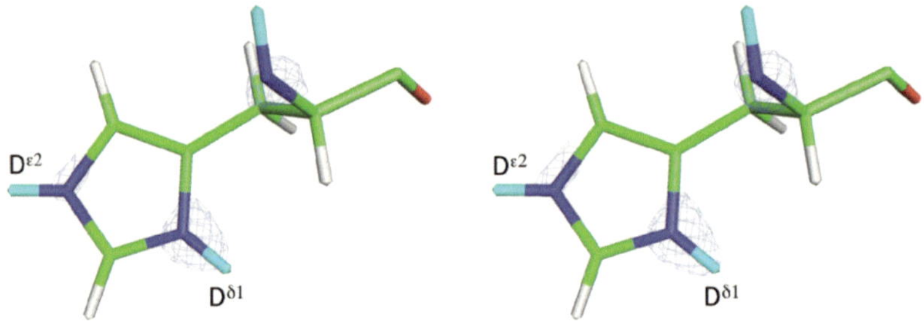

Fig. 7. A stereo view of an omit map showing the His12 residue of RNase A. (Blue contours show positive peaks at 10σ.)

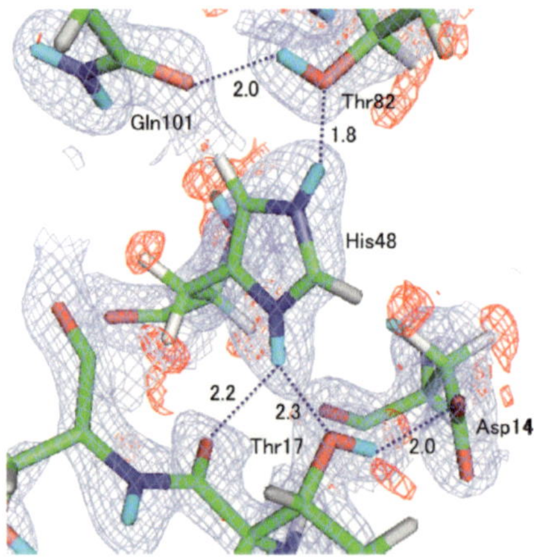

Fig. 8. A 2Fo-Fc nuclear density map around the His48 residue of RNase A.. (Blue contours are at 1.5σ, and red contours are at -2.5σ.) H-bonds are indicated by broken lines. Gln101 is hydrogen bonded from atom donates a hydrogen bond to the carbonyl oxygen atom of the Gln101 side-chain (H...O distance 2.0 Å). The $N^{\delta1}$ atom of the His48 imidazole acts as a donor in a bifurcated manner, donating a H-bond not only to the $O^{\gamma1}$ atom of Thr17 (H...$O^{\gamma1}$ 2.3 Å), but also to its backbone carbonyl oxygen atom (H...O 2.2 Å). Finally, the $O^{\gamma1}$ hydroxyl group of Thr17 acts as a H-bond donor to the $O^{\delta2}$ atom of Asp14 (H...$O^{\delta2}$ 2.0 Å).

The details of the hydrogen bonding network around His48 at various pHs were speculated upon - without knowing the actual positions of the H atoms - in an earlier X-ray analysis. This estimation was done in order to explain the movement of the Gln101 residue at alkaline pH (Berisio et al. 2002). In the neutron diffraction results, this speculation was confirmed not only by determining the H atom positions but also by defining their donor and/or acceptor characteristics (Figure 8)The $N^{\varepsilon 2}$ atom of the His48 imidazole ring donates a hydrogen bond to $O^{\gamma 1}$ of Thr82 (with an $H \ldots O^{\gamma 1}$ distance of 1.8 Å), and the $O^{\gamma 1}$

His48 through the Thr82 residue. Note that this histidine residue is doubly-protonated and that one of the two N-H bonds forms a bifurcated H-bond with two oxygen atoms.

2.4 Trypsin-BPTI complex

The target enzyme of BPTI, bovine trypsin, is a well-known serine protease. This protein is a member of the chymotrypsin family of proteases, and its active residues are His57, Ser195 and Asp102 (based on the numbering of residues in chymotrypsin). The catalytic mechanism of BPTI has been extensively studied, and thus the reaction mechanism of this protease is suggested in many previous reports. Trypsin is specific to basic amino acids—lysine and arginine—and the substrate recognition site is Asp189 (Voet *et al.*, 1999; Hedstrom, 2002; Bode & Huber, 1992).

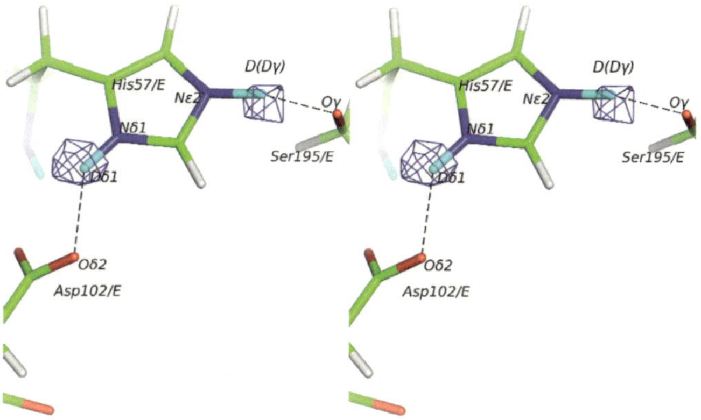

Fig. 9. The F_o-F_c nuclear density maps (in stereo view) of the catalytic triad in the Trypsin-BPTI complex. The maps have been calculated without the $D^{\delta 1}$ and $D(D^{\gamma})$ deuterium atoms and are contoured at 5.5 σ positive density. The broken line indicates a hydrogen bond between the hydrogen atom and the acceptor atom.

The specific roles of the $HisN^{\varepsilon 2}$ within the catalytic triad of serine proteases are as follows: to act as a base and extract the H from the -OH of the hydroxyl group of serine; to act as an acid and transfer H to the scissile peptide amide; to act as a base and extract the H from a hydrolytic H_2O molecule; and to act as an acid and transfer H to the deprotonated $-O^-$ of the hydroxyl group of serine. The protonation state of the $His57N^{\varepsilon 2}$ is crucial to understanding the catalytic mechanism of trypsin, because the $N^{\varepsilon 2}$ serves as both a base and an acid for activating the nucleophilic species of Ser195 and for transferring a proton to the leaving group-NH in the substrate polypeptide, respectively. In this study, we used neutron crystal-structure analysis to examine the protonation (deuteration) states of the catalytic triad (Asp102, His57 and Ser195 of trypsin) of trypsin in the complex state with BPTI, which is in a Michaelis complex state. Figure 9 shows the F_o-F_c nuclear-scattering map of both His57D$^{\delta1}$ and Ser195D$^{\gamma}$ in the trypsin–BPTI complex (Kawamura et al. 2011); this map was made at the 5.5σ contour level and was calculated without incorporation of the D atoms. Using this method, we found two positive peaks, with one between Asp102O$^{\delta2}$ and His57N$^{\delta1}$ and the other between His57N$^{\varepsilon2}$ and Ser195O$^{\gamma}$. The distances between Asp102O$^{\delta2}$ and D$^{\delta1}$ and between D$^{\delta1}$ and His57N$^{\delta1}$ were determined to be 2.0 Å and 1.0 Å, respectively. The distance between Asp102O$^{\delta2}$ and His57N$^{\delta1}$ was 2.8 Å, and the angle of O$^{\delta2}$-D$^{\delta1}$-N$^{\delta1}$ was ~140°.The distances between His57N$^{\varepsilon2}$ and D$^{\gamma}$ and between D$^{\gamma}$ and Ser195O$^{\gamma}$ were determined to be 1.3 Å and 1.7 Å, respectively. The distance between His57N$^{\varepsilon2}$ and Ser195O$^{\gamma}$ was 2.8 Å, and the angle of N$^{\varepsilon2}$-D$^{\gamma}$-O$^{\gamma}$ was ~144°.

3. Carboxylate groups of acidic residues

3.1 Lysozyme

The enzymatic activity of lysozyme, a saccharide-cleaving enzyme, is maximal at pH 5 and diminishes at pH 8. It has been postulated that, at pH 5, the carboxyl group of Glu35 is protonated and that it is this proton that is transferred to the oxygen atom of the bound substrate (an oligosaccharide) during the hydrolysis process. During the reaction, another acidic residue, Asp52, is postulated to remain in its dissociated (anionic) state (D. C. Phillips 1966). To elucidate the role of hydrogen atoms in this reaction, neutron diffraction experiments of hen egg-white lysozyme were carried out, using crystals that were grown at different acidities, specifically, pD = 4.9 on BIX and pD = 8.0 on LADI (Maeda et al. 2001; N. Niimura et al. 1997b; N. Niimura et al. 1997a). The $(2|Fo|-|Fc|)$ nuclear density maps around the carboxyl group of Glu35 are shown at pD 4.9 in (a) and at pD 8.0 in (b). At pD 4.9, the Fourier map shows a positive region (at the arrow in Figure 10 (a)) extending beyond (i.e., attached to) the position of the O atom of the carboxyl group labeled E35 O$^{\varepsilon1}$, suggesting that this carboxyl oxygen atom (circled) is protonated at pD 4.9. On the other hand, at pD 8.0, it can be seen that this residue is deprotonated (the circled atom in Figure 10 (b)), and hydrogen-bonded to a water molecule. The observation that the Glu35 catalytic site is deprotonated at pD 8.0 rationalizes why lysozyme has significantly reduced activity at neutral conditions. Thus, these results suggest that Glu35 is the site of the enzymatically active proton that is subsequently transferred to the oxygen atom of the carbohydrate substrate during the hydrolysis process. Incidentally, these results at the less acidic pH value of 8.0 (N. Niimura et al. 1997a) are consistent with the conclusions of an earlier neutron investigation, as described in a paper by Mason and co-workers many years ago (Mason et al. 1984).

4. How to grow a large single crystal under different pH

One of the highest barriers of neutron protein crystallography experiments is the supply of large single crystals. Single crystals larger than 1mm^3 at least are necessary for neutron diffraction experiments because of the weak incident neutron beams so far. The crystallization phase diagram is very helpful to grow a large single crystal. The crystallization of HEWL at wide range of pH from 2.5 to 8.0 with 0.5 step was carried out as follows: HEWL was crystallized in a buffer of 50 mM NaH_2PO_4, 33.5 mg/ml protein concentration, 3.5 w/v% NaCl as the crystallization agent and T= 20°C at each pH by batch method. The condition is indicated by a yellow filled circle in the crystallization phase diagram as shown in Figure 11 and this condition was fixed in the following preliminary crystallization experiment at wide range of pH (Iwai et al. 2008). The pHs of solutions at pH

2.5 - 4.5 and at pH 5.0 - 8.0 were adjusted in 50 mM NaH_2PO_4 buffer with H_3PO_4 and with Na_2HPO_4, respectively, in order to prepare solutions of wide range of pH. Crystallization at pH 2.5 - 6.5 was carried out in the designed pH solution directly. Crystals did not appear beyond pH 7.0 and therefore the crystals at pH 7.0 - 8.0 were prepared by soaking crystals (initially grown at pH 5.5) in the solution of designed pH in the meta-stable region. Crystallization phase diagrams at pH 2.5, 6.0, 7.5 were determined. (Figure 12) At pH less than 4.5 the border between meta-stable region and nucleation region shifted to the left (lower precipitant concentration) and at larger than 4.5 the border shifted to the right (higher precipitant concentration) in the phase diagram. The qualities of these crystals were characterized by Wilson plot method.(Arai et al. 2002) The qualities of all crystals at different pH were more or less equivalent (B-factor values are within 25 - 40). It is expected that the neutron diffraction from these crystals of different pH provides equivalent data in quality for discussion of protein pH titration in the crystalline state of HEWL.

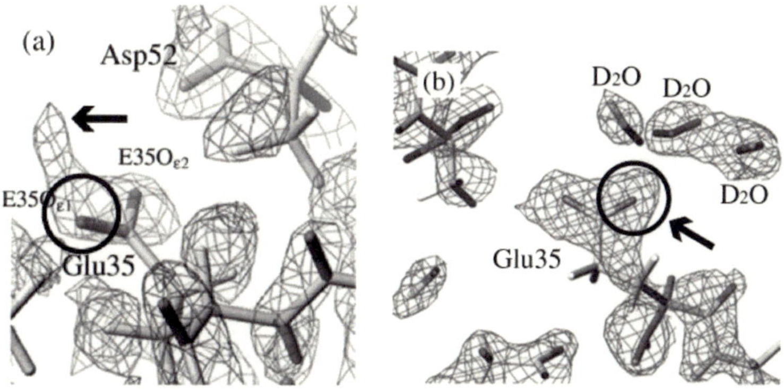

Fig. 10. The 2 Fo–Fc nuclear density maps around the carboxyl group of Glu35 of HEWL (a) at pD 4.9 and (b) at pD 8.0. The Fourier map at pD 4.9 shows a positive region extending beyond (i.e., attached to) the position of the O atom of the carboxyl group labeled E35 Oε1.

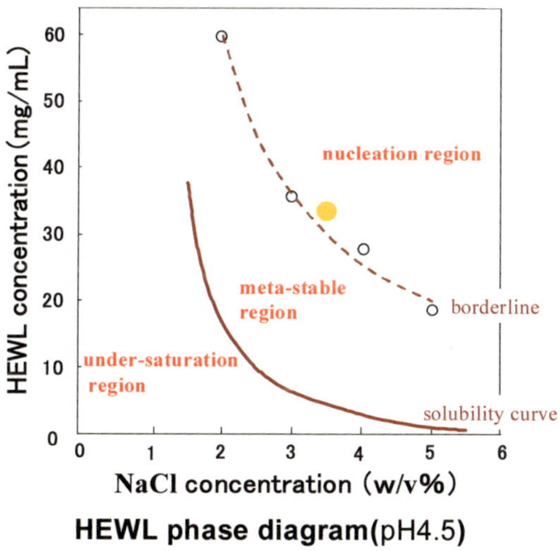

Fig. 11. Crystallization phase diagram of HEWL at pH 4.5 and at 20°C. The plots of circles (o) are experimental data. A solid line is solubility curve and a broken line is a borderline, guide for eyes respectively.

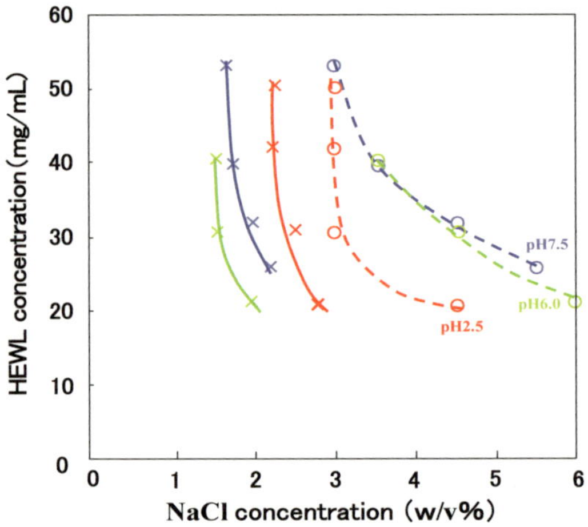

HEWL phase diagram

Fig. 12. The results of crystallization phase diagrams of HEWL at 20°C. The cross (x) and circle (o) are experimental data. Solid lines are solubility curves and broken lines are borderlines, guide for eyes respectively. Results at pH 2.5, 6.0 and 7.5 are drawn in red, green and blue, respectively.

5. Conclusion

The determination of protonation states has historically been a major application of neutron diffraction. Biological mechanism such as enzymatic reactions depend on these protonation states, but they cannot always be obtained from X-ray diffraction data. On the other hand, since neutrons scatter from nucleus in proteins, neutron interacts with protons and identify the protonation states in proteins. These were certificated by several examples, such as insulin, RNase A, Trypsin-BPTI complex and lysozyme.

One of the most difficult probles in NPC is to obtain a single crystal that is large enough to obtain diffraction results with the low flux of a neutron beam. Often, a crystal with a volume of several cubic millimeters is required. To grow such a large single crystal of a proein, a seed crystal is placed into a metastable growth environment, in which new crystals do not nucleate but protein molecules stillcrystallize on the seed crystal. This method was demonstrated successfully in case of the crystallization of lysozyme under different pH.

6. References

Adams, M. J., et al. (1969), 'Structure of Rhombohedral 2 Zinc Insulin Crystals', *Nature,* 224 (5218), 491-95.

Arai, S., et al. (2002), 'Crystallization of a large single crystal of a B-DNA decamer for a neutron diffraction experiment by the phase-diagram technique', *Acta Crystallogr D Biol Crystallogr,* 58 (Pt 1), 151-3.

Avey, H. P., et al. (1967), 'Structure of ribonuclease', *Nature,* 213 (5076), 557-62.

Baker, E. N., et al. (1988), 'The structure of 2Zn pig insulin crystals at 1.5 A resolution', *Philos Trans R Soc Lond B Biol Sci,* 319 (1195), 369-456.

Berisio, R., et al. (2002), 'Atomic resolution structures of ribonuclease A at six pH values', *Acta Crystallogr D Biol Crystallogr,* 58 (Pt 3), 441-50.

Blundell, Tom, et al. (1972), 'Insulin: The Structure in the Crystal and its Reflection in Chemistry and Biology by', in Jr John T. Edsall C.B. Anfinsen and M. Richards Frederic (eds.), *Advances in Protein Chemistry* (Volume 26: Academic Press), 279-86, 86a, 87-402.

Blundell, T. L., et al. (1971), 'X-ray analysis and the structure of insulin', *Recent Prog Horm Res,* 27, 1-40.

Bode, W. & Huber, R. (1992). 'Natural protein proteinase inhibitors and their interaction with proteinases. ' *Eur J Biochem.* 204, 433–451.

Chatake, T., et al. (2003), 'Hydration in proteins observed by high-resolution neutron crystallography', *Proteins,* 50 (3), 516-23.

Diao, J. (2003), 'Crystallographic titration of cubic insulin crystals: pH affects GluB13 switching and sulfate binding', *Acta Crystallogr D Biol Crystallogr,* 59 (Pt 4), 670-6.

Findlay, D., et al. (1962), 'The active site and mechanism of action of bovine pancreatic ribonuclease. 7. The catalytic mechanism', *Biochem J,* 85 (1), 152-3.

Hedstrom, L. (2002). 'Serine protease mechanism and specificity', *Chem Rev.* 102, 4501–4524.

Hill, C. P., et al. (1991), 'X-ray structure of an unusual Ca2+ site and the roles of Zn2+ and Ca2+ in the assembly, stability, and storage of the insulin hexamer', *Biochemistry,* 30 (4), 917-24.

Ishikawa, Takuya, et al. (2008a), 'An abnormal pKa value of internal histidine of the insulin molecule revealed by neutron crystallographic analysis', *Biochemical and Biophysical Research Communications,* 376 (1), 32-35.

--- (2008b), 'A neutron crystallographic analysis of a cubic porcine insulin at pD 6.6', *Chemical Physics,* 345 (2-3), 152-58.

Iwai, W., et al. (2008) 'Crystallization and evaluation of Hen Egg-White Lysozyme (HEWL) crystals for protein pH titration in the crystalline state.', *J. Synchr. Rad.* , 58, 312-315

Iwai, W., et al. (2009), 'A neutron crystallographic analysis of T6 porcine insulin at 2.1 A resolution', *Acta Crystallogr D Biol Crystallogr,* 65 (Pt 10), 1042-50.

Kawamura,K. , et al: (2011) 'X-ray and neutron protein crystallographic analysis of the trypsin-BPTI complex', *Acta Crystallogr D Biol Crystallogr,* 67, 140-148

Maeda, M., et al. (2001), 'Neutron structure analysis of Hen Egg-White Lysozyme at pH 4.9', *J. Phys. Soc. Jpn. Suppl. A,* 70, 403-05.

Maeda, M., et al. (2001), 'Neutron structure analysis of Hen Egg-White Lysozyme at pH 4.9', *J. Phys. Soc. Jpn. Suppl. A,* 70, 403-05.

--- (2004), 'Crystallization of a large single crystal of cubic insulin for neutron protein crystallography', *J Synchrotron Radiat,* 11 (Pt 1), 41-4.

Mason, S. A., Bentley, G. A., and McIntyre, G. J. (1984), 'Deuterium exchange in lysozyme at 1.4-A resolution', *Basic Life Sci,* 27, 323-34.

Niimura, N., et al. (2004), 'Hydrogen and hydration in proteins', *Cell Biochem Biophys,* 40 (3), 351-69.

--- (2006), 'Recent results on hydrogen and hydration in biology studied by neu tron macromolecular crystallography', *Cell Mol Life Sci,* 63 (3), 285-300.

--- (1997a), 'Neutron Laue diffractometry with an imaging plate provides an effective data collection regime for neutron protein crystallography', *Nat Struct Biol,* 4 (11), 909-14.

--- (1997b), 'Neutron Laue diffractometry with an imaging plate provides an effective data collection for neutron protein crystallography', *Physica B: Condensed Matter,* 241-243, 1162-65.

Niimura, N. and Bau, R. (2008), 'Neutron protein crystallography: beyond the folding structure of biological macromolecules', *Acta Crystallogr A,* 64, 12-22.

Niimura,N., & Podjarny, A., (2011) 'Neutron Protein Crystallography. Hydrogen, Protons, and Hydration in Bio-macromolecules.' ISBN 978-0-19-957886-3, Oxford University Press.

Ostermann, A. and Parak, F. G. (2005), 'Hydrogen Atoms and Hydration in Myoglobin: Results from High Resolution Protein Neutron Crystallography', in N. Niimura, et al. (eds.), *Hydrogen- and Hydration-Sensitive Structural Biology* (Tokyo: KubaPro), 87-102.

Park, C., Schultz, L. W., and Raines, R. T. (2001), 'Contribution of the active site histidine residues of ribonuclease A to nucleic acid binding', *Biochemistry,* 40 (16), 4949-56.

Phillips, D. C. (1966), 'The three-dimensional structure of an enzyme molecule', *Sci Am,* 215 (5), 78-90.

Smith, G. D. and Blessing, R. H. (2003), 'Lessons from an aged, dried crystal of T(6) human insulin', *Acta Crystallogr D Biol Crystallogr,* 59 (Pt 8), 1384-94.

Smith, G. D., Pangborn, W. A., and Blessing, R. H. (2003), 'The structure of T6 human insulin at 1.0 A resolution', *Acta Crystallogr D Biol Crystallogr,* 59 (Pt 3), 474-82.

Usher, D. A., Erenrich, E. S., and Eckstein, F. (1972), 'Geometry of the first step in the action of ribonuclease-A (in-line geometry-uridine2',3'-cyclic thiophosphate- 31 P NMR)', *Proc Natl Acad Sci U S A,* 69 (1), 115-8.

Voet, D., Voet, J. G. & Pratt, C. W. (2008). *Fundamentals of Biochemistry.* (3rd Edition.) New York: John Wiley & Sons.

Wlodawer, A. (1980), 'Studies of ribonuclease-A by X-ray and neutron diffraction', *Acta Crystallographica Section B,* 36 (8), 1826-31.

Wlodawer, A., Miller, M., and Sjolin, L. (1983), 'Active site of RNase: neutron diffraction study of a complex with uridine vanadate, a transition-state analog', *Proc Natl Acad Sci U S A,* 80 (12), 3628-31.

Wlodawer, A., et al. (1986), 'Comparison of two independently refined models of ribonuclease-A', *Acta Crystallographica Section B,* 42 (4), 379-87.

Yagi, D., et al. (2009), 'A neutron crystallographic analysis of phosphate-free ribonuclease A at 1.7 A resolution', *Acta Crystallogr D Biol Crystallogr,* 65 (Pt 9), 892-9.

13

Determination of Internal Stresses in Lightweight Metal Matrix Composites

Guillermo Requena[1], Gerardo Garcés[2],
Ricardo Fernández[2] and Michael Schöbel[1]
[1]*Institute of Materials Science and Technology, Vienna University of Technology*
[2]*Department of Physical Metallurgy, National Center for Metallurgical Research-C.S.I.C.*
[1]*Austria*
[2]*Spain*

1. Introduction

Internal stresses are those stresses found in a body when this is stationary and in equilibrium with its surroundings (Withers & Badeshia, 2001). These stresses can arise at different length scales within a microstructure ranging from the size of the analysed body down to the atomic scale. Multiphase materials are prone to develop internal stresses due to the different mechanical and physical properties usually found between the phases that form these materials. This is essential for composites because the distribution and magnitude of the internal stresses may determine their mechanical/physical behaviour. Neutron diffraction has become an essential tool to determine internal stresses non-destructively in metal-based composite materials. The present chapter gives a thorough description of the state of the art of the technique and its use to determine internal stresses developed in lightweight metal matrix composites under mechanical, thermal and thermo-mechanical loading.

2. Metal matrix composites

The term composite material refers, in a first approach, to all kinds of materials systems composed of more than one solid phase. However, if we follow this definition, any alloy with a miscibility gap would have to be considered as a composite material since it is composed of two phases. To avoid this confusion, a more general and currently accepted definition of composite is the one provided in (mmc-assess.tuwien.ac.at):

A composite (or composite material) is defined as a material that consists of at least two constituents (distinct phases or combinations of phases) which are bonded together along the interface in the composite, each of which originates from a separate ingredient material which pre-exists the composite.

Following this definition, the composite is produced via a physical combination of at least two pre-existing ingredient materials. The so called in-situ composite materials are those where the pre-existing ingredients change their chemical composition or their shape during mixing. Those produced by reactions between elements or between compounds within the material system (Tjong 2000) fall outside this definition. Although their properties can be described using composite theory (e.g. grey cast iron) they will not be considered as composite materials in the present work. As already stated, a composite presents at least two constituents, one of them may act as the matrix and the other one/s as the reinforcement/s. To complete our previous definition we define the components of the composite as follows:

The matrix of a composite refers to the phase which presents a continuous structure and usually, but not exclusively, is the phase with highest volume fraction. The other phases of the composite are called the reinforcement.

Composites can now be classified by means of the type of matrix they have. The matrices of technological interest are polymeric, metallic and ceramic matrices. Thus, we have three different types of composites:

- MMC: *Metal Matrix Composites*
- CMC: *Ceramic Matrix Composites*
- PMC: *Polymer Matrix Composites*

Now, we are able to give a consistent definition of metal matrix composites:

A Metal Matrix Composite (MMC) is a composite material in which one constituent, the matrix, is a metal or alloy forming at least one percolating network. The other pre-existing constituent or their derivatives are embedded in this metal matrix and play the role of reinforcement phases. In addition, it is convenient to limit the volume fractions of the constituents to be considered as MMC (mmc-assess.tuwien.ac.at): matrix volume fraction > 5% and reinforcement volume fraction > 5%.

2.1 Types of MMC

MMC are classified into two subgroups according to the architecture of the reinforcement:

1. Continuously reinforced MMC:
    - Monofilament reinforced metals (MFRM) (Fig. 1 a): monofilaments of approximately 0.1 mm in diameter (e.g. SiC) are embedded uni- or bidirectionally within a metallic matrix by diffusion bonding (Leyens et al., 2003; Brendel et al., 2007; Peters et al., 2010).
    - Continuous fibre reinforced metals (CFRM) (Fig. 1 b): uniaxial or multiaxial multifilament tows of at least some hundreds of fibres of about 0.01 mm in diameter (e.g. C, Al_2O_3) are embedded within a metallic matrix, usually, by infiltration methods.
    - Interpenetrating composite metals (ICM) (Fig. 1 c): percolating ceramic structures can also be used as continuous reinforcement resulting in an MMC in which both the matrix and the reinforcement are continuous phases forming an interconnected 3D structure throughout the volume.
2. Discontinuously reinforced MMC:
    - Particle reinforced metals (PRM) (Fig. 1 d): irregular or spherical particulates or platelets (e.g. SiC, Al_2O_3, B_4C) are embedded within the metallic matrix. Stir casting, infiltration and powder metallurgy (PM) are the usual processing routes used to produce PRM. Dispersion strengthened metals produced by powder metallurgy incorporating particles <1μm in size are frequently called nano-

composites, but contain usually <5 vol% of particles placing them outside of the definition of MMC.

- Short fibre reinforced metals (SFRM) (Fig. 1 e): short fibres (usually Al_2O_3) of high aspect ratio (length-to-diameter ratio) or whiskers (e.g. SiC) are embedded within the metallic matrix. SFRM are usually produced by infiltration of randomly distributed short fibre preforms. Strengthening is only provided by over-critical volume fractions of short fibers, usually >5 vol%. Whiskers have been forbidden in Europe due to their asbestos-like toxicity (Birchal et al., 1988).). The harmfulness of carbon nano tubes (CNT) is still in discussion, but it is hardly possible to achieve a uniform distribution of > 5 vol% of CNT within a metal matrix. The verification of an increase in stiffness by CNT needs further research, whereas the strength increase is mainly due to dispersion strengthening.

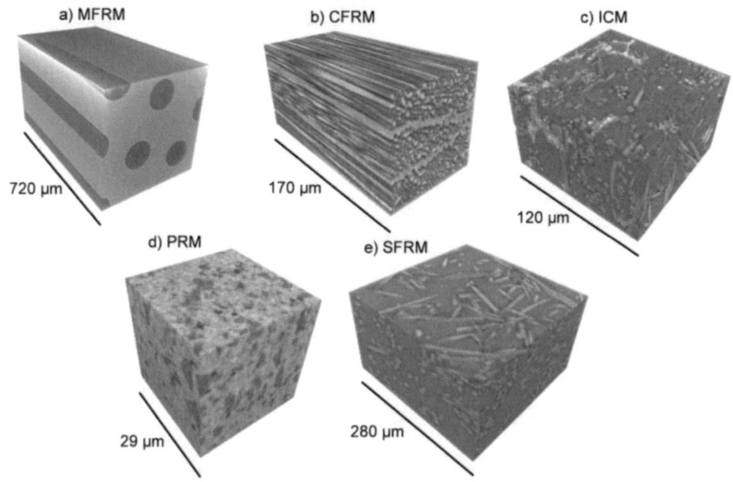

Fig. 1. Microstructures of different types of MMC revealed by synchrotron microtomography: a) SiC monofilament reinforced Cu (MFRM), b) Al matrix reinforced with continuous C fibres (CFRM), c) ICM formed by an Al matrix reinforced with an interpenetrating network of Al_2O_3 short fibres (light grey), eutectic Si (dark grey) and Ni- and Fe-rich aluminides (white), d) SiC particle reinforced 2124 Al-alloy (PRM) and e) Al_2O_3 short fibre reinforced Al matrix.

The possibility to combine light alloy matrices, i.e. Al-, Mg-, and Ti-alloys, with ceramic reinforcement has been a matter of technical interest during the last decades due to the increase in mechanical properties that lightweight MMC offer (Degischer, 1997), particularly regarding specific stiffness. This can be understood from the Ashby map (Ashby, 2005) in Fig. 2, where the Young's modulus of engineering metals and ceramics is shown as a function of their density. Guide lines for minimum mass design are introduced for the cases of bars under tensile / compressive loading, bending bars and for membranes. The E/ρ factor (specific modulus) for light alloys and steels is essentially the same (red dashed line). The only chance to increase the specific moduli of metals is by moving it towards the upper left corner (indicated by the arrow). The engineering ceramics are located in a region with a higher E/ρ but these materials are usually too brittle to be used for structural parts. Lightweight MMC appear as an innovative solution for this problem since they tailor the properties of ceramic and metallic materials and can increase the specific stiffness of their matrices, as it is shown in Fig. 2 for the case of Al-based composites.

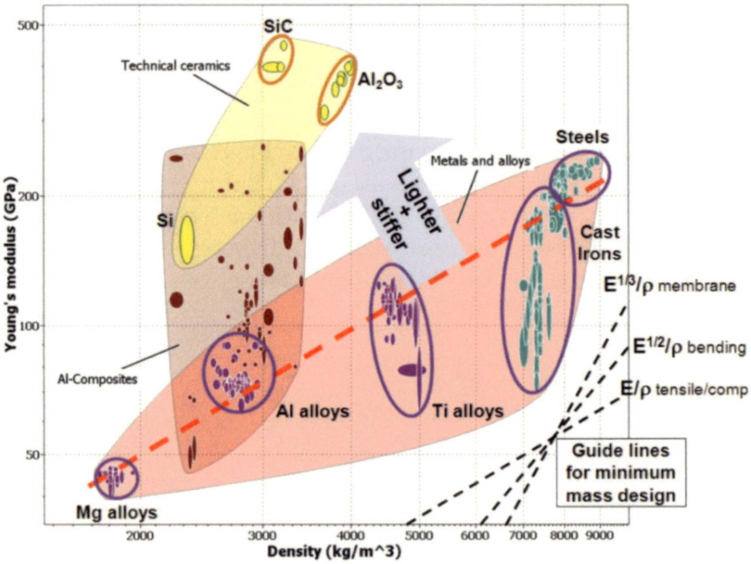

Fig. 2. Young's modulus vs density for engineering metals and some ceramics. The technical lightweight alloys are indicated together with the steels and cast irons. Guide lines for minimum mass design are included (CES Edupack, 2010)

3. Origins of internal stresses in materials

Internal stresses are those arisen in materials from shape misfits between different parts, regions or phases of materials and/or components. If no external load is acting on the considered body they are called residual stresses. Shape misfits are caused by different thermo-mechanical treatments. It is well established that almost every processing step during materials manufacture introduces residual stresses. For instance, machining, usually exemplified by a cold hole cutting in a sheet, strongly modifies the residual stress state of materials. Other cold deformation steps that introduce residual/internal stresses by mechanical working apart from cutting are bending, peening, cold laminating and cold forging. The resultant internal/residual stresses will depend on sample geometry, microstructure and process parameters such as magnitude and type of external applied stress, time, etc.

Other sources of residual stresses are temperature changes and temperature gradients. Thus, residual stresses may be introduced when materials or components are subjected to thermal treatments. These are very common in industrial processes, achieving its maximum significance for heat treatable alloys such as aluminum alloys of the 2xxx, 5xxx or 6xxx series. Thermal stresses can be divided in two main groups. The first one includes those treatments that produce strong residual stress profiles, e.g. quenching. They are usually introduced by sudden temperature variations on the sample that can result in the occurrence of plastic gradients within the body. The second group includes those heat treatments that relieve residual stresses, as for example annealing treatments.

Both thermal and mechanical effects usually appear simultaneously. The reason is that most of the manufacturing steps applied to metallic materials and components are composed of a combination of thermal and mechanical steps. Typical examples are hot forming and shape casting with mechanical constraints exerted by the tools. In these situations, it is not possible to separate the contributions of each individual process parameter and the resulting internal/residual stress distribution is usually very complex.

Furthermore, during service, internal stresses include residual stresses produced by thermal exposures of components, e.g. thermal cycling of multi-phase or multi-material systems.

3.1 Types and origin of internal stresses relevant to MMC

Internal stresses are normally divided in three categories depending on their length scale:

Type I - Macrostresses: these stresses are those acting over a length scale of the size of the considered body and are usually developed during processing and/or thermo-mechanical processing of the material. Among the thermal treatments, quenching processes are a paradigmatic example for the generation of this kind of stresses. In relation to mechanical processing, macrostresses appear in those steps that introduce plastic strain gradients such as shot-peening or bending. In this size range, anisotropy in the residual stress profile may be introduced depending on the geometry of the body. Fig. 3 illustrates the different types of internal stresses that can arise during extrusion of PRM. The cooling temperature profile generated after extrusion results in the development of a macrostress profile that goes from tensile stresses in the centre of the bar to compressive in the surface (Noyan and Cohen, 1987, Hutchings et al., 2005). It is important to point out that macrostresses are homogenous at least along one direction and that they must be balanced throughout the body, i.e.

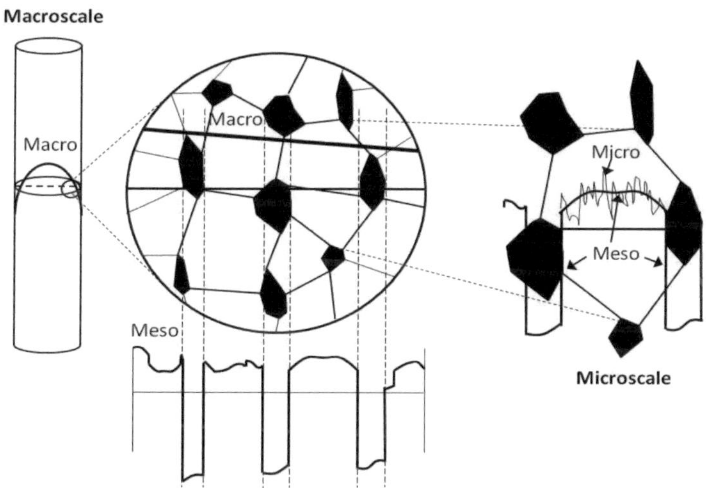

Fig. 3. Schematic view of the residual stresses that can arise in the different structural scales of a PRM. Ceramic particles are indicated as black regions.

$$\int_V \sigma_{ij} dV = 0 \qquad (1)$$

where σ_{ij} represents the stress in direction ij and V is the volume of the body.

Macrostresses can be relieved by relaxation heat treatments or by plastic deformation.

Type II - Microstresses (also known as mesostresses): this kind of residual or internal stresses is originated from inhomogeneities in a material and thus is related to a length scale of the heterogeneity, e.g. the grain size, reinforcement size, interspacing between different phases, etc. They can arise due to several reasons, e.g. inhomogeneous distribution of plastic deformation on a micro scale, thermal/elastic mismatch between phases, crystallographic texture, etc. In the case of MMC, the different mechanical and physical properties of the component phases are sources of type II microstresses. Thus, the different thermal expansion coefficients between the metallic matrix and the ceramic reinforcement result in the generation of type II thermal microstresses during temperature variations. Furthermore, the load partitioning between matrix and reinforcement, typical of MMC, is in fact a type II microstresss. The following equation indicates the condition for the balance of type II microstresses in the absence of external loads,

$$\sum_{i=1}^{n} f_i \, \sigma_{jki} = 0 \qquad (2)$$

where f is the volume fraction of the phase i and $<\sigma_{jk}>$ represents the mean microstress in the direction jk for the phase i. If an external load is applied the sum must be equal to this and equation (2) must be modified accordingly. Type II microstresses can be reduced by annealing or by plastic deformation. However, they can be more or less re-established after cooling or loading. For instance, thermal cycling of MMC is accompanied by reversible cyclic changes of mesostresses.

Type III - microstresses: these stresses are found in the smallest length scale, e.g. stress fields around dislocations or within dislocation entangles, intergranular stresses or stress field around solutes, vacancies or coherent precipitates. For the case of MMC, high dislocation density regions, called work hardened zones, can generate in the vicinity of the reinforcement during plastic deformation and temperature changes.

4. Neutron diffraction for non-destructive stress analysis

4.1 Why neutrons

Neutrons are nucleons without electric charge and with magnetic moment (s = ½). They interact strongly with the nuclei of atoms and with the magnetic moment of uncoupled electrons (Brückel et al. 2005; Pynn, 2011). Their deep penetration in solids (no Coulomb interaction) and their scattering potential give neutrons superior properties for diffraction (Fitzpatrick & Lodini, 2003) compared to electrons and photons. For thermal neutrons (~ 25 meV) the scattering centers ($Ø_{nucleus} \sim 10^{-15}$ m) can be assumed as infinitely small point potentials compared to the lattice distances ($d \sim 10^{-10}$ m) and wavelengths used (Behrens, 2011). Geometrical effects of the scattering center can be neglected. The scattering cross sections, which are independent of the atomic number Z, can be useful and restrictive as well, especially for some engineering materials with a negative scattering length (high incoherent background) such as Li, Ti and Mn. Besides their sophisticated radiation properties, the main problem with neutrons is their availability in terms of sources (Pynn,

2011). In fission reactors thermal neutrons are extracted from a "cloud" source with a diameter adequate to their slowing-down length ($L_s \sim 30$ cm) in the moderator (D_2O). These continuous sources use only a small fraction from the initial intensity of $\sim 10^{15}$ n cm^{-2} s^{-1}. Primary monochromators reduce the flux on the sample to $\sim 10^{7}$ n cm^{-2} s^{-1}. Spallation sources (Carpenter, 2008) generate polychromatic neutron beams by proton acceleration and collision with a neutron emissive target reaching intensities of $\sim 10^{17}$ n cm^{-2} s^{-1}. These pulsed sources, which are the most sophisticated neutron sources now available, are still not comparable to novel undulator or wiggler insertion devices at synchrotrons reaching photon fluxes of $\sim 10^{30}$ ph cm^{-2} s^{-1}.

Neutron diffraction is particularly well suited to tackle the problem of three-dimensional strain measurement in unreinforced and reinforced materials for several reasons: due to their large penetration depth neutrons allow lattice strain measurements in the bulk of the specimen under investigation, thus avoiding unwanted surface effects; the orientation of the scattering vector with respect to the specimen axes can be changed arbitrarily by simple specimen manipulation, which is a prerequisite for three-dimensional stress analysis; the gauge volume can be adjusted to specific needs in the range from a few cubic millimetres to about a cubic centimetre without compromising the strain resolution. This is an important fact in materials with coarse grain size, since other non-destructive techniques (in particular synchrotron radiation diffraction) suffer from graininess.

4.2 Diffraction techniques

4.2.1 Angle dispersive neutron diffraction

Continuous neutron sources (fission reactors) emit thermal neutrons for diffraction with kinetic energies of ~ 25 meV (Brückel et al. 2005). Their wavelengths of $\lambda \sim 1 - 2$ Å is further restricted to a monochromatic neutron beam with a well defined λ. Si and Ge crystal arrays or bended single crystals are used as focusing monochromators which increase the intensity of the overall diffuse neutron radiation (thermal neutrons) to a high flux neutron beam directed towards the sample. Setups with fixed as well as with variable monochromator angle/wavelength can be used. An angle dispersive neutron diffraction experiment is shown schematically in Fig. 4. The neutron beam is diffracted by the polycrystalline sample producing Debye-Scherrer cones. A position sensitive detector system (PSD), usually a 2D ^{3}He detector array, acquires sections of these cones at preselected angles and distances (Brückel et al. 2005). The gauge volume in the sample is defined by primary and secondary slits. An angle dispersive diffraction measurement on an AlSi7/SiC/70p PRM (Schöbel et al., 2011) is shown in Fig. 5. The wavelength was set to 1.67 Å and the image in the detector shows the sections of the Debye-Scherrer cones of the Al(311) and SiC_{cubic}(311) crystallographic planes. The peak intensities are then vertically integrated over the PSD image and the d-spacing of the corresponding crystallographic plane can be calculated with the well-known Bragg's law,

$$d_{hkl} = \frac{n \cdot \lambda}{2 \cdot \sin\theta} \tag{3}$$

where n is a natural integer, and θ the diffraction angle (Fig. 5).

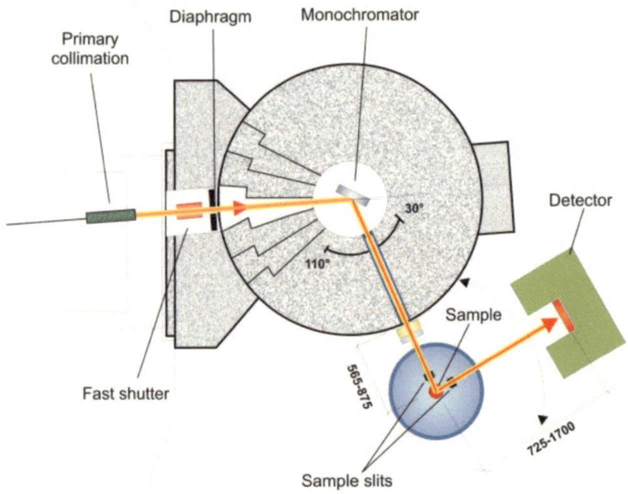

Fig. 4. Example of an angle dispersive neutron diffraction experiment using the Stress Spec instrument at the FRM II neutron source, Garching, Germany (www.frm2.tum.de).

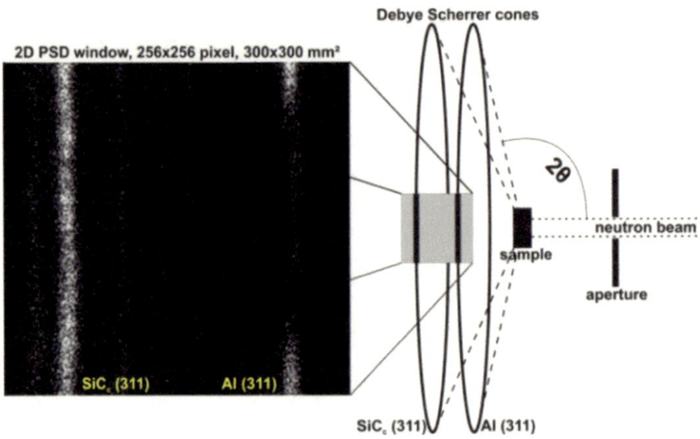

Fig. 5. Angle dispersive diffraction geometry and section of the Debye-Scherrer cones obtained by a monochromatic neutron beam scattered from an AlSi7/SiC/70p.

4.2.2 Time Of Flight neutron diffraction (TOF)

On pulsed neutron sources (spallation sources) a white neutron beam which covers a wide energy range (Santisteban, 2008) is used for diffraction. A polycrystalline sample diffracts all neutron wavelengths on all {h k l} lattice planes in space. Detector banks at two fixed positions from the sample (± 90°) cover two main strain axes of the sample in q_1 and q_2 direction (Fig. 6).

In a TOF strain scanner several neutron velocities are defined using a chopper in the primary beam. The wavelength of a neutron is given by the de Broglie relation $h{\cdot}\lambda = m{\cdot}v$, where h is the Planck's constant, m is the mass of the neutron and v is the velocity of the neutron. Furthermore, $v = t{\cdot}(L_1 + L_2)$, where t is the travelling time along well-known path lengths L_1 (distance source-sample) and L_2 (distance sample-detector). The lattice distances can be determined using the Bragg's law depending on the travelling time at constant diffraction angle θ,

$$d_{hkl} = \frac{h}{2 \cdot \sin\theta \cdot m \cdot (L_1 + L_2)} \cdot t_{hkl} \tag{4}$$

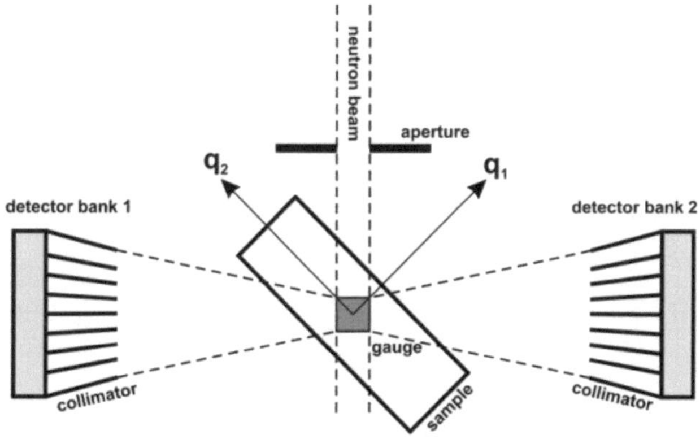

Fig. 6. TOF setup with two detector banks at fixed θ positions.

4.3 Neutron diffractometers for strain measurements

Table 1 shows an overview of the most important angle dispersive and time of flight neutron diffractometers available together with their main characteristics.

4.4 Some experimental issues

4.4.1 The d_0-problem

Neutron diffraction is a phase-sensitive strain measurement technique since it allows the determination of the average elastic strain acting in each microstructural phase independently. As a consequence, a reliable interpretation of the strain data depends greatly on the determination of a stress-free reference d_0 for each phase (see e.g. Fitzpatrick & Lodini, 2003). In principle, this can be achieved by investigating a stress-free reference sample (e.g. powders) using the same diffraction set-up. However, this can be affected by differences in composition between the reference and the studied materials due to different precipitation kinetics (this is typical for MMC (Dutta et al., 1991)), segregation during casting, etc. A description of several strategies to obtain d_0 is given for several cases in (Fitzpatrick & Lodini, 2003) but it must be kept in mind that this is a matter that presents serious difficulties for experimentalists. Some researchers have also reported methodologies that are independent of the measurement of a stress-free reference sample (Garcés et al., 2006; Young et al., 2007).

4.4.2 Systematic errors during strain measurement experiments

Some geometrical effects can influence the diffraction signal in angle dispersive diffraction. The intensity and position of the diffraction peak will depend not only on the internal stress state but also on absorption effects and the number and position of contributing crystals within the gauge volume that fulfil the Bragg condition.

Beamline / Facility	Flux on the sample	Sample size	Stage Type	In situ tests rigs	Experimental environment	Detector
Angle dispersive neutron diffraction						
STRESS-SPEC / FRM II (FR) (Hofmann et al., 2006; FRM II. STRESS-SPEC , 2011)	≤9.2×10^7 n s^{-1} cm^{-2} (graphite monocromator) ≤2×10^7 n s^{-1} cm^{-2} (Gemono-cromator)	max 300 kg	- Rotation table. - X Y Z translation table: x,y = ± 110 mm, z = 300 mm - 6 axis robot. - Eulerian cradles.	- 50 kN tension/compression rig (heatable to 1000 °C). -50 kN tension/comprssion + 100 Nm torsion rig.	- Variable take-off angle allows λ from 1.2Å– .4 Å - Mirror furnaces up to 1100 °C; Vacuum furnaces up to 1900 °C; Cryostats down to 30 mK.	Mirrotron MK-300-1; Gas mixture 3 bar ^3He + 2 bar CF_4; 300×300 mm^2 (256×256 pixel)
BT-8 / NIST NCNR (USA) (NIST, 2011)	~ 1 × 10^6 n $s^{-1}cm^{-2}$	- 0.6 m (h) x 0.6 m (w) x 1.0 m (l) - 100 kg	- Rotation table - Table with X Y Z -translation	- 10 kNrig, with table rotation in and tilt in Ψ, - 100 kNrig with rotation, combinable with 10 kNrigfor biaxial load.	Variable take-off angle allows λ= 0.9 Å – 3.0 Å withpossible $2\theta_{sample}\approx$ [40°,120°]	3He PSDactive area 10 cm x 5 cm (h)corresponding to approximately 8° width in 2θ
HB-2B/HFIR Oak Ridge National Laboratory (USA) (Oak Ridge National Laboratory, 2011)	≤ 3 × 10^7 n s^{-1} cm^{-2}	max. 450 kg	- XY translation / rotation table - Sample 400mm x 200mm; ≤ 1000 kg. - Z_1= 400 mm / 50 kg at 5 cm from stage;Z_2= 200 mm / 500 kg lift capacity from below. - ϕ rotation axis; Huber χ-ϕ full circle orienter,	- 22 kN Tension/compression load frame. - Tension/compression/torque - 4-point bending	- Furnaces (T). - User apparatus such as High Voltage field, gear loading apparatus.	Seven ORDELA 1155N PSD detectors covering 5° in the horizontal plane and +/- 15° from horizontal plane

Table 1.

Beamline / Facility	Flux on the sample	Sample size	Stage Type	In situ tests rigs	Experimental environment	Detector
G5-2 DIANE / LLB (FR) (LLB, 2011) (instr. being rebuild)	now~6.1x10^6 n cm^{-2} s^{-1}; future ~ 7.1x10^6 n cm^{-2} s^{-1}	max 300 kg max 1 m^3	- X Y Z translation table: - Eurelian cradle	- ± 25kNuni-axial tensile rig		100 x 100 mm^2 EMBL ^{5}He PSD
SALSA / ILL (FR) (Pirling et al., 2006; ILL, 2011)	Double focusing bent Si (400) monochromator, 2.5×10^7 n s^{-1} cm^{-2} at 1.6 Å, 0.25% Δd/d	- max length 1.5m - max -weight 700 kg	- Hexapod (Stewart platform, max tilt ±25°, x,y = ± 300 mm, z = ± 170 mm) - 90 Degree Eulerian cradle	- 10 kN vertical/horiz.mo untingrig for small samples; 50 kN horizontally mounting rig for static / dynamic tests. - 3kNthermo-mechanical rig; resistive heating T ≤ 1100 °C.	- Variable take-off angle allows λ from 1.3Å–2.4 Å - Variable optics (slits and radial collimators) Camera assisted metrology system	- 80 x 80 mm^2 2-dimensional Position Sensitive microstrip Detector.
E3 / HZB (D) (Wimpory et al., 2008; HZB, 2011)	~ 5 x 10^6 n cm^{-2} s^{-1}	max 250 kg	- -rotation table equipped with X Y Z - translation table - Eulerian cradles	Tensile, compressive and torsionmulti-axial load frame.	- cryostats (1.4K to 550 K); - furnaces (400K to 2000K); - Gas atmospheres	PSD ^{3}He Detector 30 x 30 cm^2
SMARTS / LUJAN LANL (USA) (Los Alamos, 2011)	n.a.	1 mm^3 - 1 m^3. max 1500 kg	- Rotation table (370°) - Table with X Y Z –translation (theodolites to position sample within 0.01 mm)	- 250 kN and at extreme T ≤ 1500°C. - Uniaxial 2 GPaloading on samples φ≤1 cm; with lower stresses T ≤ 1500°C.	- Cryogenics 200K to room temperature. - Up to 1800°C in stand alone configuration.	n.a.

Table 1.

Beamline/ Facility	Flux on the sample	Sample size	Stage Type	In situ tests rigs	Experimental environment	Detector
Time of flight neutron diffraction						
ENGIN-X / ISIS (UK) (ISIS, 2011)	3×10^6 n $cm^{-2}s^{-1}$	- Capable for complex shape, - max 1000 kg	- X Y Z rotation table	- Stress rigs (up to 100 kN) can be equipped with furnce or cryostats (-200 °C – 1000 °C).	- Three types of furnaces: air or inert gas furnace can be equipped with stressrig (≤1100 °C), furnacewith ceramic heating pad and vacuum resistance furnace (≤1800 °C). - Cryostat down to -200°C	- For diffr. (2-θ at 90°): λ - shifting fibre coded Zn detector. - For Bragg edge transmission (2θ of 180°): 2D area detector / 10×10 scintillation detectors, each 2×2 mm^2 with 0.5 mm pitch.
SIRIUS / KENS (JP) (KENS, 2011)	n.a.	n.a.	n.a.	- 10 kN tensile stress machine	- Cryostat (7 K – RT). - Furnaces: RT-1000°C (vacuum, gas); 300 °C.	- 90° bank with 500 and 144 1D PSD (3He ½"x 60 cm)
POLDI / SINQ (CH) (PSI, 2011;Stuhr, 2005a; Stuhr, 2005b)	~10×10^7 n $cm^{-2}s^{-1}$ Thermal spectrum (1.1Å - 5Å)	- ≤ 10 ton - φ<450mm height<2000mm - Gauge vol.: $(0.6mm)^3$-$3.8 \times 30 \times 3.8mm^3$	- 6 axis sample manipulator: X Y Z + rotation (360°) + 2-axis quarter cradle; Translation range vertical: 570mm Translation range horiz.:150mm<X, Y<150mm	- Tension-Compression, 20 kN, RT, mounting: vertical and horizontal - Rig under construction: 100 kN, multi-axial, RT to 1100 °C, LCF	standard furnaces (up to 1800 K); standard cryostats	He wire chamber; 2θ-range:30° ΔQ/Q: better 2×10^{-3} (Angular resolution: 0.075°; Time resolution: 1 μs) Radial collimators: 0.6mm; 1.5mm; 3.8mm

Table 1.

If large gauge volumes are used (≥ sample size) an absorption effect on the diffracted intensity must be taken into account (Fig. 7 a). The absorption gradient over the gauge volume causes a shift of the diffraction centre from the centre of gravity towards the primary beam. The consequence is an asymmetric peak shape. This effect differs for materials with different absorption characteristics as it is the case for the different phases in a MMC. Thus, relative peak shifts which are not caused by microstresses between the phases but by different absorption of the constituents may be observed.

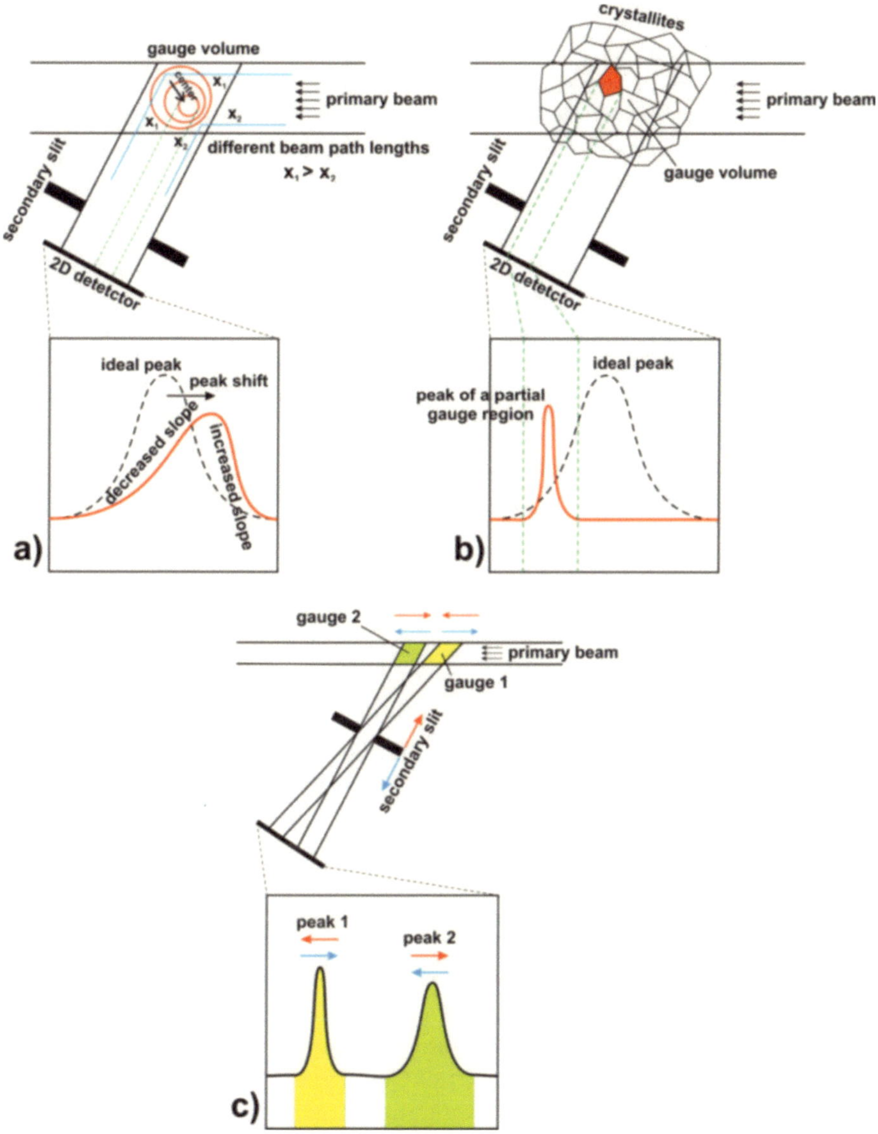

Fig. 7. Systematic errors that can occur using an angle dispersive diffraction setup (Schöbel, 2011a): a) displacement of the diffraction centre from the centre of the gauge volume due to absorption effects, b) coarse grains and c) pseudo micro strains generated by a change of the secondary-slit-to-sample distance.

Other systematic errors come from coarse grained materials or small samples. Here, only partial regions in the gauge volume contribute to the diffracted signal resulting in a peak shape not representative for the stress condition (Fig. 7 b). Low grain statistics can be improved by measurements of crystallographic planes with high multiplicity and rotation of the sample to increase the number of grains fulfilling the Bragg condition.

Diffraction patterns of multiphase materials usually include several peaks from different microstructural phases. In this case the secondary slit can produce a phase-sensitive systematic problem. The diffraction peaks of two hypothetical phases result in a "splitting" of the gauge volume dependent on the diffraction angle as shown in Fig. 7 c). Relative peak shifts occur if the secondary-slit-to-sample distance is changed, resulting in the measurement of pseudo micro stresses. This problem can be overcome using samples smaller than the gauge volume, with necessarily equally dimensioned reference samples, or by collimation of the diffracted beam.

In TOF neutron diffraction experiments the path length differences in the gauge volume cause comparable absorption and coarse grain influences on the peak positions but these can be neglected due to the long sample-detector path length in comparison with the gauge volume (Santisteban, 2008). The biggest advantage of TOF diffraction is the acquisition of complete diffraction patterns in one scan. However, the asymmetric peak shape (moderator effect) and long counting periods are the limiting factors for this method.

5. Evolution of internal stresses during heat treatment and thermal cycling of MMC

Examples of ex situ and in situ investigations to determine the evolution of internal stresses during heat treatment and thermal cycling of MMC are presented in this chapter.

5.1 Ex situ investigations

5.1.1 Macro-stresses

Annealing processes are known as one of the simplest ways to reduce the level of macrostresses in metallic materials and MMC. When the temperature is high enough to allow thermal relaxation processes, the plastic deformation gradients generated during thermo-mechanical processing, e.g. after quenching, decrease following a exponential decay with time. This is shown in Fig. 8 a) and b) for a cylindrical sample of 6061/SiC/15w with an aspect ratio of about two and a Ti6Al4V/TiB+TiC/8p disc of 15 mm diameter and 3 mm height, respectively. The stress relaxation in 6061/SiC/15w starts with a high tensile stress, while the initial stress condition is compressive in Ti6Al4V/TiB+TiC/8p. The magnitude of stress relaxation depends mainly on the microstructure of the materials and the temperature. However, the progress of the stress decay is independent from the internal architecture of the MMC and from the sign of the stress. For comparison, the stress state reduction after annealing is shown for an unreinforced 6061 Al alloy in Fig. 8 a).

Other authors (Chowdhury et al., 2010) divide this behaviour into two different stages: the first one (initial strong relaxation) is related to a mechanical transient loss and the second one (progressive relaxation) to the microstructural evolution. Although the stress decay is observed for both the MMC and the unreinforced matrix, the presence of the reinforcement reduces the magnitude of the relaxation of the macrostresses (Fig. 8). On the one hand, the higher dislocation density in the matrix acts as a barrier for dislocation movement. These extra dislocations are mainly the geometrical ones necessary to accommodate the matrix around reinforcement when deformation progresses. This effect is accompanied by the elastic strain fields introduced around the reinforcement. The load transfer from matrix to the stiffer (ceramic or intermetallic) discontinuous reinforcement decreases the amount of stress in the metallic matrix that can act as the driven force for relaxation during annealing.

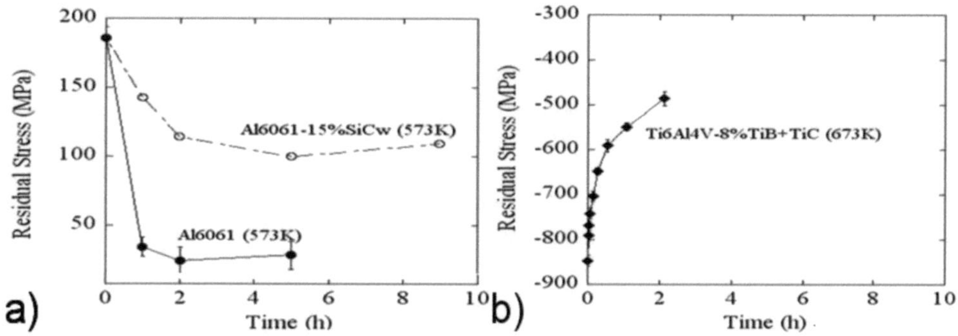

Fig. 8. Axial macrostress decay with annealing time for: a) PM Al alloys 6061 and 6061/SiC/15w at 300 °C (P. Fernández et al., 2005), and b) a disc shape sample of a Ti6Al4V/TiB+TiC/8p (Xie et al., 2005) at 400 °C.

As described above, the macrostress relaxation in lightweight MMC presents a similar stress decay. The relaxation of deviatoric and hydrostatic stress components usually differs in discontinuously reinforced composites. In particular, the hydrostatic component in the matrix of the MMC relaxes slower than in the corresponding unreinforced alloy (Fig. 9). This trend is similar to that found when the creep effects in these materials are compared (Bruno et al., 2004). This fact has not been completely explained yet and more work is needed to understand the correlation between this difference and material's microstructure.

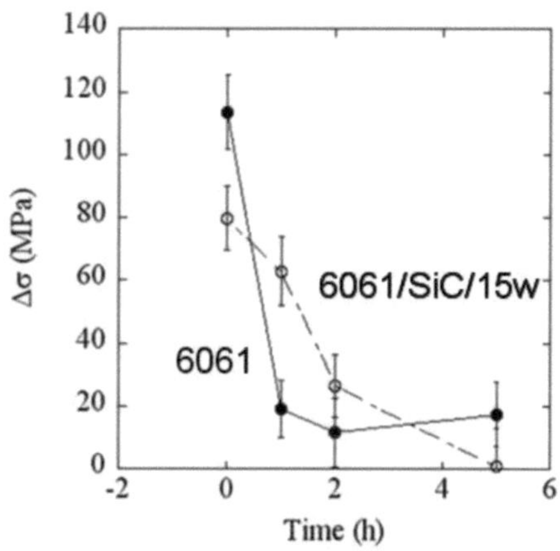

Fig. 9. Evolution of the stress difference between deviatoric and hydrostatic residual stress components during annealing for a 6061/SiC/15w and its corresponding unreinforced 6061 aluminium alloy matrix.

5.1.2 Type II microstresses

Type II microstresses in lightweight MMCs are directly related to their microstructure. The reinforcement architecture (content, size, morphology, distribution, etc.) determines the magnitude of internal stresses in these materials. (Fernández et al., 2011) showed that for PRM with a small interparticle distance (typically 5 μm) the original stress state is conserved almost unaltered after heat treatments. Small interparticle distance can be obtained either by a high reinforcement content (typically > 25vol%) and/or by small particle size (typically >2 μm). The relaxation of internal microstresses is compared in Fig. 10 for materials with different interparticle distances subjected to an overageing heat treatment. For the PRM summarized in Fig. 10, both ingot metallurgy (IM) materials, 2014/Al_2O_3/15p (=W2A15A) and 6061/Al_2O_3/15p (=W6A15A), present an interparticle distance of about 50 μm. The PM 6061/SiC/15w=E219 exhibits a mean interparticle distance of around 5 μm. For the IM MMC (W2A15A and W6A15A), with large interparticle distances, an almost complete relaxation is achieved (Fig. 10 a) and b). However, there is no relaxation in the PM 6061/SiC/15w because dislocation movement is hindered by nanometric Al_2O_3 dispersoids introduced during the PM process (Fig. 10 c).

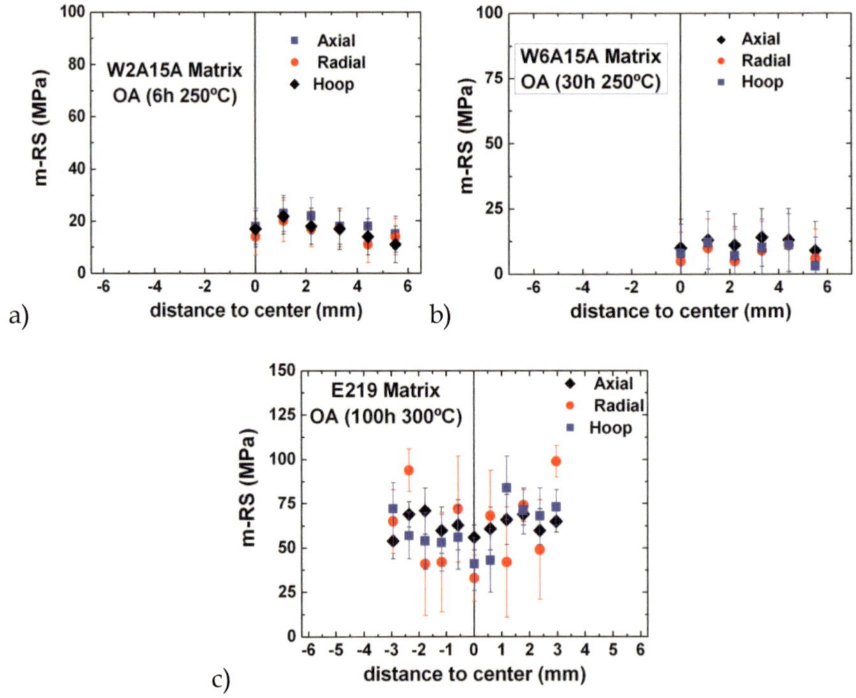

Fig. 10. Relaxation of microstresses after thermal treatment for: a) IM 2014/Al_2O_3/15p, b) IM 6061/Al_2O_3/15p, c) PM 6061/SiC/15w (Fernández et al., 2011).

It is worthy to mention that there are some materials that present an evolution of the internal stresses at room temperature. A typical example is the 7075 Al alloy that undergoes relaxation of internal stresses at room temperature during periods as long as 2 months (Linton et al., 2008). This phenomenon is explained by the ageing of this alloy at room temperature. In other cases, is the application of some specific thermomechanical processes, such as severe plastic deformation, that allows this room temperature relaxation process. For example, there exists a sigmoidal stresses decay much less pronounced than in the case of the small cylindrical samples shown in Fig. 8 a) (Woo et al., 2009). Since the time-dependent decrease of internal stresses was only observed in the nugget of a friction stir welded part, this evolution must be explained by the microstructural characteristics of this particular zone of the material undergoing natural aging. The presence of high concentrations of vacancies, atoms in solid solution and a large numbers of grain boundaries, may explain this process (Woo et al., 2009).

5.2 In situ determination of microstresses during thermal cycling of lightweight MMC

Insulated gate bipolar transistors (IGBT) are developed for high power applications in hybrid vehicles or railway traction (Baliga, 1979). These sealed IGBT modules consist of semiconducting chips mounted on a ceramic insulator which is soldered onto the baseplate.

The power reaches tens of megawatts concentrated in the small Si chips from which the heat produced must be dissipated through the baseplate into a heat sink. Therefore, base plate materials with high thermal conductivity (TC) and low coefficient of thermal expansion (CTE) are required to avoid delamination and thermal fatigue damage. This can be achieved using MMC formed by an aluminium matrix reinforced with high volume fractions of SiC particles or diamond (CD) particles (> 60 vol.%) to combine the thermal properties of Al (TC_{Al} ~ 250 W/mK, CTE_{Al} ~ 25 ppm/K) with those of SiC (TC_{SiC} ~ 140 W/mK, CTE_{SiC} ~ 5 ppm/K) or CD (TC_{CD} ~ 1000 - 2000 W/mK, CTE_{CD} ~ 1 ppm/K). Such PRM can be produced by liquid metal infiltration (gas pressure infiltration or squeeze casting), where a densely packed particle preform with mono-, bi-, trimodal particle size distributions (Ø ~ 5 - 200 µm) is infiltrated with the melt (Al, AlSi7, AlSi7Mg) (Huber et al., 2006). An AlSi7Mg/SiC/70p composite with a trimodal particle size distribution and voids in between the large SiC particles is shown in Fig. 11 a). In this MMC, with an AlSi7 matrix, the SiC particles are interconnected by eutectic Si bridges (Schöbel et al., 2010). These bridges result in the formation of an ICM with a 3D Si-SiC network embedded in the α-Al matrix.

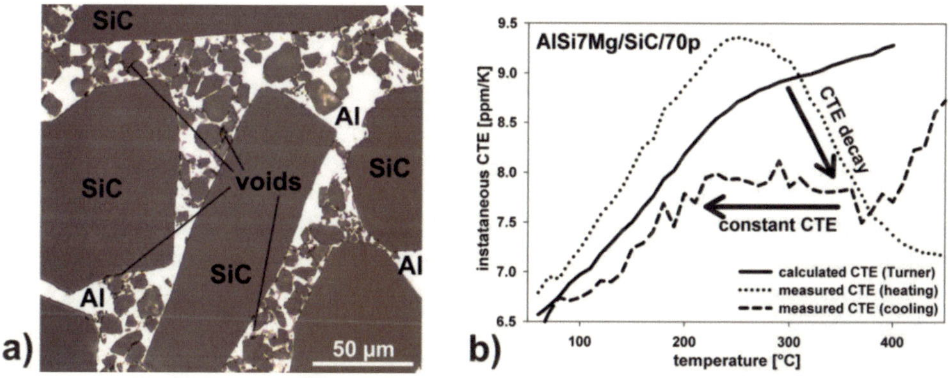

Fig. 11. a) Light optical micrograph of AlSi7Mg/SiC/70p with voids (dark) in the Al matrix (bright) between the SiC particles and b) measured instantaneous CTE(T) compared with that calculated by the thermo-elastic Turner-model (Huber, 2006).

The instantaneous CTE of AlSi7Mg/SiC/70p was investigated in (Huber et al. 2006), where an anomalous behaviour was observed above 200 °C which did not fit to the predictions of classical thermo-elastic models (Turner, 1946) (Fig. 11 b). A combination of in situ angle dispersive neutron diffraction and in situ synchrotron tomography experiments were performed (Schöbel et al., 2011) to explain this behaviour in terms of internal stresses generated during thermal cycling of the ICM. As explained previously, internal stresses can relax at elevated temperatures. Therefore, short acquisition times in the range of minutes are required for an overall heating / cooling rate comparable to the service conditions of the MMC. Another restrictive property for diffraction studies of these materials is presented by the coarse grains of the α-Al that requires the use of large gauge volumes (> 100 mm^2) to increase the grain statistics (see section 3.4.2).

The microstresses obtained by in situ neutron diffraction in an AlSi7/SiC/60p PRM during two thermal cycles (RT - 350°C) are shown in Fig. 12. High hydrostatic compressive stresses up to - 120±80 MPa are generated in the α-Al matrix during heating to 350°C. These stresses invert at ~ 100°C during cooling and become tensile down to room temperature (~ 50°C). The initial tensile stress level at RT ~ 100 MPa increases after the first cycle to ~ 170±70 MPa.

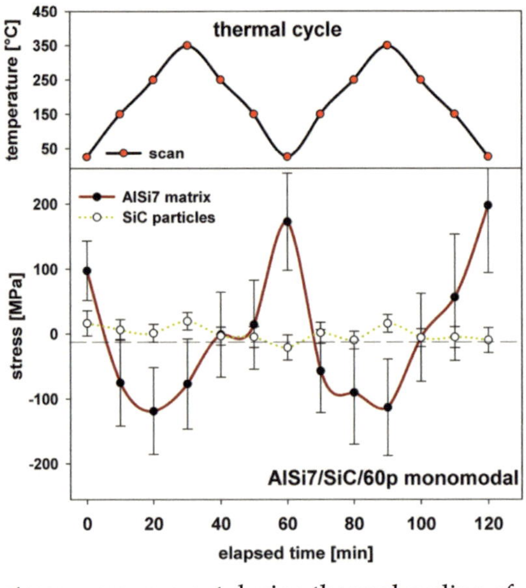

Fig. 12. In situ neutron stress measurement during thermal cycling of AlSi7/SiC/60p (RT - 350°C) showing the stress changes in the α-Al and in the particle stresses (Schöbel et al. 2011).

Synchrotron tomography with $(1.4\mu m)^3$/voxel of an AlSi7Mg/SiC/70p composite gives information on the evolution of the void volume fraction during thermal cycling (RT - 400°C). Fig. 13 a) shows voids in the AlSi7Mg matrix between large SiC particles. Voids are formed in MMC with a large CTE mismatch between the components and high reinforcement volume fractions during cooling after infiltration even if perfectly infiltrated (Schöbel et al., 2011). Fig. 13 b) shows that the voids close during heating and reopen after cooling. Only voids > $(\sim 5\mu m)^3$ are resolved by the produced tomographies, therefore the given volume fractions only indicate the relative change reliably. During thermal cycling large microstresses between the particles and the matrix (CTE mismatch) change the void volume fraction by visco-plastic matrix deformation (Nam et al., 2008) as indicated by the 2D finite element analysis presented in Fig. 14. The matrix embedded in a 3D reinforcement

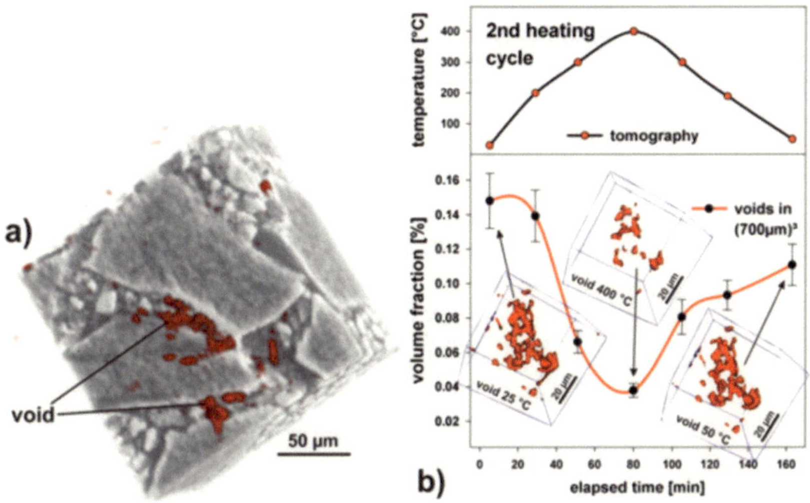

Fig. 13. Voids in an AlSi7Mg/SiC/70p composite during thermal cycling (RT – 400 °C). a) A single void between large SiC particles compared to b) evolution of the void volume fraction during thermal cycling (Schöbel et al. 2011).

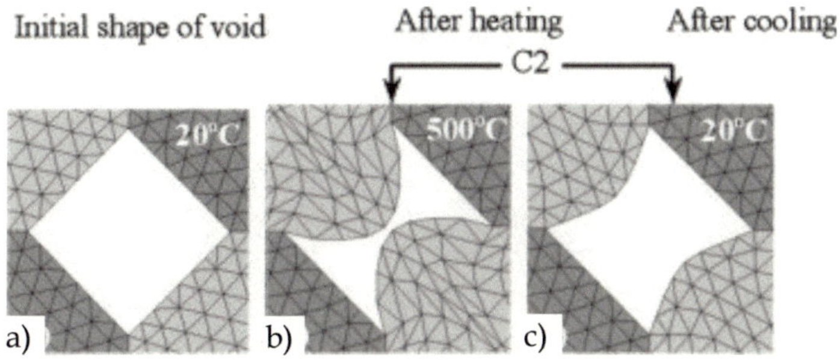

Fig. 14. The change of volume fraction and shape of a void (white) in an ICM formed by SiC (dark grey) and Al (light grey) obtained by finite element analysis: a) 0.25 vol.% void at initial zero stress shrinks to b) 0.103 vol.% after first heating and c) increases to 0.199 vol.% after the following cooling (Nam et al., 2008).

structure expands during heating and closes the voids (CTE decay in Fig. 11 b) which reopen during cooling (constant CTE region in Fig. 11 b) accommodating the volume mismatch.

In the case of a pure Al matrix without Si content, lower stress amplitudes and inverse void kinetics are observed. This was shown using the same experimental procedure, namely in situ neutron diffraction with synchrotron tomography during thermal cycling, for Al/CD/60p and AlSi7/CD/60p composites. The results are shown in Fig. 14. AlSi7/CD/60p behaves similar to AlSi7/SiC/60p with hydrostatic compression up to ~ -120±20 MPa closing the voids during heating, going into to ~ 120±20 MPa tensile stress at RT in which the voids reopen again. In Al/CD/60p, without eutectic Si connecting the diamond particles, low micro stress amplitudes were observed (Schöbel et al., 2010) together with an increase in void volume fraction during heating and decrease during cooling owing to the capability of the matrix to expand in a composite with isolated particles only touching each other.

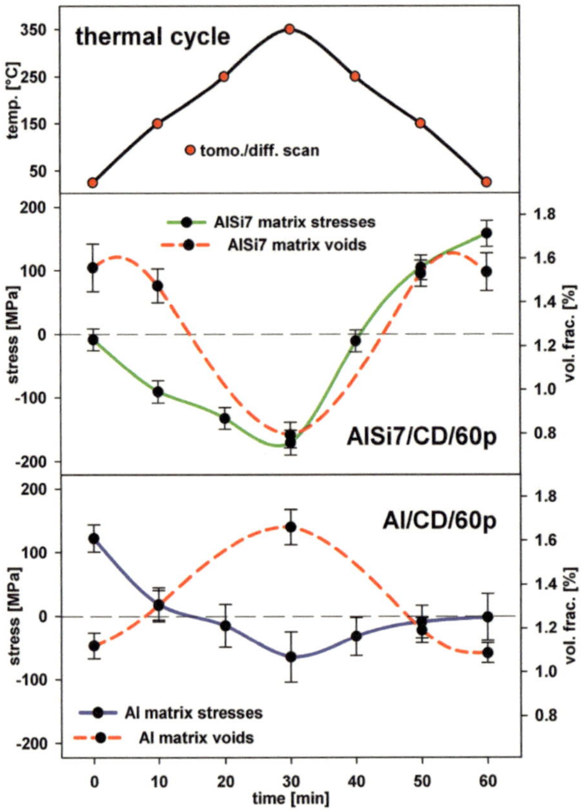

Fig. 15. Evolution of the microstresses in α-Al and of the matrix voids during thermal cycling of Al/CD/60p and AlSi7/CD/60p composites (Schöbel et al. 2010).

6. Evolution of internal stresses during deformation of lightweight MMC

In the present chapter, the evolution of internal stresses during plastic deformation of MMC is presented. Some examples of ex situ and in situ investigations will be summarized in order to describe the main features of this topic.

6.1 Macro-stresses

Similarly to the case of annealing heat treatments, when a material containing residual stresses is deformed, internal stresseses opposite to the applied stress can relax following a exponential decay as deformation progresses. This phenomenon has been extensively studied during the last decade (Levy-Tubiana et al., 2003). It is well establish that during early stages of plastic deformation, typically below 2%, the initial residual stresses are almost totally relaxed. After this initial stage the stress relaxation become less pronounced, like in the case of annealing (see section 4.1). However, after some degree of cumulative deformation, the internal structure of the material is changed by dislocation multiplication, forest dislocations, subgrains, etc. imposing new strain gradients increasing the general residual stress level (R. Fernández et al., 2005). As in the case of thermal relaxation, the behaviour of the deviatoric and hydrostatic components of the macrostresses evolves in a different manner depending on the microstructure. The results obtained for some discontinuously whisker reinforced MMC with the whiskers oriented preferably along the extrusion direction are shown in Fig. 16. Here, the deviatoric component relaxes completely after a certain value of deformation, while the hydrostatic component increases due to the new plastic strain gradients (P. Fernández et al., 2005). The relaxation behaviour of a conventional unreinforced 2024 IM Al alloy is shown for comparison. It is shown that the presence of reinforcement particles, introduce a higher residual stress level and a higher increase of the hydrostatic component progressing with deformation, mainly due to microstresses induced by geometrical necessary dislocations.

Apart from the effect of microstructural features as reinforcement size, aspect ratio and distribution, imposing slight modifications to the exponential decay of the residual stresses (Fig. 16), the most important difference between the described materials can be related to the processing route. Thus, in the case of the IM materials, the IM 6061/ Al_2O_3/15p MMC and the unreinforced 2024 IM Al alloy, the stress relaxation is more pronounced than in the PM 6061/SiC/15w (C38, C45).

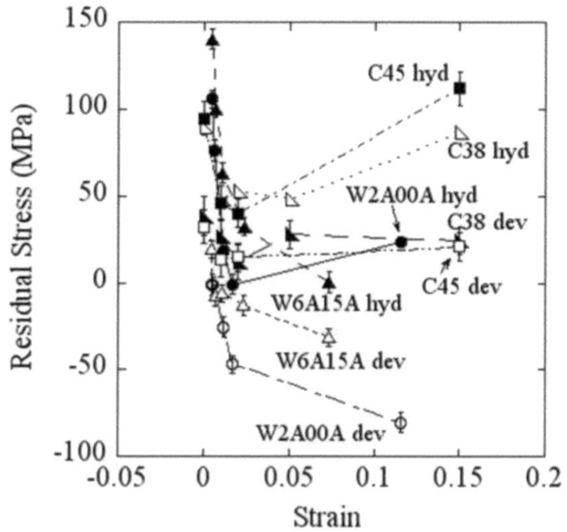

Fig. 16. Residual stress evolution (hyd = hydrostatic, dev = deviatoric) in the matrix of PM 6061/SiC/15w (C38, C45), IM 6061/ Al_2O_3/15p (W6A15A) and 2024 IM alloy (W2A00A) (R. Fernández et al., 2005).

6.2 Microstresses

The evolution of the microstresses with plastic deformation is different from that of the macrostresses. In particular, in PM discontinuously reinforced 6061/SiC/15w with the extrusion direction coinciding with the axial residual stress component, the hydrostatic component of the microstresses continuously relaxes to strain values as high as 15% (Fig. 17). However, the deviatoric microstresses remain constant, within the error bars, as it is shown in Fig. 17.

The fact that microstresses become totally deviatoric after some degree of plastic deformation accounts for the relevance of the activity of geometrically necessary dislocations (GND). In the PM PRM, a part of the reinforcement population is aligned along

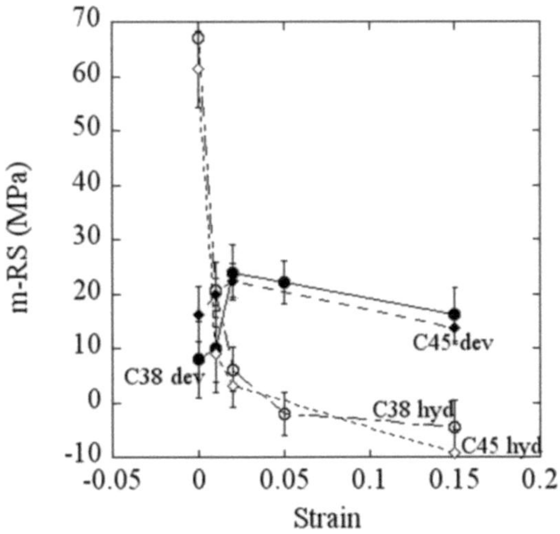

Fig. 17. Hydrostatic (hyd) and deviatoric (dev) microstress evolution in the aluminium phase with plastic deformation for two PM 6061/SiC/15w MMC (C38, C45) (R. Fernández et al., 2005).

the extrusion axis. Therefore, as the GND are mainly at the ends of the whiskers, the microstresses become predominantly deviatoric. This is the case for fibre reinforcements, particularly for short fibres with aligned fibre orientation. The relaxation processes are especially interesting in the case of metal matrix composite with random planar short fibre distribution. Such composites with a relatively low volume fraction of reinforcement (< 30 vol%) are in many cases more attractive, from a mechanical point of view, than composites with a high content of aligned whiskers or continuous fibres because they can be used under multiaxial loading conditions (Hutchinson et al., 1993; Dragone et al., 1991). It has been demonstrated that the mechanical behaviour and the creep properties depend on the direction of the stress with respect to the fibre plane (parallel or transverse to the fibre plane) and the sign of the stress (tension or compression) (Garces et al., 2006b; Garces et al, 2007). During plastic deformation, the MMC stores elastic strain energy within the fibres and when the load is removed the matrix is deformed in the opposite direction. In the case where the fibres were loaded in tension/compression, the matrix surrounding the fibre rapidly unloads elastically and goes into elastic compression/tension. At this stage, the fibre is still in elastic tension/compression. From this point onwards, fibres can only be further unloaded when backflow occurs in the matrix.

The plastic deformation of short-fibre reinforced metals is strongly affected by two competitive mechanisms which occur at the same time: i) the load transfer from the matrix to stiff short fibres under an applied load and ii) the internal damage reducing the load-bearing capacity of the fibres. The internal damage mechanisms reported in the literature are fibre fragmentation, buckling and debonding, as well as void formation at the fibre / matrix interface. The damage mechanisms reduce the capacity to store load by the fibre and, therefore, the internal stresses change depending on the direction and sign of the stress. Fig. 18 shows the evolution of the von Mises internal stress in an Al/Al_2O_3/15s SFRM for three cases: tensile (T) / compressive (C) stress in the direction parallel (P) to the fibre plane (TP and CP, respectively) and compressive stress in a direction perpendicular to the fibre plane

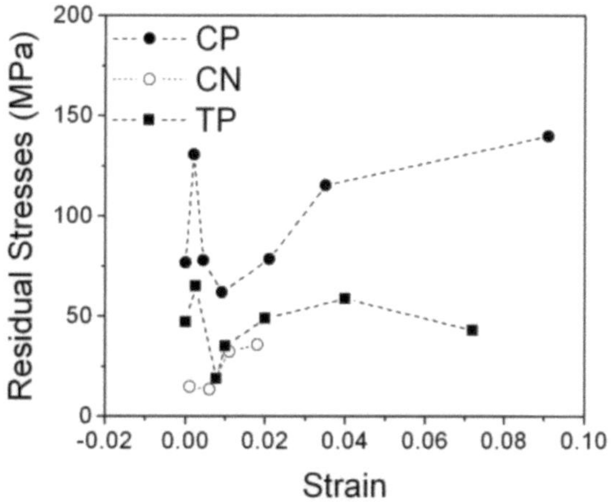

Fig. 18. Von Mises stress in an Al/Al_2O_3/15s SFRM as a function of pre-strain accumulated in the composite for the three loading modes: CP (compression in plane), CN (compression out of plane) and TP (tension in plane) (Garcés et al., 2006b).

(CN). It is important to point out that the von Mises stress in all the cases does not start from zero stress since initial residual stresses generated during the fabrication step pre-exist in the composite. Furthermore, the residual stresses in the CP plane are higher than the CN and TP cases, which are similar. This fact is caused by internal damage, which is less in the case of the compressive direction (CP). On the one hand, in the case of the TP mode, the fibres are mainly loaded in tension, and break earlier or suffer debonding. On the other hand, fibres in the CP and CN modes are slightly buckled, which is not as damaging as tension.

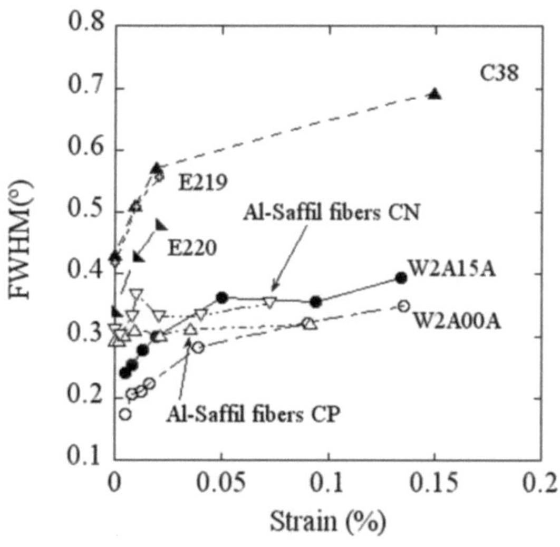

Fig. 19. Evolution of the microstresses represented by FWHM of the neutron diffraction peaks during plastic deformation of some Al matrix composites and their corresponding unreinforced alloys: PM 6061/SiC/15w extruded using a conical (C38) and a flat-shaped extrusion tool (E219); PM unreinforced 6061 alloy (E220); IM 6061/Al_2O_3/15p (W6A15A) and IM 2024 alloy (W2A00A) (R. Fernández et al., 2005).

As the central position of the diffracted peak is related to the lattice parameter, the Full Width at Half Maximum (FWHM) of a diffraction peak quantifies the scatter introduced in the diffraction process by fine grains and type III microstresses, mainly by the dislocation density. The FWHM is shown in Fig. 19 for some MMC as a function of applied strain (Garcés et al., 2006b; R. Fernández et al., 2005). The FWHM increases with the dislocation multiplication process as plastic deformation proceeds. In particular, PM materials present a more pronounced evolution of the FWHM. In the IM 2124/Al_2O_3/15p PRM and Al/Al_2O_3/15s SFRM this increment is weaker. It is interesting to point out that this fact seems to be independent of the fibre aspect ratio and it is more related with the fabrication route. Nanometric Al_2O_3 dispersoids introduced during the PM process anchor dislocations suppressing annihilation processes as plastic deformation advances.

6.3 Load partition in lightweight MMC with three-dimensional interpenetrating structures

Neutrons have been extensively used to follow in situ the evolution of internal stresses in MMC during deformation (see e.g. Withers et al., 1989; Shi et al., 1997; Daymond et al., 1999). Recently, there has been a great interest in ICM due to a more rigid and efficient reinforcement effect that these composites offer, especially at high temperatures (e.g. Roy et al., 2011; Long et al., 2011; Requena et al., 2011). The microstructure of these interconnected systems is usually characterized by a random distribution of ceramic and/or intermetallic phases, but infiltrated preforms of random planar oriented short fibres can present preferred orientations of the reinforcement (Roy et al., 2011; Requena et al., 2009). Within this group of materials, the composites reinforced with short fibres interconnected with eutectic Si and aluminides (see Fig. 1 c) are of technological interest and have been used to locally reinforce diesel pistons. As we commented briefly above, such composites exhibit a more balanced property profile, are less anisotropic and present a better mechanical response under multiaxial loading conditions.

The strength of an AlSi12/Al_2O_3/20s ICM was investigated in (Requena et al., 2009a) by in situ neutron diffraction during compression tests. Besides the high interconnectivity that is usually obtained in the eutectic Si in the as cast (AC) condition (Lasagni et al., 2007) this phase forms bridges between the ceramic short fibres (Requena & Degischer, 2006). Thus, an IMC with an interconnected 3D reinforcement of eutectic Si and Al_2O_3 short fibres is obtained. The strength, stability and degree of interconnectivity of this 3D structure are sensitive to the amount and size of the Si bridges, which can be modified by heat treatment (Lasagni et al, 2008). The composite was therefore investigated in AC condition and after a solution treatment (ST) at 540°C during 4 hours to study the effect of different Si-short-fibres architectures on the strength.

The Si bridges connecting the short fibres can be seen in Fig. 20 a) for AlSi12/Al_2O_3/20s-AC. The large amount of Si bridges results in a 3D structure formed by the Al_2O_3 short fibres and the eutectic Si that remains together after leaching out the Al matrix. After the spheroidisation treatment, the Si forms round particles, most of which stick to the interface of the short fibres (see Fig. 20 b) but the number of Si bridges decreases as well as the connectivity of the 3D structure. The highest strength is exhibited by the AlSi12/Al_2O_3/20s-AC composite (Requena et al., 2009a). The stress directions considered for the stress partition analysis as well as the load direction are shown schematically in Fig. 21 taking into

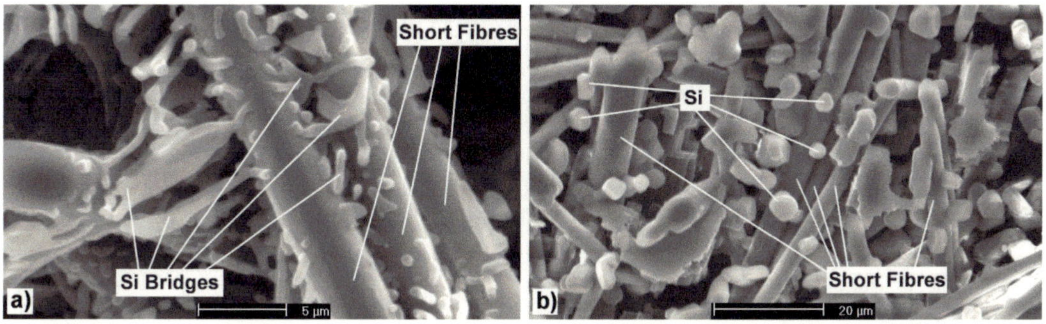

Fig. 20. Morphology of the Si-short-fibres structure in an $AlSi12/Al_2O_3/20s$ after deep etching: a) AC condition, where fine elongated eutectic Si bridges connect the short fibres and b) ST condition (lower magnification), where Si forms round particles, which stick to the interface of the ceramic short fibres (Requena et al., 2009a).

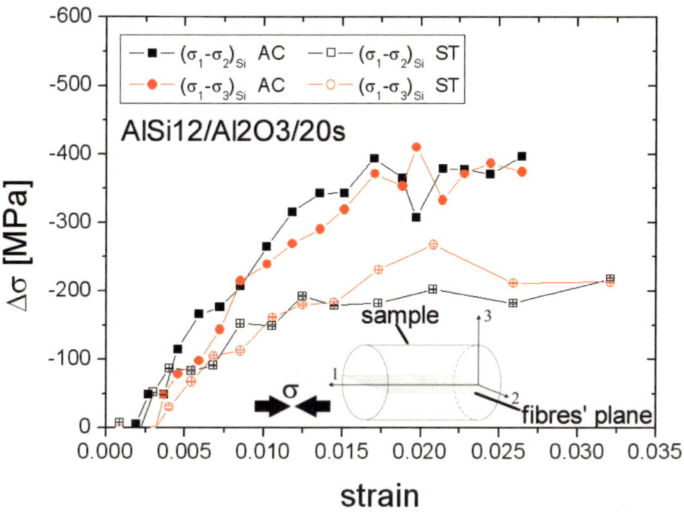

Fig. 21. Evolution of the stress differences (σ_1-σ_3) and (σ_1-σ_2) in $AlSi12/Al_2O_3/20s$ determined by in situ neutron diffraction for the eutectic Si in the AC and 540°C/4h conditions as a function of the applied compressive strain. The geometry of the sample and the orientation of the fibres' plane (1-2) with respect to the applied stress σ (1) are indicated (Requena et al. 2009a).

account the shape of the samples used for the in situ tests and the orientation of the fibres. Furthermore, Fig. 21 shows the evolution of the stress differences (σ_1-σ_3) and (σ_1-σ_2) obtained during the in situ neutron diffraction experimentes as a function of the applied strain for the eutectic Si in the AC and 540°C/4h conditions. It is observed, that the eutectic Si in the $AlSi12/Al_2O_3/20s$-AC ICM presents a higher load bearing capacity than the eutectic Si after 540°C/4h. However, the eutectic Si in the ICM carries a smaller portion of load than in the case of the unreinforced AlSi12 alloy (Requena et al, 2009b) although the composite exhibits a higher strength. This is due to the fact that the load is carried by the network formed by short fibres and the eutectic Si in the case of the ICM. A mean field model proposed rationalized the evolution of the principal stress differences and the von-

Mises stress in the aluminium matrix, the eutectic Si and the short fibres (Requena et al., 2009a). The reinforcing effect of the fibres and of the eutectic Si were reflected as well as the reinforcing potential of isolated elongated Si particles. However, this model does not take into account the interconnectivity between the short fibres and the eutectic Si underestimating the load transfer from the matrix to the reinforcement architecture. These results support the need to take into account the interconnectivity of the reinforcing structure to understand the high strength exhibited by these composites, especially at high temperatures.

7. Acknowledgments

M. Hofmann (FRM II), J. Repper (PSI), S. Van Petegem (PSI), A. Evans (ILL), T. Pirling (ILL), R. Wimpory (HZB), C. R. Hubbard (Oak Ridge Nat. Lab.), T. H. Gnäupel-Herold (Nat. Inst. of Stand. and Techn.), S. Y. Zhang (ISIS), A. Paradowska (ISIS), V. Klosek (LLB).

G. Requena and M. Schöbel would like to acknowledge the "K-Project for Non-Destructive Testing and Tomography" supported by the COMET-Program of the Austrian Research Promotion Agency (FFG) and the Province of Upper Austria (LOÖ), Grant No. 820492.

8. Conclusions

The contribution of the neutron diffraction technique to the understanding of the generation and evolution of thermo-mechanically induced residual and internal stresses in lightweight metal matrix composites has been reviewed in the frame of relevant literature. Besides the conclusions corresponding to the specific works described, the following conclusions are withdrawn:

- The neutron diffraction technique has evolved during the last decades to become a well-established technique to measure residual / internal stresses both by ex situ and in situ experiments in crystalline multiphase materials. The considerable amount of dedicated state of the art instruments available around the world to external users shows the importance of the neutron diffraction technique for materials scientists.
- Neutron diffraction still presents some advantages to determine internal/residual stresses which are decisive under certain experimental conditions in comparison with the modern high brilliance synchrotron radiation sources, e.g. a larger penetration depth and the possibility to use larger gauge volumes. Furthermore, the modern neutron sources allow now acquisition times of only a few minutes to acquire reliable data, which is crucial for in situ experiments.

9. References

Ashby, M. *Materials Selection in Mechanical Design* (3rd edition), Butterworth-Heinemann, ISBN 075061682, Oxford.

Baliga B.J. (1979). Enhancement and Depletion Mode Vertical Channel MOS Gated Thyristors. *Electronic Letters*, Vol. 15, pp. 645-647.

Behrens, M. (2011). Powder X-ray and neutron diffraction, Lecture series: Modern Methods in Heterogeneous Catalyst Research.

Birchall, J.; Stanley, D. Mockford, M.; Pigott, G. & Pinto, P. (1988). The Toxicity of Silicon Carbide Whiskers, *J Mater Sci Letters*, Vol. 7, 350-352.

Brendel, A.; Popescu, C.; Köck, T. & Bolt, H. (2007). Promising composite heat sink material for divertor of future fusion reactors, *J Nuclear Mater,* Vol. 367-370 B, pp. 1476-1480.

Bruno, G.; Fernández, R. & González-Doncel, G. (2004). Relaxation of the residual stress in 6061Al-15 vol.% SiCw composites by isothermal annealing *Mat Sci Eng A,* Vol. 382, No. 1-2, pp. 188-197.

Brückel, T.; Heger, G.; Richter, D. & Zorn, R. (2005). *Neutron Scattering, Matter and Materials* 28, Jülich.

Buffa, G.; Fratini, L.; Pasta, S.; Shivpuri, R. (2008). *CIRP annals*, Vol. 1, No. 57, 287.

Carpenter, JM. (2008) *Neutron Sources for Materials Research.* Tenth international School on Neutron and X-ray Scattering, University of Chicago.

Chowdhury,A.; Mari,D. & Schaller,R. (2010).Thermal stress relaxation in magnesium matrix composites controlled by dislocation breakaway. *Comp Sci Tech,* Vol.70, pp. 136-142.

Daymond, M.R.; Lund, C.; Bourke, M.A.M. & Dunand, D.C. (1999). Elastic phase-strain distribution in a particulate-reinforced metal-matrix composite deforming by slip or creep *Metall Mater Trans A*, Vol. 30, pp. 2989-2997.

Degischer, H. P. (1997). Innovative light metals: metal matrix composites and foamed aluminium, *Materials & Design*, Vol. 18, pp. 221-226.

Dragone, T.L.; Schlauchtmann, J.J. & Nix, W.D. (1991). Processing and creep characterization of a model metal matrix composite: Lead reinforced with nickel fibers. *Metall Trans. A* 22, No. 5, pp. 1029-1036.

Dutta, I.; Allen, S. & Hafley, J. (1991). Effect of reinforcement on the aging response of cast 6061 Al-Al_2O_3 particulate composites *Metall Trans A*, Vol. 22, No. 11, pp. 2553-2563.

CES Edupack Aerospace Edition, October 2011, Available from: http://www.grantadesign.com/education/content.htm.

Fernández, P. ; Fernández, R. ; González-Doncel, G. & Bruno, G. (2005). Correlation between matrix residual stress and composite yield strength in PM 6061Al–15 vol% SiCw. *Scripta Mater,* Vol. 52, No. 8, pp. 793-797.

Fernández, R.; Bruno, G. & González-Doncel, G. (2005). Residual stress evolution with compressive plastic deformation in 6061Al–15 vol.% SiCw composites studied by neutron diffraction *Mat Sci Eng A*, Vol. 403, No. 1-2, pp. 260-268.

Fernández, P.; Bruno, G.; Fernández & R. González-Doncel G. (2011). Fifth AUSE user meeting. Valencia 7-9 September 2011.

Fitzpatrick, M.E. & Lodini, A. (2003) *Analysis of Residual Stress by Diffraction using Neutron and Synchrotron Radiation*, Taylor & Francis, London.

MMC-ASSESS, October 2011, Available from: http://mmc-assess.tuwien.ac.at.

FRM II. STRESS-SPEC Eigenspannungs- und Texturdiffraktometer, October 2011, Available from: http://www.frm2.tum.de/wissenschaftliche-nutzung/diffraktion/stress-spec/

Garcés, G.; Bruno, G. & Wanner, A. (2006). Residual stresses in deformed random-planar aluminium/Saffil short-fibre composites *Mat Sci Eng A*, Vol. 417, No. 1-2, pp. 73-81.

Garcés, G.; Bruno, G. & Wanner, A. (2006) Residual stresses in deformed random-planar aluminium/Saffil® short-fibre composites deformed in different modes *Int J Mat Res,* Vol. 97, No. 10, pp. 1312-1319.

Garcés, G. Bruno & G. Wanner, A. (2007) Load transfer in short fibre reinforced metal matrix composites *Acta Mater,* Vol. 55, No. 16, pp. 5389-5400.

Hofmann, M.; Seidl, G.A.; Rebelo-Kornmeier, J.; Garbe, U.; Schneider, R.; Wimpory, R.C.; Wasmuth, U. & Noster, U. (2006). The new materials science diffractometer STRESS-SPEC at FRM-II. *Mat Sci Forum*, Vol. 524-525, pp. 211-216.

Huber, T.; Degischer, H.P.; Lefranc, G. & Schmitt, T. (2006). Thermal expansion studies on aluminium-matrix composites with different reinforcement architecture of SiC particles *Comp Sci Tech*, Vol. 66, No. 13, pp. 2206-2217.

Hutchinson, J. & McMeeking, R. (1993). *Continuous models for deformation: discontinuous reinforcement*. Suresh, S. Mortensen, A. Needleman, A. editors. Fundamentals of metal matrix composites. Butterworth-Heinemann.

Hutching, M.T.; Withers, P.J.; Holden, T.M. & Lorentzen T. (2005). *Introduction of the characterization of residual stress by neutron diffraction*. Taylor and Francis Group.

HZB. Neutron Residual Stress/Materials diffractometer, October 2011, Available from: http://www.helmholtz-berlin.de/forschung/funkma/werkstoffe/methoden/neutronendiffraktion_en.html

ILL. Strain imager for engineering applications SALSA, October 2011, Available from: http://www.ill.eu/instruments-support/instruments-groups/instruments/salsa/

ISIS. Engin-X Engineering materials beamline, October 2011, Available from: http://www.isis.stfc.ac.uk/instruments/engin-x/

KENS. SIRIUS (High resolution and high intensity powder diffractometer), October 2011, Available from:
http://neutron-www.kek.jp/kens2/kens_e/spectrometer/sirius.html

Lasagni, F.; Lasagni, A.; Marks, E.; Holzapfel, C.; Mücklich, F. & Degischer, H.P. (2007). Three-dimensional characterization of 'as-cast' and solution-treated AlSi12(Sr) alloys by high-resolution FIB tomography. *Acta Mater*, Vol. 55, pp. 3875-3882.

Lasagni, F.; Acuña, J. & Degischer, H.P. (2008). Interpenetrating Hybrid Reinforcement in Al_2O_3 Short Fiber Preforms Infiltrated by Al-Si Alloys, *Met Mat Trans A*, Vol. *39*, No. 6, pp. 1466-74.

Levy-Tubiana, R.; Baczmanski, A. & Lodini, A. (2003). Relaxation of thermal mismatch stress due to plastic deformation in an Al/SiCp metal matrix composite. *Mat Sci Eng A*, Vol. 341, No. 1-2, pp. 74-86.

Leyens, C.; Kocian, F.; Hausmann & J. Kaysser, W.A. (2003). Materials and design concepts for high performance compressor components, *Aero Sci Tech*, Vol. 7, pp. 201-210.

Linton, V. & Ripley, M.I. (2008). Influence of time on residual stresses in friction stir welds in agehardenable 7xxx aluminium alloys *Acta Mater*, Vol. 56, No. 16, pp. 4319-4327.

LLB. Two-Axis Strain Diffractometer "DIANE", October 2011, Available from:
http://www-llb.cea.fr/en/fr-en/pdf/g52-llb.pdf

Long, J.; Li, W.; Chen, S.; Lin, J. & Zeng, Y. (2011). Interface Study of Short Mullite Fiber Reinforced Al-Cu-Si Alloy Composites. *Adv Mater Res*, Vol. 150-151, pp. 1574-1579.

Los Alamos National Laboratory. Spectrometer for Materials Research at Temperature and Stress, October 2011, Available from:
http://lansce.lanl.gov/lujan/instruments/SMARTS/

Nam, T.H.; Requena, G. & Degischer, P. (2008). Thermal expansion behaviour of aluminum matrix composites with densely packed SiC particles. *Compos Part A*, Vol. 39, No. 5, pp. 856-865.

NIST. The BT8 Residual Stress Diffractometer, October 2011, Available from: http://www.ncnr.nist.gov/instruments/darts/

Oak Ridge National Laboratory. Neutron Residual Stress Mapping Facility (HB-2B), October 2011, Available from: http://neutrons.ornl.gov/instruments/HFIR/HB2B/

Noyan, I.C. & Cohen, J.B. (1987). *Residual stress. Measurement by diffraction and Interpretation* Springer-Verlag New York Inc.

Peters, P.; Hemptenmacher, J. & Schurmann, H. (2010). The fibre/matrix interface and its influence on mechanical and physical properties of Cu-MMC. *Comp Sci Tech,* Vol. 70, No. 9, pp. 1321-1329.

Pirling. T.; Bruno, G. & Withers, P.J. (2006). SALSA, a new concept for strain mapping at the ILL. *Mat Sci Eng A,* Vol. 437, pp. 139-144.

PSI. POLDI: Pulse Overlap time-of-flight Diffractometer, October 2011, Available from: http://poldi.web.psi.ch/

Pynn R. (2011). Neutron Scattering, Lectures, Los Alamos National Laboratory.

Requena, G. & Degischer, H.P. (2006). Creep behaviour of unreinforced and short fibre reinforced AlSi12CuMgNi alloy. *Mat Sci Eng A,* Vol. 420, No. 1-2, pp. 265-275.

Requena, G.; Garcés, G.; Danko, S.; Pirling, T. % Boller, E. (2009). The effect of eutectic Si on the strength of short-fibre-reinforced Al. *Acta Mater,* Vol.57, No. 11, pp. 3199-3210.

Requena, G.; Garcés, G.; Rodríguez, M.; Pirling, T. & Cloetens, P. (2009). 3D architecture and load partition in eutectic Al-Si alloys. *Adv Eng Mater,* Vol. 11, No. 12, pp.1007-1014.

Requena, G.; Garcés, G; Asghar, Z.; Marks, E.; Staron, P. & Cloetens, P. (2011). The effect of the connectivity of rigid phases on strength of Al-Si alloys. *Adv Eng Mater.,* Vol. 13, No. 8, pp. 674-684.

Roy, S.; Gibmeier, J.; Kostov, V.; Weidenmann, K.A.; Nagel, & A. Wanner, A. (2011) Internal load transfer in a metal matrix composite with a three-dimensional interpenetrating structure *Acta Mater.,* Vol. 59, No. 4, pp. 1424-1435.

Santisteban J.R. et al. Neutrons and Synchrotron Radiation in Engineering Materials Science, WILEY-VCH, Weinheim, 2008.

Schöbel, M.; Degischer, H.P.; Vaucher, S.; Hofmann, M. & Cloetens, P. (2010). Reinforcement architectures and thermal fatigue in diamond particle-reinforced aluminum. *Acta Mater,* Vol. 58, No. 19, pp. 6421-6430.

Schöbel, M.; Altendorfer, W.; Degischer, P.; Vaucher, S.; Buslaps, T.; Di Michiel, M. & Hofmann, M. (2011). Internal stresses and voids in SiC particle reinforced aluminum composites for heat sink applications. *Comp Sci Tech,* Vol. 71, No. 5, pp. 724-733.

Shi, N.; Bourke, M.A.M.; Roberts, J.A. & Allison, J.E. (1997). Phase-stress partition during uniaxial tensile loading of a TiC-particulate-reinforced Al composite *Metall Mater Tran. A,* Vol. 28A, No. 12, pp. 2741-2743.

Stuhr, U. (2005). Time-of-flight diffraction with multiple pulse overlap. Part I: The concept. *Nucl Instrum Meth A,* Vol. 545, pp. 319-329.

Stuhr, U.; Spitzer, H.;Egger, J.; Hofer, A.; Rasmussen, P.; Graf, D.; Bollhalder, A.; Schild, M. Bauer, G. & Wagner W. (2005). Time-of-flight diffraction with multiple frame overlap Part II: The strain scanner POLDI at PSI. *Nucl Instrum Meth A,* Vol. 545, pp. 330-338.

Tjong S. C. & Ma Z.Y. (2000), Microstructural and Mechanical Characteristics of In Situ Metal Matrix Composites *Mat Sci Eng A,* Vol. 29, pp. 49-113.

Wimpory, R.C.; Mikula, P.; Šaroun, J.; Poeste, T.; Li, J.; Hofmann, M. & Schneider, R. (2008). Efficiency boost of the materials science diffractometer E3 at BENSC: one order of magnitude due to a horizontally and vertically focusing monochromator. *Neutron News*, Vol.19, No. 1, pp. 16-19.

Withers, P.J.; Stobbs, W.M. & Pedersen, O.B. (1989). The application of the Eshelby method of internal stress determination to short fibre metal matrix composites *Acta Metal,* Vol. 37, No. 11, pp. 3061-3084.

Wither, P.J.; Badeshia, H. (2001). Residual stress. Part 1 – Measurement techniques. Mat Sci Tech, Vol. 17, pp. 355-365.

Woo, W.; Feng, Z.; Wang, X.-L. & Hubbard, C.R. (2009). Neutron diffraction measurements of time-dependent residual stresses generated by severe thermomechanical deformation. *Scripta Mater*, Vol. 61, No. 6, pp. 624-627.

Xie, L.; Jiang, C. & Ji, V. (2005). Thermal relaxation of residual stresses in shot peened surface layer of (TiB + TiC)/Ti–6Al–4V composite at elevated temperatures. *Mat Sci Eng A,* Vol. 528, pp. 6478-6483

Young, M.L.; DeFouw, J.; Almer, J.D. & Dunand D.C. (2007). Load partitioning during compressive loading of a Mg/MgB_2 composite. *Acta Mater,* Vol. 55, No. 10, pp. 3467-3478.